Introduction to the Relativity Principle

G. Barton

Centre for Theoretical Physics,
University of Sussex, Brighton

JOHN WILEY & SONS, LTD
Chichester · New York · Weinheim · Brisbane · Singapore · Toronto

Other Wiley Editorial Offices

John Wiley & Sons, Inc., 605 Third Avenue,
New York, NY 10158-0012, USA

WILEY-VCH Verlag GmbH, Pappelallee 3,
D-69469 Weinheim, Germany

Jacaranda Wiley Ltd, 33 Park Road, Milton,
Queensland 4064, Australia

John Wiley & Sons (Asia) Pte Ltd, 2 Clementi Loop #02-01,
Jin Xing Distripark, Singapore 129809

John Wiley & Sons (Canada) Ltd, 22 Worcester Road,
Rexdale, Ontario M9W 1L1, Canada

Library of Congress Cataloging-in-Publication Data

Barton, Gabriel.
    Introduction to the relativity principle / G. Barton.
      p. cm.
    Includes bibliographical references and index.
    ISBN 0-471-99895-8 (cloth : alk. paper). – ISBN 0-471-99896-6
  (pbk. : alk. paper)
    1. Relativity (Physics) I. Title.
  QC173.55.B377 1999   99-17294
  530.11–dc21   CIP

British Library Cataloguing in Publication Data

A catalogue record for this book is available from the British Library

ISBN 0-471-99895-8 (Hardback)
0-471-99896-6 (Paperback)

Typeset in 10/12 pt Times by C.K.M. Typesetting, Salisbury, Wiltshire.

Contents

Preface   ix

Part I   Introductory   1

1   Preliminaries   3
    1.1 Aims   4
    1.2 Prerequisites   4
    1.3 Organization   5
    1.4 History   7
    1.5 Further Reading   7
    1.6 Context   8
    1.7 Terminology   9
    1.8 Notes   10

2   The Relativity Principle, and its Applications in Newtonian Physics   11
    2.1 Reference Frames and Transformation Rules   11
    2.2 Galilean Transformations: Kinematics   13
         2.2.1 Time Coordinates   13
         2.2.2 Observers in Uniform Relative Motion   13
    2.3 Inertial Frames and the Galilean Relativity Principle   16
         2.3.1 Inertial Frames   16
         2.3.2 The Galilean Relativity Principle   17
         2.3.3 Form-Invariance   18
         2.3.4 Active and Passive Transformations   20
    2.4 Galilean Transformations: Dynamics   21
         2.4.1 Mass   21
         2.4.2 Momentum   22
         2.4.3 Force   24
         2.4.4 Energy   24
         2.4.5 The Relativity Principle and the Conservation Laws   26

2.5   Plane Waves   26
      2.5.1   Introductory   26
      2.5.2   Galilean Transformations: $\omega, \mathbf{k}, \mathbf{u}_g$, and $\mathbf{u}_p$ 27
      2.5.3   Galilean Transformations: The Dispersion Relation   30
      2.5.4   The Doppler Effect for Sound   31
2.6   Notes   35

3   **Einstein's Relativity Principle**   **37**
3.1   The Existence of an Invariant Speed c 37
3.2   The Relativity Principle   39
3.3   Curtain-Raiser: Light Clocks   40
      3.3.1   Transverse Light Clock: Time Dilation   40
      3.3.2   Longitudinal Light Clock: Lorentz Contraction   42
3.4   Evidence   43
      3.4.1   Preamble   43
      3.4.2   Testing Ballistic Models: Does Light Speed Depend on the
              Source Velocity? (Low Velocity)   44
      3.4.3   Testing Ballistic Models: Does Light Speed Depend on the
              Source Velocity? (High Velocity)   46
      3.4.4   Testing Aether Theories: Does Light Speed Depend on the
              Velocity of the Apparatus Relative to the Fixed Stars?   47
3.5   Notes   49

Part II   Kinematics   **51**

4   **Lorentz Transformations**   **53**
4.1   Statement of the Transformations   53
4.2   Some Simple Consequences   55
      4.2.1   Coordinate Differences (Intervals)   55
      4.2.2   Time Dilation   55
      4.2.3   Lorentz Contraction   55
      4.2.4   Particle Velocity   57
      4.2.5   Particle Acceleration   60
4.3   Derivation of the Transformations   62
4.4   Evidence: Time Dilation   64
4.5   Notes   65

5   **Invariant Intervals and Space–Time Diagrams**   **67**
5.1   Intervals Timelike, Spacelike, and Lightlike   67
5.2   Space-time Diagrams and the Light Cone   69
5.3   Note   73

6   **Proper Time and Nonuniform Motion**   **74**
6.1   Preamble: Nonuniform Motion   74
6.2   The Proper Time τ 75

6.3   Effects of Acceleration Stresses   78
      6.3.1   Dimensional Estimates   78
      6.3.2   Three Models   79
6.4   Evidence   81
6.5   Notes   82

7   Four-Vectors   **83**
7.1   Definitions   83
7.2   Particle Four-Velocity   86
7.3   Lorentz Transformations in $1+1$ Dimensions   89
7.4   Covariance   90
7.5   Notes   91

8   Four-Acceleration   **93**

Part III   Momentum and Energy   **99**

9   Particle Dynamics: Momentum and Energy   **101**
9.1   The Form-Invariant Conservation Laws: Redefinition of
      Momentum and Energy   101
9.2   Basics of the Energy-Momentum Four-Vector   105
      9.2.1   Finite-Mass Particles   105
      9.2.2   Zero-Mass Particles   107
      9.2.3   Lorentz Transformations   109
9.3   Evidence   110
      9.3.1   Energy as a Function of Speed   110
      9.3.2   The Cockroft–Walton Reaction   111
      9.3.3   Positron Annihilation   113
      9.3.4   Why Was Mass Thought to be Conserved in Chemistry?   115
9.4   Notes   115

10   Natural Units, and the Prevalence of MeV   **117**
     Notes   121

11   Systems of Particles: Four-Momentum Conservation using Invariants   **122**
11.1 The Simplest Example: Two-Body Decays   123
      11.1.1   The Obvious Method   123
      11.1.2   The Obvious Method Improved   124
      11.1.3   Four-Vectors: Segregate and Square   124
11.2 Several Free Particles Treated as One   125
11.3 Production Thresholds   126
      11.3.1   Stationary Target   127
      11.3.2   Colliding Beams   128
11.4 The Compton Effect   129
      11.4.1   Theory   129
      11.4.2   Evidence   131
11.5 Elastic Scattering   132

11.6 Discrete Masses and Continuous Masses   136
    11.6.1 Discrete Masses: Bound States and Elementary Particles   136
    11.6.2 Continuously Variable Mass   136
    11.6.3 The Relativistic Rocket   137
11.7 Notes   140

Part IV   Waves   **143**

12   Plane Waves   **146**
12.1 Introduction   146
12.2 Lorentz Transformations: The Four-Vector $\vec{K} = (\omega/c, \mathbf{k})$ 146
12.3 Group Velocity   148
12.4 Phase Velocity   150
    12.4.1 Collinear Motion   150
    12.4.2 Light Speed   150
    12.4.3 Phase Velocity and Refractive Index in the General Case 151
12.5 Evidence from the Fizeau Effect: The Refractive Index of a
      Moving Medium   153
    12.5.1 Introduction   153
    12.5.2 The Experiment   153
    12.5.3 Why is the Fizeau Effect not Second-Order Small?   156
12.6 Quantum Mechanics: The Wave Functions of Free Particles   157
12.7 Notes   159

13   Light Waves in Empty Space: Aberration and Doppler Effect **160**
13.1 Notation and Agenda   160
13.2 Aberration, from First Principles   161
    13.2.1 Theory   161
    13.2.2 Evidence: Stellar Aberration   161
    13.2.3 The Galilean View: Euler's Paradox   164
13.3 The Collinear Doppler Effect from First Principles   164
13.4 The Method of Invariants   167
13.5 Evidence: Doppler Shifts of Spectral Lines   170
    13.5.1 First-Order Shift   170
    13.5.2 Second-Order Shift   172
13.6 Notes   174

Appendices
A   Lorentz Transformations with Arbitrary Relative Velocity   **175**
    Note   177

B   Vectors, Four-Vectors, and Transformation Matrices   **178**
    B1   What is a Vector? (Definition and Representations)   178
    B2   Rotation Matrices   181
    B.3 Four-Vectors   182
    B.4 Three Space Dimensions   185
        B.4.1   Rotations   185
        B.4.2   Lorentz Transformations and Thomas Precessions   186
    B.5 Notes   187

C   Motion Under Given Forces   **180**
   C.1 Newton's Second Law   180
   C.2 Examples   189
       C.2.1 Motion Parallel to a Constant Homogeneous E Field   189
       C.2.2 Motion in a Constant Homogeneous B Field   190
       C.2.3 Motion in a Coulomb Field   191
   C.3 Notes   193

D   Wave Equations   **195**
   Notes   197

E   Black-Body Radiation: The Lorentz Transformation of Planck's Law **198**
   E.1 The Problem and its Solution   198
   E.2 The Proof that $f'(\mathbf{p}') = f(\mathbf{p})$ 201
   E.3 Applications   202
   E.4 Notes   204

Problems   **205**

Index   **226**

*This book could not have evolved
but for refuge and courtesy in England
and the company of Jenifer*

Preface

Too often, the relativity principle is introduced as if it were a recent discovery, yet with details and applications strongly coloured by the circumstances of its revival almost a century ago. I take the contrary view on both counts: namely that the principle and its implications are better taught as a classic thread running right through physics, but that modern evidence is generally preferable to old, because it is more direct and more precise. Indeed I believe that ideally one should not treat elementary relativity as a free-standing topic at all. This book proposes a compromise: if practicalities mandate such a course, then it should at least observe various decencies, narrow and wide.

First, the relativistic ideas most productive in the real world centre on form-invariance; it would be crass to ignore them, or to obscure them through lack of exercise. Their application requires four-vectors and four-scalar products, amounting to a nontrivial but clearly bounded minimum of formalism, perfectly manageable if introduced early enough and then exploited as a matter of course. Second, no idea can be assimilated without much critical reflection, here least of all, and it would be disingenuous to pretend that mere instruction can short-circuit this; but many of the essentials can be thought through in the familiar context of Newtonian physics, well below light speed. Third, for the philosophically vulnerable one should keep in plain view that relativity was made for physics, not physics for relativity. This is another reason for a leavening of examples fit to stimulate curiosity in their own right: like other principles, the relativity principle is difficult to appreciate if perceived as largely decorative, rather than as answering real needs. Conversely, the indispensable modicum of hard work should require only tools already mastered; this rules out electromagnetism, which as a relativistic field theory demands sophistication that reasonable syllabuses would not expect until later. Fourth, in relativity very short-term strategies are unproductive: logical distinctions are better explained, and untruths are better avoided, even at the cost of marginally retarding the argument here and there.

Like other actors, teachers are accountable both to the play and to their audience, with its requirements present and future; it is fortunate that we have also, in season, a theatre and the rest of the cast. My own debts by now are untraceable: countless questions, discussions, and printed pages, pro and con, have coalesced beyond individual acknowledgement. They add up to an obligation impossible to meet, and barely capable of diminution through yet another textbook. If I make no apology for this one, it is only because there cannot at any level be a definitive text on relativity, any more than there can be a definitive performance of, say, *Love's Labour's Lost*.

Gabriel Barton
Kingston, Lewes
December 1998

Part I
Introductory

1 Preliminaries

The relativity principle highlights certain simple features evident most readily in mechanics, but found to be characteristic of physics as a whole. It concerns the special class of *inertial reference frames*,[1] namely those where a body free of external influences moves with constant velocity, conformably to Newton's first law of motion, also known as the law of inertia. Roughly speaking, the principle asserts that the velocity of one such frame with respect to another can be determined only by direct comparison between them, but not by observations internal to any one such frame. Thus the principle denies physical significance to uniform motion in any absolute sense, a denial to which it presumably owes its name. Alternatively but equivalently, it asserts that the laws of nature are described by equations of exactly the same form, equally simple or equally complicated, with respect to all inertial frames, privileging no such frame over another.

One must keep in mind that the relativity principle is not a principle of logic or a mathematical deduction, but a summary of observed facts. What might have been debatable is how much importance one should attach to the particular regularity in the facts that is emphasized by the principle: should one choose to view it as a secondary consequence of the various equations governing the various branches of physics, or as one of the axioms on which it is convenient to erect physical theory in general? In Newtonian physics[2] the principle was long disregarded, or counted as a liability more than an asset;[3] in any case it lay largely fallow until brought into sharp focus by an apparent threat from the fact that Maxwell's electrodynamics predicts light to travel through vacuum at a definite speed, call it $c = 2.9979 \times 10^8$ m/s, yet fails to specify the reference frame relative to which the speed should have this value. What observations show is that signals moving with speed c relative to one inertial observer move with the same speed relative to all, an impossibility given the Newtonian/Galilean notions of space and time. Thus the facts and the Maxwellian prediction obey the relativity principle; but they force a revision of the earlier views that physics had taken of kinematics, i.e. of space, time, and velocity. It then turns out that comparably drastic changes are needed also in dynamics. In systems with typical velocities $v \ll c$, the differences between the old "nonrelativistic" and the new "relativistic" predictions are as a rule only of relative order $(v/c)^2$; but they become overwhelming when speeds become comparable to c.

The revision, the work of Einstein nearly a hundred years ago, is a well-understood and undisputed part of classical physics: it constitutes what is commonly called the (special) theory of relativity, and is the subject of this book. For the glamour so widely attached to it there are probably two main reasons. First, it refines our everyday notions of time, showing in particular that two events simultaneous with respect to one observer may not be so with respect to another. Second, it dominates the dynamics of elementary particles in modern high-energy physics, and the energy changes in nuclear reactions, with their applications military and civil.

1.1 Aims

This book is meant to enable you to:

- appreciate the privileges of inertial reference frames, the Galilean transformations between observations referred to different inertial frames, and how the transformations are used to exploit the relativity principle in Newtonian physics;
- understand how the principle, when combined with the observed invariance of c, forces one to replace the Galilean by the Lorentz transformations;
- handle the kinematic implications regarding time-dilation, the Lorentz contraction, and the velocity-combination rules;
- understand the consequent changes in the definitions of momentum and energy;
- apply these ideas to solve simple problems on the collisions, decay, and production of particles; and
- understand and apply the Lorentz transformation rules for plane waves generally, and more specifically for waves travelling at speed c through empty space, electromagnetic waves being the prime example.

We do not consider electromagnetic theory, even though the Einsteinian revision was born of electromagnetism, and even though Maxwell's theory is fully relativistic, and by far the most powerful such theory yet known. Electromagnetism is omitted because this introduction is meant to be accessible well before students acquire the techniques that can show them the advantages of taking an explicitly relativistic view of electromagnetic fields. One can live with the omission, because the fact central to the theory is that a finite invariant speed c exists, rather than that, as far as we can tell, electromagnetic waves in empty space happen to travel at this speed.

1.2 Prerequisites

The prerequisites from physics are (a) elementary mechanics, from Newton's laws of motion to two-particle systems; (b) energy and momentum conservation applied to two-particle and to many-particle systems; and (c) plane waves (sound or light), up to the notions of amplitude, phase, phase-velocity, refractive index, and group velocity. For evaluating some of the evidence, one needs also (d) the rudiments of standing waves and interference.

The prerequisites from mathematics are minimal: (e) vectors and their scalar products; (f) linear and quadratic equations; and (g) very occasionally an integral and an elementary ordinary differential equation. In the important Chapter 6, on particle trajectories and proper times, one needs to understand what a line integral is. Matrices and matrix products are needed only in Appendix B.

However, the student of relativity should be alerted from the outset to the notorious contrast between the simplicity of its mathematics and the difficulty of assimilating its concepts. Experience suggests that once the ideas have been stated, learners gain little from repetition or paraphrase: the ideas can be mastered only by prolonged reflection, self-questioning, and problem-solving. On the other hand, once mastered, the ideas are easy to apply. *If your attempt on a problem from this book runs into long or difficult algebra, the fault is likely to be insufficient thought in choosing your approach rather than insufficient mathematical skill.*

1.3 Organization

The material is assembled as follows.

Chapter 2 formulates the relativity principle and discusses its status; identifies the Galilean transformations whereby the principle can be exploited in Newtonian physics; and illustrates its uses there, (a) because this is the best way to familiarize oneself with many of the essential technicalities, which recur almost or altogether unchanged post-Einstein; and (b) because the ideas are important to Newtonian mechanics in its own right. Any topics in Chapter 2 may be skipped if they are already familiar from elementary mechanics courses.

In Chapter 3 we take note of the fact that the speed c is invariant, and of the incompatibility of this fact with Galilean transformations. Then we derive time-dilation and the Lorentz contraction directly from the fact, as a curtainraiser.

Serious study of Einsteinian relativity starts with the Lorentz transformations, in Part II. In Chapter 4, the Lorentz transformations[4] are, to begin with, merely asserted, and then applied to kinematics systematically, so as to yield in particular the three chief kinematic results, namely the two already found in Chapter 3, plus the velocity-combination rule for particles. The transformations are derived in Section 4.3, using as input Newton's first law, the relativity principle, some basic empirical properties of space, and the existence of an invariant speed c. Only right at the end does one need to say whether c is finite or infinite: the same reasoning yields the Galilean transformations if c is infinite, and the Lorentz transformations (i.e. Einsteinian physics) if c is finite. A logically self-consistent theory results in either case: the choice must be made on grounds not of self-consistency but of fact.

Chapters 5 and 6 explore general properties of the Lorentz transformations of the time and position coordinates $t, \mathbf{r}$ of an event. In particular they introduce the famous space–time diagrams for visualizing such coordinates;[5] identify the combination $c^2 t^2 - r^2$ as in invariant; and define the crucial notion of the proper time τ of an arbitrarily moving particle having position $\mathbf{r}(t)$. Chapters 7 and 8 then introduce four-vectors, starting with $\vec{X} \equiv (ct, \mathbf{r})$, and with the four-velocity $\vec{U} \equiv d\vec{X}/d\tau$. Four-vectors generalize the notion of ordinary three-dimensional vectors to relativisic kinematics, and play the same essential role of simplifying concepts and easing calculations. Vectors and four-vectors are discussed somewhat more formally in Appendix B, emphasizing what they have in common.

This far one has been concerned only with kinematics. Part III (Chapters 9 to 11) discusses particle dynamics as far as the Einsteinian redefinition of energy ε and momentum $\mathbf{p}$; the four-vector $\vec{p} \equiv (\varepsilon/c, \mathbf{p})$; and the conservation law for $\vec{p}$. This is the physics that governs the energy release in nuclear reactions, and the collisions and transmutations whereby one explores the structures of elementary particles. On the other hand, the motion of particles under given forces is discussed only briefly, in Appendix C, because a thorough treatment would need more electromagnetic theory than we wish to assume.

Finally, Part IV deals with the Lorentz transformation of plane waves, Chapter 12 in general and Chapter 13 with light waves in empty space. Waves are left till last because concepts, technicalities, and experimental tests are all more elaborate than for particles. The wave equations that some of these plane waves satisfy are considered only in appendix D, very briefly, and with an eye particularly to the wave functions of quantum mechanics. Finally Appendix E derives the Lorentz transformation of the Planckian black-body photon spectrum. This is a surprisingly delicate problem, but interesting in itself, and important because it serves to determine our motion relative to the cosmic background radiation, which is absolute motion in the cosmological sense.

Experimental evidence is of the essence thoughout, as in science it always is. But it requires particular emphasis in Einsteinian relativity where (as in quantum mechanics) input assumptions, analysis, and predictions fall well outside the range of common experience, and of the dangerously deceptive so-called common sense that it engenders. Since the theory though old is closely meshed into contemporary physics, we cite modern experiments, apart from a few exceptionally direct, elegant, or entertaining historical firsts.

However, our necessarily limited selection of evidence may mislead the beginner in another way, by obscuring the fact that major and long-established theories are assessed by their successes and failures over a wide range of experience, rather than on the strength of just a few supposedly critical tests. Conversely, one worries little about the lack of direct evidence for particular predictions however dramatic (e.g. for the Lorentz contraction), provided they are clear logical consequences of a theory well warranted by other means. Spectacular direct tests of basic features, input or output, are rarely if ever crucial to the acceptance of a theory, and play this role mainly in folklore. Nevertheless, though usually performed quite late in the day, they do have functions vital to the hygiene of any discipline.

First, they are a dialectical courtesy to nature, giving her a chance to countermand the theory through a single discrepancy beyond statistical uncertainty and experimental error. In such circumstances of course one never does find root and branch discrepancies: they would have surfaced long before the theory came to be taken seriously. That is just what brings one to the second function of such tests, that they explore the limits of validity of the theory. For instance, Newtonian physics was and remains adequate until questioned by speeds comparable to the invariant speed c: but its precise limits of validity emerge only after the Einsteinian revision. And the relativity principle itself ceases to apply unconditionally in gravitational fields, where inertial frames can be identified only in limited regions or to limited accuracy: to clarify these conditions one needs the general theory of gravitation, a point to which we return below, and again in Chapter 3. Alternatively, imagine that a future test like those described in Section 3.4 were to find a weak dependence of light speed on the speed of the source or of the laboratory, of a kind not explainable by known gravitational fields. The dependence would certainly be weak, in virtue of evidence already in; and while it would entail radical innovations, the new theory would necessarily reduce to the old in some appropriate limit, because otherwise it would be manifestly wrong. It is in just this way that quantum mechanics, once established, is seen to contain classical mechanics as a limiting case, albeit under mathematical conditions that were certainly not foreseeable and have in fact turned out to be highly sophisticated. In other words, good theories hardly ever die, nor do they even fade away: one merely learns when not to use them.

Minimal reading. The book holds more, by perhaps a third, than one would wish to absorb on a first reading. A minimal selection runs as follows: Sections 2.1–2.4; Chapters 3, 4, and 5; Sections 6.1, 6.2, and 6.4; Sections 7.1, 7.2, and 7.4; Chapters 9 and 10; Sections 11.1–11.4 and Section 11.6.3; Sections 12.1, 12.2, 12.5, and 12.6; and Chapter 13. Appendix B, Sections B.1 to B.3, has been found helpful by some and a burden by others: readers should take them or leave them, according to taste.

Small print may be skipped freely, and is needed at most for other small-print sections. As a rule it covers certain niceties, and certain mathematically more demanding arguments.

1.4 History

Historical accounts of the genesis and the repercussions of the Einsteinian revision must be treated with extreme caution. Those written before 1960 or so were almost invariably wrong; many still are; and the only safe course regarding popular accounts and marginal comments in textbooks is to ignore them. Since the sixties the record has been set straight, as far as the evidence allows. It is best traced through the following references:

G. Holton, *American Journal of Physics*, volume **28** (1960), page 627, and volume **37** (1969), page 968.

G. Holton, *Thematic origins of scientific thought*, chapters 6 to 8, Harvard University Press, Cambridge, Mass., 1988.

A. I. Miller, *Albert Einstein's theory of relativity*, Addison-Wesley, Reading, Mass., 1981.

1.5 Further Reading

There are countless textbooks and textbook chapters on relativity, good and bad: recommendations are bound to reflect personal views and preferences.

For the relativistic aspects of electromagnetism, which we do not discuss at all, see particularly

R. P. Feynman, *The Feynman lectures*, volume 2, chapters 25 to 28; Addison-Wesley, Reading, Mass., 1964.

All aspects of the theory at the next level of sophistication are lucidly covered by

R. P. Feynman, *The Feynman lectures*, volume 1, chapters 15 to 17, and volume 2 as above.

W. Rindler, *Introduction to the special theory of relativity*, Clarendon Press, Oxford, 1982.

W. Pauli, *Theory of relativity*, Macmillan, New York, 1958, sections 1 to 6, 24 to 29.

L. D. Landau and E. M. Lifshitz, *The classical theory of fields*, 4th edition, chapters 1 to 3; Pergamon, Oxford, 1975.

W. K. H. Panofsky and M. Phillips, *Classical electricity and magnetism*, second edition, chapters 15 to 17; Addison-Wesley, Reading, Mass., 1962.

(The chapters cited from last two books contain excellent and very workmanlike introductions to special relativity all-round.)

Most of Einstein's basic paper, "On the electrodynamics of moving bodies", can be studied perfectly well, and very profitably, starting from much the same level as the present book. A just adequate English translation appears in *The principle of relativity*, Dover publications, New York. Originally the paper appeared (in German) in the *Annalen der Physik*, volume **17** (1905), page 891.

1.6 Context

We are concerned with what is usually called the *special theory of relativity*, in supposed contrast to the so-called *general theory of relativity*. It can be argued that these are yet another pair of crass misnomers; that the general theory of relativity is not a theory of relativity at all, but a general theory of gravitation; and that the special theory is not a special case of the general, but the only relativity theory there is.[6] However, the general theory of gravitation is crucial to cosmology, and reveals the conditions under which the relativity principle applies.

(a) Viewed against the background of the general theory of gravitation, the theory of relativity is a *local* theory, and the relativity principle applies only locally, to an approximation valid within known limits. What this means is, roughly, as follows. In inhomogeneous gravitational fields Newton's first law holds only over limited regions of space, and the inertial frames identifiable well within one such region need not move uniformly with respect to those identifiable in another. Thus inertial frames exist only in an approximate sense, and the relativity principle, which is a statement about such frames, is an approximation in exactly the same sense, along with the theory derived from the principle. On a cosmological scale, the relativity principle fails altogether: the universe certainly supplies absolute standards of rest, with respect to the fixed stars, or to the cosmic background radiation, or to the cosmic mass distribution averaged over very large distances.[7] But measurable departures from relativity theory appear already on ordinary terrestrial scales, through the dependence of proper times on gravitational potential, additionally to their relativistic dependence on speed (see, for instance, Example 6.2).

(b) Even in the restricted regions where the relativity principle holds, it constrains only the form of the equations, but not the particular physical systems to which the equations may be applied. Two examples make the point. To the high-enegy physicists dealing with collisions between particles, such thoughts never even occur: to them, the laboratory and the centre-of-momentum frames for instance (see Sections 11.3 and 11.5) have completely equal status. But to describe the dynamics of electrons in interstellar space, a privileged role might very sensibly be assigned to the frame relative to which the cosmic-background black-body radiation is isotropic (equally intense in all directions). Although the equation of motion of the electrons knows nothing about this particular frame, the electrons certainly do: instead of moving uniformly, they will eventually be brought to rest relative to this frame (and this frame alone) by frictional forces due to the radiation. On the other hand, these forces are weak, and over limited time-scales they may be negligible. Just how much importance the relativity principle retains in such problems is evidently a question of taste rather than of dogma.

(c) Taken together, (a) and (b) reveal a major unsolved problem. Under conditions where the relativity principle does apply, it turns out that the inertial frames are those that move uniformly with respect to the fixed stars. This is an observed fact; it can hardly be a coincidence. It might be explained if the inertia of every particle were due to some interaction between that particle and all others masses in the universe, most of which are very distant. The idea that this should be the case is referred to as Mach's principle; but it can be called a principle only by courtesy, since there is neither experimental nor theoretical evidence to support it. Fortunately for us, the question has no bearing on the relativity principle, which takes Newton's first law, inertia, and the existence of inertial frames as facts, without needing or seeking to explain them.

1.7 Terminology

- The reader must remain alert to the difference between *velocity*, which is a vector, and *speed*, which is the magnitude of a velocity regardless of its direction. For instance, Lorentz transformations leave speeds c unchanged, but inevitably change velocities whatever the speeds.
- *Varieties of equality.* In the process of introducing new concepts one frequently needs to introduce new mathematical expressions; this makes it important to distinguish between equality by definition and equality resulting from proof or calculation. Equality by definition is written $a \equiv b$, and can serve to define either a or b, the other being known. For instance, given a particle of mass m and velocity $\mathbf{u}$, Newtonian physics defines its momentum as $\mathbf{p} \equiv m\mathbf{u}$. The symbol $\equiv$ reminds the reader not to spend time looking for a reason for this equality, except perhaps for the hindsight that makes such a definition of $\mathbf{p}$ convenient. By contrast, the kinetic energy of the particle is introduced by $k \equiv p^2/2m = mu^2/2$, where $\equiv$ signals a definition, and $=$ equality in virtue of the already known relation between $\mathbf{p}$ and $\mathbf{u}$.

  The symbol $a = b$ denotes exact equality between a and b. Approximate equality is written as $a \simeq b$ or $a \approx b$; the discrepancy that these symbols allow between a and b depends on the context. Usually it is obvious; for instance, for speeds v much smaller than c, one might write $\sqrt{1 - v^2/c^2} = 1 - v^2/2c^2 + \ldots$, meaning only to indicate the start of the binomial series for the square root; or one might write $\sqrt{1 - v^2/c^2} \simeq 1 - v^2/2c^2$, allowing a discrepancy of order v^4/c^4.

  Agreement guaranteed only between orders of magnitude is written as $a \sim b$, often as a rougher version of $a \simeq b$ (sometimes regardless of sign). For example, the Einsteinian rule for the resultant u of two parallel velocities u' and v reads

$u = (u' + v)/(1 + u'v/c^2)$, and one might wish to stress the fact that $(u' \sim c$ and/or $v \sim c) \Rightarrow u \sim c$.

- The double arrow in $A \Rightarrow B$ stands for logical implication: if A is true, then B is true, as in the example just quoted. (The symbol $\Rightarrow$ must not be confused with the single arrow: $a \to b$ is a limit in the elementary mathematical sense. For instance, expressing the resultant u as a function $u(u', v)$, one writes, interchangeably, $u(u', v \to c) \to c$, or $\lim_{v \to c} u(u', v) = c$.)

- The symbol ■ marks the end of an argument, of a proof, or of a worked example.

1.8 Notes

1. Perforce, this introduction uses some expressions ahead of precise explanation or definition, which will follow in later chapters. Meanwhile it suffices to think of a reference frame as a facility for recording measurements of position and time. *Inertial observers* are simply observers using inertial reference frames.

2. We speak of physics before Einstein as *Newtonian* or *Galilean*, using these names almost interchangeably. Physics taking account of Einstein's revision we call *Einsteinian*. No other historical implications should be read into these names. Both Newtonian and Einsteinian physics respect the relativity principle, though they implement it on the basis of different ideas about how the perceptions of one inertial observer relate to those of another. Unfortunately, pre- and post-Einstein physics are generally called *nonrelativistic* and *relativistic* respectively; these descriptions though utterly misleading are past remedy, and we make no sustained attempt to avoid them.

3. Presumably because it accepts and explicates a contradiction to one of the most quoted (and one of the very few readily understood) statements by Newton himself: Newton viewed the detectable motions of bodies relative to each other as merely secondary reflections of their motions referred to absolute although unascertainable space and time.

4. Most of this book is restricted, without significant loss of generality, to Lorentz transformations between reference frames with their corresponding axes parallel, and their relative velocity along the x axis. For completeness, Appendix A deals with relative velocities in other directions.

5. If nothing else has, these diagrams should dispose of the widely-touted nonsense that relativity somehow abolishes the differences between space and time.

6. Today this view is shared by, probably, a small majority of physicists; but there are and always have been many distinguished dissenters, including Einstein himself.

7. As near as makes no difference here, all these criteria are equivalent. The long-distance averages appear to be the same everywhere and in all directions: space remains, on average, homogeneous and isotropic. In other words, on these scales position remains relative while motion becomes absolute. Time of course is absolute in view of the evolution of the universe.

2 The Relativity Principle, and its Applications in Newtonian Physics

2.1 Reference Frames and Transformation Rules

To describe a physical system[1] one needs a *reference frame*, call it S, meaning a system of coordinates for determining positions, plus a clock, or several conveniently situated and mutually synchronized clocks, for determining times. Every *observer*, i.e. every physicist, is equipped with such a frame: the symbol S is used for observer and frame interchangeably. Different observers (and their frames) are distinguished by primes: much of what follows concerns observers S and S' comparing their respective descriptions of a given system.

Descriptions with respect to[2] S include the time and position coordinates $(t, \mathbf{r})$ of *events*, like the coincidence of a moving particle with a fixed benchmark, the collision between two particles, or the spontaneous disintegration of a radioactive particle into two or more different particles; equally they include the positions, velocities, and accelerations of the particles at arbitrary times t, specified by functions $\mathbf{r}(t)$, $\mathbf{u}(t) \equiv d\mathbf{r}/dt$, and $\mathbf{a}(t) \equiv d\mathbf{u}/dt$. Generally we use Cartesian coordinates $\mathbf{r} \equiv (r_1, r_2, r_3) \equiv (x, y, z)$. If the coordinates of an event relative to one frame S are (t, x, y, z), then the coordinates of the same event relative to a different frame S' are written as (t', x', y', z'), with primes used similarly for other variables. The coordinate origins are called $O \equiv (x = 0, y = 0, z = 0)$ and $O' \equiv (x' = 0, y' = 0, z' = 0)$. The "laws of nature", a jargon phrase for the rules governing the behaviour of physical systems, are equations connecting variables all referred to one frame: for example, Newton's first law, also called the law of inertia, asserts that relative to a certain special class of observers (inertial observers) to be discussed presently, a free particle, i.e. a particle not subject to any external influences, obeys

$$\mathbf{a} \equiv d^2\mathbf{r}/dt^2 = 0 \quad \Leftrightarrow \quad [\mathbf{r} = \mathbf{A} + \mathbf{B}t, \text{ with } \mathbf{A} \text{ and } \mathbf{B} \text{ constant}]. \tag{2.1}$$

Of course, different such observers S and S' will give different descriptions of any given system: for instance, to the trajectory in (2.1), S assigns constants $\mathbf{A}$ and $\mathbf{B}$, while S' will assign constants $\mathbf{A}' \neq \mathbf{A}$ and $\mathbf{B}' \neq \mathbf{B}$. (A physical quantity whose value is the same relative to all inertial frames is said to be *invariant*, or *an invariant*. The charge of the electron is a classic example.)

Moreover, arbitrarily chosen observers S and S' will express the laws of nature by means of equations assuming different *forms*. For example, suppose that the coordinate frame of S' is fixed to a turntable having constant angular velocity $\mathbf{\Omega}$ with respect to S, with their origins O and O' coinciding on the rotation axis. If for S the law of inertia indeed takes the familiar form $\mathbf{a} = \mathbf{0}$, then for S' it will take the very different form

$$\mathbf{a}' \equiv \mathrm{d}^2\mathbf{r}'/\mathrm{d}t' = -\mathbf{\Omega} \times (\mathbf{\Omega} \times \mathbf{r}') - 2\mathbf{\Omega} \times \mathbf{u}', \qquad (2.2)$$

where the first (position-dependent) term represents the centrifugal and the second (velocity-dependent) term the Coriolis effects. *Therefore in mechanics one must ask right at the outset whether certain reference frames are privileged, in the sense that with respect to them the laws of nature are simpler in form than they are with respect to other frames.* We shall see presently that the answer turns out to be yes: observations identify a privileged set of frames called *inertial frames*, with respect to all of which the law of inertia takes the form (2.1). Loosely speaking, the relativity principle asserts the further fact that there is no experimental basis for preferring any inertial frame over any other inertial frame.

To identify the class of inertial frames, and to compare different inertial frames with each other, one needs the rules, called *the transformation laws*, that allow us, from the results of observations referred to one such frame, to determine the results of the same observations referred to another. These are the tools required to formulate and to exploit the relativity principle.

This chapter introduces the principle and applies it to Newtonian physics, where the transformation rules are called *Galilean transformations*. In Chapter 3 the same strategy and essentially the same principle will be adapted to Einsteinian physics, where the Galilean are replaced by the Lorentz transformations. But it proves more productive to study the Newtonian/Galilean applications in their own right first, without referring forwards: comparisons are more illuminating retrospectively. The difference of input into the two scenarios is that speeds are invariant in Newtonian physics if they are infinite, and in Einsteinan physics if they are equal to c. This explains the traditionally different expositions: the Galilean transformations are everyday-familiar, whence many texts start with them, and relegate invariant speeds to an afterthought; whereas the invariance of c (i.e. the truth) is un-intuitive, whence Einsteinian physics always starts by stressing this invariance, and only then derives the Lorentz transformations. Though such different distributions of emphasis might tend to mislead, the danger should be obviated by Section 4.3, where Galilean and Lorentz transformations are derived not just in parallel but by practically the same reasoning.

It is natural to start with kinematics, and to deal with dynamics afterwards. *Kinematics* merely *describes* motion: it deals with position, time, and with the time-derivatives of position: roughly speaking it extends geometry to include time. By contrast, *dynamics* is concerned with *explaining* motion: it deals with momentum, force, and energy. Thus Newton's second and third laws of motion belong squarely to dynamics. On the other hand, in appearance at least the first law, equation (2.1), is purely kinematic, and we shall handle it as such; though in strictly Newtonian terms the choice is open to argument.

2.2 Galilean Transformations: Kinematics

2.2.1 Time Coordinates

It is fundamental to Newtonian physics that the times assigned to a given event by different observers are the same:

$$t = t', \tag{2.3}$$

provided of course that both observers use the same time-scale (i.e. identically constructed clocks), and choose the same time-origin (i.e. the same reference event to mark time zero). This is not as obvious as it might seem (in fact, i.e. in Einsteinian physics, it is false), because it is not obvious how an observer determines the time of an event happening somewhere else. (She has no problems with events that coincide with her in both space and time: she simply looks at her clock when they happen.) Hence (2.3) must be justified in terms of the mechanisms available for assigning times to events that happen at a distance. The argument depends on Newtonian dynamics admitting the possibility of signals that move arbitrarily (as if infinitely) fast, taking effectively zero time to cross arbitrarily long distances.[3] First, knowing that such signals leave and arrive effectively at the same time, two separated observers can verify that their clocks are and remain synchronized: a signal is emitted at each tick of, say, the clock at O, and the clock at O' is adjusted so that it ticks in step with the arrival of the successive signals. Second, if interesting events anywhere are immediately signalled in all directions (say by pre-arrangement), then each observer can assign to every event the time registered by her own clock when the signal arrives.

To the accuracies allowed by Newtonian physics, light pulses are often acceptable practical substitutes for the infinite-speed signals we have contemplated in principle: when all pertinent speeds are far below c and distances not too large, the travel times of signals that do travel at speed c are generally negligible.

2.2.2 Observers in Uniform Relative Motion

Consider now the frames S and S' shown in Figure 2.1, where O' moves with constant velocity **v** relative to O. For simplicity we take the trajectory to be such that O and O' meet, and their clocks to be set so that the time of meeting is zero for both: in other words their meeting is an event with coordinates $(t, \mathbf{r}) = (0, 0) = (t', \mathbf{r}')$. We look for the *transformation rules of event coordinates*, i.e. for the relation betwen the coordinates $(t, \mathbf{r})$ and $(t', \mathbf{r}')$ of an arbitrary event.

The time coordinates are equal as explained above. Inspection of the figure, amounting to ordinary vector addition, shows that the space coordinates are related

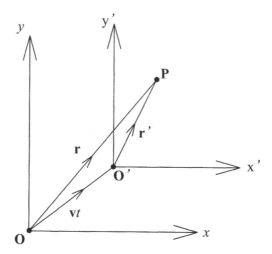

Figure 2.1　Two coordinate frames S and S', shown in the special case where (a) the origins coïncide at time zero, and (b) corresponding primed and unprimed axes are parallel. The relation between the primed and the unprimed coordinates of P can be read on sight: $\mathbf{r}' = \mathbf{r} - \mathbf{v}t$.

by $\mathbf{r}' = \mathbf{r} - \mathbf{v}t$. We display the direct and the inverse transformations in full:

$$\mathbf{r}' = \mathbf{r} - \mathbf{v}t, \qquad t' = t, \tag{2.4}$$

$$\mathbf{r} = \mathbf{r}' + \mathbf{v}t, \qquad t = t'. \tag{2.5}$$

Mathematically speaking, the inverse (2.5) is derived by treating the direct transformation (2.4) as equations to be solved for the unprimed variables, with the primed variables regarded as known. (Though here the algebraic solution is trivial, it will not always be so.) Equivalently, each transformation is obtainable from the other by the following prescription:

interchange primed and unprimed variables, and reverse the sign of $\mathbf{v}$.　(2.6)

This is obvious: (i) by definition, $\mathbf{v}$ is the velocity of S' relative to S, whence the velocity of S relative to S' is $-\mathbf{v}$; (ii) however, it is purely a matter of choice which frame carries the primes; (iii) hence we are free to reverse our choice (i.e. to transfer the primes) if we also replace $\mathbf{v}$ by $-\mathbf{v}$. For exactly the same reasons, *the prescription (2.6) applies to all other transformations we shall meet, Galilean or Lorentz.*

We say that the two frames are in *standard configuration*[4] if their corresponding axes are parallel, with the relative velocity along the x (hence also along the x') direction: see Figure 4.1. Then the Galilean transformations reduce to

$$t' = t, \quad x' = x - vt, \quad y' = y, \quad z' = z; \tag{2.7}$$

$$t = t', \quad x = x' + vt, \quad y = y', \quad z = z'. \tag{2.8}$$

It will be understood throughout the book that S, S' *are inertial frames in standard configuration, unless we say otherwise.*

The Galilean transformations between particle velocities[5] $\mathbf{u}, \mathbf{u}'$ *follow at once if we regard* $\mathbf{r}, \mathbf{r}'$ as the coordinates of a particle, and consider position changes in a small time interval. From the definition of velocity, $\delta\mathbf{r}' = \mathbf{u}'\delta t'$ and $\delta\mathbf{r} = \mathbf{u}\delta t$; from (2.4) one has $\delta t' = \delta t$ and (remembering that $\mathbf{v}$ is not a variable) $\delta\mathbf{r}' = \delta\mathbf{r} - \mathbf{v}\delta t = (\mathbf{u} - \mathbf{v})\delta t$; and taken together these relations yield $\delta\mathbf{r}' = \mathbf{u}'\delta t = (\mathbf{u} - \mathbf{v})\delta t$, whence finally

$$\mathbf{u}' = \mathbf{u} - \mathbf{v}, \qquad \mathbf{u} = \mathbf{u}' + \mathbf{v}, \tag{2.9}$$

so that in standard configuration

$$u_x' = u_x - v, \quad u_{y,z}' = u_{y,z}, \qquad u_x = u_x' + v, \quad u_{y,z} = u_{y,z}'. \tag{2.10}$$

As always, the inverse is obtainable directly from the prescription (2.6).

These transformations (particularly the inverse) can be thought of equally well as the familiar (Galilean) *velocity-combination rule*, answering the following question: the velocity of a body relative to S' is $\mathbf{u}'$; the velocity of S' relative to S is $\mathbf{v}$; what is the velocity $\mathbf{u}$ of the body relative to S? In this sense, $\mathbf{u}$ is the result of combining $\mathbf{u}'$ with $\mathbf{v}$, and from this point of view it seems obvious why $\mathbf{u}$ is symmetric in the two: the result of combining $\mathbf{u}'$ with $\mathbf{v}$ is the same as the result of combining $\mathbf{v}$ with $\mathbf{u}'$.

The velocity-combination rule so understood plays yet another role when several Galilean transformations are applied in succession. Let frame S' have velocity $\mathbf{v}_1$ relative to S, and S'' have velocity $\mathbf{v}_2$ relative to S'. To find the transformation from S to S'' one proceeds in three steps. (i) Transform from S to S' by expressing primed in terms of unprimed variables, as above. (ii) Next transform from S' to S'' by expressing double-primed in terms of primed, using the same transformation rules but with $\mathbf{v}_2$ instead of $\mathbf{v}_1$. (iii) Finally eliminate the primed variables between the results of (i) and (ii), getting the double-primed in terms of the primed. Inspection shows that the results are the same as from a single transformation where the velocity of the target frame S'' relative to the original frame S is $\mathbf{v}_{12} \equiv \mathbf{v}_1 + \mathbf{v}_2$. The reader should verify this as an exercise.[6] Knowing the velocity-combination rule one could of course have foreseen the result, and got from S to S'' with far less work.

It is evident that arbitrarily high velocities can be produced by combining sufficiently many finite velocities: for example, given n frames in standard configuration such that the velocity of each is u relative to the one before, the velocity of the last relative to the first is clearly nu. Though possibly obvious, this is important because it is a check on the internal consistency of the theory, which has assumed the possibility of arbitrarily high speeds right at the outset, in order to derive $t = t'$. It is worth stressing also that all finite velocities are indeed changed by Galilean transformations, in the sense that $\mathbf{u}' \neq \mathbf{u}$ unless $\mathbf{v} = 0$. The only velocities that fail to change are those of infinite magnitude (infinite speed): this verifies the claim we made at the outset that *only infinite speeds are invariant under Galilean transformations.*

The Galilean transformations between accelerations $\mathbf{a}, \mathbf{a}'$ *are found similarly, by considering velocity changes in a small time interval. The argument is self-explanatory: $\delta\mathbf{u}' = \mathbf{a}'\delta t' = \mathbf{a}'\delta t$ (from the definition of acceleration); the velocity-combination rule (2.9) yields $\delta\mathbf{u}' = \delta\mathbf{u} = \mathbf{a}\delta t$ (since $\mathbf{v}$ is constant); and combining these one finds

$$\mathbf{a}' = \mathbf{a}. \tag{2.11}$$

In other words acceleration is Galilean-invariant.

It is easily checked that *differences between the position coordinates of simultaneous events* are invariant too. Consider for instance the difference $(\mathbf{r}_2 - \mathbf{r}_1)$ between the simultaneous positions of two particles, or in other words the position of particle 2 relative to particle 1; each coordinate transforms according to (2.4), whence

$$\mathbf{r}_2' - \mathbf{r}_1' = (\mathbf{r}_2 - \mathbf{v}t) - (\mathbf{r}_1 - \mathbf{v}t) = \mathbf{r}_2 - \mathbf{r}_1, \tag{2.12}$$

as claimed. Here the restriction to simultaneous events is essential. To see this consider successive ticks of a clock, two events that are not simultaneous. In the rest frame S' of the clock (the frame where it stays put) the events happen at the same place, i.e. $(\mathbf{r}_2' - \mathbf{r}_1') = \mathbf{0}$. But if S' moves relative to S with velocity $\mathbf{v}$, then evidently $(\mathbf{r}_2' - \mathbf{r}_1') = \mathbf{v} \cdot (1 \text{ second}) \neq \mathbf{0}$.

Differences between velocities are likewise invariant. The proof for $(\mathbf{u}_2 - \mathbf{u}_1)$ is left as an exercise; the result is valuable because it opens a quicker way to (2.11).

Remarkably, there is no such thing as a Galilean transformation rule for angles: how angles transform depends on what they measure.

Example 2.1a.   A rod is at rest relative to S, at an angle θ to the x axis. Find the angle θ' that it makes with the x' axis of S'.

Solution. The angles are determined by the coordinates of the two ends of the rod. Suppose for instance that the rod lies in the xy plane, with its ends at (x_1, y_1) and (x_2, y_2); then $\theta = \tan^{-1}[(y_2 - y_1)/(x_2 - x_1)]$. By the same token, $\theta' = \tan^{-1}[(y_2' - y_1')/(x_2' - x_1')]$. But these coordinate differences are invariants,[7] so that $(x_2' - x_1') = (x_2 - x_1)$, $(y_2' - y_1') = (y_2 - y_1)$, whence $\theta' = \theta$. ∎

Example 2.1b.   Relative to S, a particle moves with speed u in the xy plane, at an angle $\phi = 60°$ to the x axis. What angle ϕ' does its velocity $\mathbf{u}'$ make with the x' axis of S', if $v = u$?

Solution. We start by expressing the angle in terms of the Cartesian components of the velocity, which we already know how to transform. Using (2.10) we first obtain the general formula

$$\phi' = \tan^{-1}\left[\frac{u_y'}{u_x'}\right] = \tan^{-1}\left[\frac{u_y}{u_x - v}\right] = \tan^{-1}\left[\frac{u \sin \phi}{u \cos \phi - v}\right]. \tag{2.13}$$

Then $u = v$, $\sin 60 = \sqrt{3}/2$, and $\cos 60 = 1/2$ yield

$$\phi' = \tan^{-1}\left[\frac{v \sin \phi}{v \cos \phi - v}\right] = \tan^{-1}(-\sqrt{3}) = -\tan^{-1}(\sqrt{3}).$$

Since u_y' is positive and u_x' negative, the angle is in the second quadrant: $\phi' = 180 - 60 = 120°$. ∎

2.3   Inertial Frames and the Galilean Relativity Principle

2.3.1   Inertial Frames

An inertial frame is defined to be a reference frame with respect to which Newton's first law is true. That some such frames exist is an experimental fact, and we now have the tools to find all of them. Clearly, if S is an inertial frame, so that (2.1) is true with respect to S, then the Galilean transformation rule (2.11) shows that $\mathbf{a}' = \mathbf{0}$ with respect to any other frame S' moving with constant velocity **v** relative to S. Hence all such frames are likewise inertial: having found one inertial frame we have found infinitely many, one for every value of **v**. The converse is also true:[8] if S is an inertial frame, then S' cannot be unless it has constant velocity with respect to S.

It is worth spelling out precisely how physics (observation) and mathematics (pure logic) contribute to the identification of inertial frames. If one starts from scratch, then only experiment can decide whether a given frame S is inertial or not; this is done (at least in principle) by observing particles one has reason to think are free, expressing the results in terms of the coordinates $(t, \mathbf{r})$ employed in S, and checking whether they satisfy (2.1). If they do, then S is inertial; if they do not, then it is not. On the other hand, if at least one inertial frame S' has been identified already, then no further experiments are needed: S is inertial if and only if it has constant velocity with respect to S'.

2.3.2 The Galilean Relativity Principle

The relativity principle is a two-part assertion about inertial frames. Part 1 is exactly the same in Newtonian and in Einsteinian physics; part 2 is different. Here we state part 2 in the form appropriate to Newtonian physics:

1. *(a) All true equations in physics (i.e. all "laws of nature", and not only Newton's first law) assume the same mathematical form relative to all inertial frames. Equivalently, (b) no experiment performed wholly within one inertial frame can detect its motion relative to any other inertial frame.*

2. *Only infinite speeds are invariant under transformations between inertial frames.*

We call 1 and 2 jointly the Galilean relativity principle, because 2 leads to the Galilean transformations, as we have seen. The rest of this chapter deals with various consequences of the principle, and with ways of exploiting it to solve problems. Some immediate comments are in order.

- That the formulations 1(a) and 1(b) are indeed equivalent will emerge from the comments on Example 2.4 below. It is left to readers so inclined to construct a general proof once this example has been digested. According to 1 all inertial frames are equivalent; hence it never makes physical sense to say that a particular inertial frame is at rest, or that it is moving. All that can be determined is the motion of one inertial frame relative to another; and this can be done only by direct comparison between them, for instance by looking at one from the other, or by comparing their separate observations of an object they can both see.

- Though called a "principle", the relativity principle stems not from logic or from mathematics but from experiment. Within a well-defined domain of physics (see below) no contraventions are known, and one adopts the working hypothesis that within this domain the principle is obeyed by any laws yet to be discovered, just as it is obeyed by all laws already known. In this respect it is on exactly the same footing as other similarly general assertions, like the second law of thermodynamics: it could be falsified by future observations, but it is not expected to be, and meanwhile it proves useful.

- As outlined in the introduction, the principle fails on a cosmological scale, and in strong enough gravitational fields. For instance, it fails whenever one needs to take into account that in a gravitational potential Φ clocks run at a rate proportional to $1/\sqrt{1 + 2\Phi/c^2}$. These failures invalidate already part 1; but to explore them one must first progress to the Einsteinian version of part 2, and thence to the general theory of gravitation, which is beyond our scope.

- What the principle asserts to be invariant is the form of the *equations*. Particular *solutions* (e.g. particular trajectories) are specified by numerical parameters that do assume different values in different inertial frames. Equation (2.1) affords an example. Suppose that with respect to one frame S, a free particle follows the trajectory $\mathbf{r} = \mathbf{A} + \mathbf{B}t$. Then the Galilean transformation rules show that with respect to another frame S' the same trajectory is described by

$$\mathbf{r}' \equiv \mathbf{A}' + \mathbf{B}'t' = \mathbf{A} + (\mathbf{B} - \mathbf{v})t' \;\Rightarrow\; \mathbf{A}' = \mathbf{A}, \text{but } \mathbf{B}' = (\mathbf{B} - \mathbf{v}) \neq \mathbf{B}. \quad (2.14)$$

- Just as the laws of nature take the same form with respect to all inertial frames, which differ from each other only by their (uniform) relative motions, so it may be worth spelling out that the laws take the same form also with respect to such frames at relative rest, but differing from each other (i) by a translation, i.e. by a time-independent displacement of the origin of coordinates, or (ii) by a rotation, in the sense of a time-independent change in the directions of the coordinate axes; or in both these ways. (i) If the origin of S' is at the point $\mathbf{b}$ relative to S, with $\mathbf{b}$ constant, then obviously $\mathbf{r}' = \mathbf{r} - \mathbf{b}$; the other rules are left to be derived as an exercise. Form-invariance under translations reflects the experimental fact that space is homogeneous, i.e. that in physics there is no absolute (but only a relative) distinction between different positions. (ii) The transformation rules between rotated frames are discussed in appendix B. Form-invariance under rotations reflects the fact that space is isotropic, i.e. that in physics there is no absolute distinction between different directions. Clearly, both these invariances are purely geometric properties of space (they do not involve time); as such they are simpler and more familiar than the relativity principle, and in strict logic they should perhaps have been discussed first.

2.3.3 *Form-Invariance*

If an equation does assume the same form with respect to all inertial frames, then we say that it is *form-invariant*, or equivalently *covariant*, under the appropriate transformations (Galilean transformations, in this chapter). Evidently it is useful to be able to recognize whether an equation proposed by some theory is form-invariant, i.e. whether it complies with the relativity principle. The testing procedure is simple and explicit, and runs as follows.

(i) *Assume* for the present that the equation, expressed in terms of unprimed variables $t, \mathbf{r}, \ldots$, is true.

(ii) In this equation, express $t, \mathbf{r}, \ldots$ in terms of primed variables, referred to S', using the transformation rules.

(iii) The resulting "primed" equation is true by assumption, because it asserts the same as did the original unprimed equation in (i), though using different variables (i.e. a different language).

(iv) We are interested here not in its truth, but in its *form*. Ask: does the primed equation have the same *form* as the unprimed? In other words: if we simply drop the primes, does the result coincide with the original equation in (i)?

(v) If *yes*, then we say that the equation is form-invariant. Then it has *a chance* of being true, and if true in one inertial frame is true in all. It may be worth checking by experiment.

(vi) If *no*, then the equation contravenes the relativity principle, cannot be true (not in any inertial frame), and is not worth a test: it is consigned to the waste-paper basket, just as we would not spend time and money on testing designs for perpetual motion machines.

Conversely, *it follows from the relativity principle that if an equation is true in any one inertial frame, and if it is form-invariant, then it is true generally, i.e. in all inertial frames.*

All this applies equally to Einsteinian physics, where one simply replaces the Galilean by the Lorentz transformations introduced in the next chapter.

Example 2.2. It is claimed that, with respect to some inertial frame S, the equations of motion for two mutually interacting particles are

$$\mathbf{a}_1 = g_1(\mathbf{r}_2 - \mathbf{r}_1), \quad \mathbf{a}_2 = g_2(\mathbf{r}_1 - \mathbf{r}_2),$$

where $g_{1,2}$ are constants. Could this be true?

Solution. Check whether the equations are form-invariant.

Since the transformation rules give $\mathbf{a}_1 = \mathbf{a}_1'$ and $\mathbf{r}_2 - \mathbf{r}_1 = \mathbf{r}_2' - \mathbf{r}_1'$, the first equation re-expressed in terms of primed variables reads $\mathbf{a}_1' = g_1(\mathbf{r}_2' - \mathbf{r}_1')$; on dropping the primes this reverts to the original equation, showing it to be indeed form-invariant; and similarly for the second equation. Hence the claim might be true. (In fact the equations describe two particles coupled by a Hooke's-law spring.) ∎

Example 2.3. Two bodies confined to the x axis collide. Before the collision their velocities are $w_{1,2}$; afterwards they are $u_{1,2}$. Newton's rule for inelastic collisions reads $(u_1 - u_2) = -e(w_1 - w_2)$, where e is a constant characteristic of the materials, with $0 \le e \le 1$. The rule is based on experiment, and is generally quite accurate; however, it has the unmistakeable air of an approximation linear in the velocities, liable to amendment when the velocities are high. It is proposed that a more accurate version might read

$$u_1 - u_2 = -e(w_1 - w_2) - (g/s^2)(w_1^3 - w_2^3), \tag{a}$$

where s is the speed of sound in the bodies, and g another constant. Is this proposal reasonable?[9]

Solution. Check whether the equation is form-invariant. Step (ii): on substitution from the (inverse) Galilean transformations between velocities (i.e. from $u_1 = u_1' + v$, etc.) it becomes

$$u_1' - u_2' = -e(w_1' - w_2') - (g/s^2)\big([w_1' + v]^3 - [w_2' + v]^3\big)$$

$$= -(e + 3gv^2/s^2)(w_1' - w_2') - 3(gv/s)(w_1'^2 - w_2'^2) - (g/s^2)(w_1'^3 - w_2'^3). \tag{b}$$

Step (iv): we drop the primes, obtaining

$$u_1 - u_2 = -(e + 3gv^2/s^2)(w_1 - w_2) - 3gv/s(w_1^2 - w_2^2) - (g/s^2)(w_1^3 - w_2^3), \tag{c}$$

and ask whether this is the same as the original equation (a). Step (vi): obviously (c) is not the same as (a), because the coefficient of the first term on the right is different, and because of the extra term in the middle. Therefore the proposed equation (a) cannot be a true general rule for collisions.

It is easy to see how the proposal should be amended so as to make it reasonable: since the left-hand side is Galilean invariant, the right-hand side must also be invariant, whence it might be sensible to consider

$$u_1 - u_2 = -e(w_1 - w_2) - (g/s^2)(w_1 - w_2)^3. \quad \blacksquare \qquad \text{(d)}$$

The precise manner in which (a) fails the test for form-invariance illustrates very instructively the equivalence between the two versions (a) and (b) of part 1 of the relativity principle. For suppose that (a) did hold with respect to the frame S, i.e. that it did correctly describe the results of measurements on collisions when the results are expressed in terms of $(t, \mathbf{r}, \mathbf{u}, \ldots)$. Then the results of the same measurements referred to S', i.e. expressed in terms of $(t', \mathbf{r}', \mathbf{u}', \ldots)$, would be described by (b). Now in (b), the coefficient $(e + 3gv^2/s^2)$ of the linear term depends on v; since this coefficient can be determined experimentally, we would then have a method, workable wholly within S', of finding the velocity v of S' relative to S; in other words we could find this relative velocity without ever looking directly from either frame at the other. But this is exactly what the formulation (b) says is impossible.

2.3.4 Active and Passive Transformations

The transformation rules have been introduced in order to relate descriptions of a given process (or system) viewed from two different frames having relative velocity $\mathbf{v}$: same process but different descriptions. These are called *passive transformations*. Since the two frames are on a perfectly equal footing, each description fits a possible process. At first this statement might appear too obvious to be worth making, but it has far-reaching consequences: for it follows that one is entitled to regard the two descriptions as fitting two different processes both viewed from the same frame: different descriptions for different processes. By varying the parameter $\mathbf{v}$, one can be led in the same way to a wide variety of processes all guaranteed to be possible, merely from knowing that the original process was possible. The relations connecting such processes are called *active transformations*. Corresponding active and passive transformations are identical in form, but their physical significance is different and they tend to be exploited for different purposes: passive transformations serve primarily to check (possibly dubious) equations for compatibility with the relativity principle, while active transformations serve to apply correct but at first sight very partial information in a much wider context. The elementary method for doing so transforms explicitly; but often it is quicker and more elegant simply to re-express the original information in manifestly invariant form.

Example 2.4.   A particle with velocity u_i hitting a fixed buffer head-on is reflected so that afterwards its velocity is

$$u_r = -e u_i. \qquad \text{(a)}$$

Determine u_r in terms of u_i if the buffer is constrained to recede with velocity w (unaffected by the collision). Obviously we take $u_i > w$. (The question might perhaps be answered by appeal to Newton's empirical rule quoted in Example 2.3, unless one worried about the buffer being constrained. But the point is that we can manage with far less information, obtainable without experiments on moving buffers.)

Solution using the explicit method. Consider the transformation between the rest frame S of the buffer (where the information is given, i.e. where (a) applies), and the laboratory frame S', where the question is asked. Evidently the velocity of S relative to S' is w. (Thus we use the formulae for standard configuration, but with $v = -w$.) Hence $u_{i,r} = u'_{i,r} - w$, and substitution into (a) yields

$$u'_r - w = -e(u'_i - w) \quad \Rightarrow \quad u'_r = -eu'_i + (1 + e)w. \tag{b}$$

On dropping or simply ignoring the primes one obtains the relation we want.

Solution using invariants. What we have been told, namely equation (a), refers to the rest frame S of the buffer, but we choose to think of it as the special case $w = 0$ of an arbitrary scenario, where w need not vanish. To put the information into invariant form, we recall that, as velocity differences, the combinations $(u_i - w)$ and $(u_r - w)$ are Galilean invariants. Hence (a) can be expressed as

$$u_r - w = -e(u_i - w) \quad \Rightarrow \quad u_r = -eu_i + (1 + e)w. \tag{c}$$

This is the equation we require, because (i) it is form-invariant, so that it is either false for all w or true for all w; and (ii) it is true for $w = 0$. It agrees with (b) as it must. ∎

In practice little would be gained by labelling every transformation as active or passive, in spite of the differences in principle, and we shall follow general usage by using the two kinds indiscriminately. Alert readers will identify at least two crucial appeals to active transformations: to generalize the dispersion relations for waves from stationary to moving media (Sections 2.5.3 and 12.2), and to identify the proper time of an arbitrarily moving particle (Chapter 6).

2.4 Galilean Transformations: Dynamics

Much of dynamics deals with the motion of bodies under given forces, or under forces that they exert on each other. But, except briefly in Appendix C, this book considers only the conservation laws holding for systems of particles isolated from the outside world: studying the effects of forces in any detail is rather sterile until one has a theory of the forces; the simplest such theory valid beyond Newtonian/Galilean physics is electromagnetism; and electromagnetic theory is beyond our present scope.

Accordingly, this section is concerned largely with mass, momentum, and energy, and we must start by finding out how they transform from one inertial frame to another.

2.4.1 Mass

In Newtonian physics, the mass m of a particle is independent of its velocity or speed, and has the same value with respect to all frames: in other words it is an invariant under Galilean transformations. Formally one would write this as

$$m' = m;$$

but having said so once, we never bother to put primes on m (nor as a rule on any invariants once they have been identified as such). Also, in any transmutation of particles, the total mass is conserved, meaning that the joint total mass of the particles present before the reaction is the same as the joint total mass of the particles present afterwards. For example, in the chemical reaction $H_2 + Cl_2 \rightarrow 2HCl$ one has, in an obvious notation, $m(H_2) + m(Cl_2) = 2m(HCl)$. The conservation law for mass is an experimental fact, established with the chemical balance. The general formulation of this fact reads

$$\sum_i m_i \Big|_{\text{after}} = \sum_j m_j \Big|_{\text{before}}, \tag{2.15}$$

where m_i is the mass of particle number i, and the sums on the left and right run over all the particles present before and after any reaction. Different summation indices have been used on the two sides, to avoid obscuring the fact that at least some of the entries in the two sums are different.

For brevity, we shall write the rule (2.15) as

$$\Delta \sum_i m_i = 0, \tag{2.16}$$

defining, for any summand $(\ldots)$,

$$\Delta \sum_i (\ldots) \equiv \sum_i (\ldots) \Big|_{\text{after}} - \sum_j (\ldots) \Big|_{\text{before}}, \tag{2.17}$$

so that positive Δ signifies an increase.

Notice that, unless a reaction does occur, mass conservation tells one nothing: since the mass of every individual particle is constant, there is no mechanism whereby total mass might change unless there is some difference in number and/or type between the particle present before and after. Notice also that we treat molecules and atoms as particles all on the same footing: for instance, it is irrelevant here that the chemist views the reaction as a rearrangement of hydrogen and chlorine atoms. We shall return to this point in Section 11.6.1.

To sum up, mass is an invariant under Galilean transformations, and it is conserved. These are two independent experimental facts.

2.4.2 Momentum

The momentum of a particle with mass m and velocity $\mathbf{u}$ is a vector defined as

$$\mathbf{p} \equiv m\mathbf{u}. \tag{2.18}$$

It follows from Newton's second and third laws, and experiment confirms, that in a system of particles possibly interacting with each other but free of any other (so called "external") forces, the total momentum is conserved in the sense that it does not change in time even if the individual momenta do change:[10]

$$\frac{d}{dt} \sum_i m_i \mathbf{u}_i(t) = \frac{d}{dt} \sum_i \mathbf{p}_i(t) = 0. \tag{2.19}$$

We wish to study systems where between reactions the particles move freely, and where reactions happen fast enough for the time they take to be negligible or of no interest. Reactions can be induced by collisions, or it can happen that a particle disintegrates spontaneously into particles of different types, without needing to be stimulated by collisions with particles already present. For instance, in a gas mixture the reaction quoted above would result from a collision between a hydrogen and a chlorine molecule, while an appropriate excited state of a complex enough molecule might disintegrate spontaneously into other smaller molecules. For such abrupt reactions affecting particles that between reactions move freely, the conservation law (2.19) reduces to the same form as (2.16):

$$\Delta \sum_i \mathbf{p}_i = 0. \tag{2.20}$$

The Galilean transformation of $\mathbf{p}$ follows from the rule for $\mathbf{u}$ and from the invariance of m:

$$\mathbf{p}' \equiv m\mathbf{u}' = m(\mathbf{u} - \mathbf{v}) \implies \mathbf{p}' = \mathbf{p} - m\mathbf{v}, \quad \mathbf{p} = \mathbf{p}' + m\mathbf{v}. \tag{2.21}$$

Often one needs names for the total mass and the total momentum of a system of particles not interacting except through collisions. Hence we define

$$M \equiv \sum_i m_i, \qquad \mathbf{P} \equiv \sum_i \mathbf{p}_i, \tag{2.22}$$

with the sums evaluated at any time when no collision is taking place. Then the conservation rules reduce to

$$\Delta M = 0, \qquad \Delta \mathbf{P} = \mathbf{0}. \tag{2.23}$$

Example 2.5. Derive the transformation rule for the total momentum.

Solution. From (2.22) and (2.21) we find

$$\mathbf{P}' \equiv \sum_i \mathbf{p}_i' = \sum_i (\mathbf{p}_i - m_i\mathbf{v}) = \sum_i \mathbf{p}_i - \mathbf{v} \sum_i m_i = \mathbf{P} - M\mathbf{v}, \tag{2.24}$$

where the second equality follows because $\mathbf{v}$ is a common factor that can be brought outside the summation. In other words, the total momentum transforms just as if the entire system constituted a single particle with mass and momentum equal to the total mass and total momentum of the system. ∎

The point of view suggested by Example 2.5 proves useful time and again: one looks at a system of particles as it were from the outside, temporarily ignores anything one might know about its internal structure, and treats it as if it were a single particle with mass M, momentum $\mathbf{P}$, and eventually with the kinetic and total energies K and E to be introduced presently. For example, the frame S' where $\mathbf{P}' = \mathbf{0}$ is sometimes called the centre-of-mass frame of the system, though it is better referred to as *the centre-of-momentum frame*, a concept easier to generalize later on.

2.4.3 Force

By Newton's second law, force is mass times acceleration, $\mathbf{f} \equiv m\mathbf{a}$. Since mass and, as we have seen, acceleration are invariant, so is force:

$$\mathbf{f}' \equiv m\mathbf{a}' = m\mathbf{a} = \mathbf{f}. \tag{2.25}$$

The same follows from the alternative but equivalent definition of force as rate of change of momentum: since $t = t'$ we have $d/dt = d/dt'$, whence (recalling that $\mathbf{v}$ is not a variable),

$$\mathbf{f}' \equiv \frac{d\mathbf{p}'}{dt'} = \frac{d(\mathbf{p} - m\mathbf{v})}{dt} = \frac{d\mathbf{p}}{dt} = \mathbf{f}. \tag{2.26}$$

2.4.4 Energy

We consider in turn kinetic, potential, and total energies.
  The *kinetic energy of a particle is*[11]

$$k \equiv p^2/2m = mu^2/2. \tag{2.27}$$

In view of (2.21) it transforms as follows:

$$k' \equiv p'^2/2m = (\mathbf{p} - m\mathbf{v})^2/2m = p^2/2m - \mathbf{v}\cdot\mathbf{p} + mv^2/2,$$

$$k' = k - \mathbf{v}\cdot\mathbf{p} + mv^2/2, \qquad k = k' + \mathbf{v}\cdot\mathbf{p}' + mv^2/2. \tag{2.28}$$

The second transformation follows from the standard prescription (2.6) for the inverse; as an exercise it should be rederived algebraically from the direct transformation plus (2.21). In terms of velocities instead of momenta the rule reads

$$k' = k - m\mathbf{v}\cdot\mathbf{u} + mv^2/2, \qquad k = k' + m\mathbf{v}\cdot\mathbf{u}' + mv^2/2. \tag{2.29}$$

Note that kinetic energy transforms into a combination of kinetic energy and momentum: this link between energy and momentum is interesting, and will become even more so under Lorentz transformations.
  It is easily shown, on the lines of Problem 2.5, that *the total kinetic energy of a system of particles*, namely

$$K_{\text{tot}} \equiv \sum_i k_i = \sum_i p_i^2/2m_i = \sum_i m_i u_i^2/2, \tag{2.30}$$

also transforms as if it belonged to a single particle of mass M:

$$K'_{\text{tot}} = K_{\text{tot}} - \mathbf{v}\cdot\mathbf{P} + Mv^2/2. \tag{2.31}$$

Recall from mechanics that

$$K_{\text{tot}} = K + K_{\text{int}}, \qquad K \equiv P^2/2M. \tag{2.32}$$

Here K is due to the motion of the system as a whole; we have exercised hindsight by choosing for it a symbol without a suffix. It transforms according to

$$K' \equiv P'^2/2M = (\mathbf{P} - M\mathbf{v})^2/2M = K - \mathbf{v} \cdot \mathbf{P} + Mv^2/2, \qquad (2.33)$$

that is to say exactly like K_{tot}.

By contrast, K_{int} is the internal kinetic energy due to the motion of the constituents relative to each other, defined formally as their joint kinetic energy in the centre-of-momentum frame. By virtue of its definition K_{int} is an invariant:

$$K'_{int} \equiv K'_{tot} - K' = (K_{tot} - \mathbf{v} \cdot \mathbf{P} + Mv^2/2) - (K - \mathbf{v} \cdot \mathbf{P} + Mv^2/2)$$
$$= K_{tot} - K = K_{int}. \quad \blacksquare$$

As an illustration consider a two-body system, where

$$\mathbf{P} \equiv \mathbf{p}_1 + \mathbf{p}_2, \quad \mathbf{p} \equiv \{m_2\mathbf{p}_1 - m_1\mathbf{p}_2\}/M \Rightarrow \mathbf{p}_1 = m_1\mathbf{P}/M + \mathbf{p}, \quad \mathbf{p}_2 = m_2\mathbf{P}/M - \mathbf{p},$$

whence

$$K_{int} = p^2 M/2m_1m_2.$$

It is easily seen that $\mathbf{p}$ and thereby K_{int} are indeed invariants:

$$\mathbf{p}' \equiv \{m_2\mathbf{p}'_1 - m_1\mathbf{p}'_2\}/M = \{m_2(\mathbf{p}_1 - m_1\mathbf{v}) - m_1(\mathbf{p}_2 - m_2\mathbf{v})\}/M$$
$$= \{m_2\mathbf{p}_1 - m_1\mathbf{p}_2\}/M = \mathbf{p}. \quad \blacksquare$$

The total potential energy, call it V, is a function of the instantaneous coordinate differences between the particles, and therefore an invariant: $V' = V$.

Now consider *the total energy E of a particle*, like the molecule HCl, which can be thought of as a bound complex of others. By virtue of (2.32) we may write

$$E = K_{tot} + V = K + K_{int} + V = K + U, \qquad U \equiv K_{int} + V, \qquad (2.34)$$

where the total internal energy U is evidently invariant, because K_{int} and V are. In other words $U' = U$. Hence the transformation rule for the total energy of the complex reads

$$E' \equiv K' + U' = K - \mathbf{v} \cdot \mathbf{P} + Mv^2/2 + U = E - \mathbf{v} \cdot \mathbf{P} + Mv^2/2. \qquad (2.35)$$

Remarkably, this has exactly the same form as the rules (2.31) for K_{tot} and (2.33) for K: the invariant internal energy is just a passenger.

Finally we recall from Newtonian mechanics that total energy is conserved:[12]

$$\Delta E = 0. \qquad (2.36)$$

In elastic collisions where nothing changes except the individual momenta $\mathbf{p}_i$ this should be a familiar constraint, often exploited to good effect jointly with (2.22); while in reactions it governs the conversion of internal energy of the reagents into kinetic energy of the products.

Example 2.6. A seagull sits on the ground. The wind-velocity is $\mathbf{v}$. How high can the gull rise without doing any work?

Solution. The trick is (i) to identify the most convenient reference frame, (ii) transform the problem to that frame, (iii) solve the problem, and (iv) transform the result back again so that it is expressed in the original frame.

Here the original frame is the rest frame of the ground, S. However, (i) the most convenient frame is the rest frame of the air, S', moving horizontally with velocity $\mathbf{v}$ relative to S. Then (ii) with respect to S' the gull starts with a horizontal velocity $\mathbf{u}' = -\mathbf{v}$. So (iii) by converting all its kinetic energy $mu'^2/2$ into potential energy mgy', it can rise to a height $y' = \sqrt{u'^2/2g} = \sqrt{v^2/2g}$. But (iv) because Galilean transformations have no effect on coordinates perpendicular to the relative motion of the two frames, we have $y = y'$. Hence the gull can rise to a maximum height $y = \sqrt{v^2/2g}$. ■

2.4.5 The Relativity Principle and the Conservation Laws

We have just met five separate conservation laws[13] for isolated systems of particles: one for the total mass M; three for (the three components of) the total momentum $\mathbf{P}$; and one for the total energy E. Mass conservation is an experimental fact; the other four are predictions of Newtonian mechanics. Thus it would appear that a full check on the theory requires at least five separate experiments. However, it turns out that, by virtue of the relativity principle, energy conservation by itself already implies the other four conservation laws, reducing the number of logically necessary tests from five to just one. Equivalently, starting with Newton's laws, the theorist needs to derive only energy conservation; the others then follow directly from the relativity principle.

We proceed to show this, partly as an instructive exercise in exploiting the principle, and partly because similar (but different) relations between conservation laws feature prominently in Einsteinian dynamics later on. Although the argument is perfectly general, it might be visualized more easily by thinking of the differences Δ as changes in a chemical reaction, like the one cited in the preceding section.

- Experiment, or theory, gives us energy conservation, equation (2.36).
- The relativity principle tells us that if an equation of this form holds in S, then an equation of the same form holds also in S':

$$\Delta E' = 0.$$

- By means of the Galilean transformation (2.35) we express this in terms of the unprimed variables:

$$\Delta E - \mathbf{v} \cdot \Delta \mathbf{P} + (v^2/2)\Delta M = 0. \qquad (*)$$

- The crucial point is that $(*)$ is an *identity*, valid for *any* value of the relative velocity $\mathbf{v}$, whence the coefficients of $\mathbf{v}$ and of v^2 must vanish separately: otherwise the equality could be upset by changing $\mathbf{v}$. Therefore indeed $\Delta \mathbf{P} = 0$ and $\Delta M = 0$. ■

2.5 Plane Waves

2.5.1 Introductory

Consider a harmonic plane wave described with respect to an inertial frame S by

$$\psi = A\cos(\phi), \qquad \phi = \omega t - \mathbf{k} \cdot \mathbf{r}, \qquad (2.37)$$

where A is the amplitude and ϕ the phase. In the phase, $\omega = 2\pi f$ is the circular frequency, and $\mathbf{k} = \hat{\mathbf{k}}k$ the wave-vector, with $k = 2\pi/\lambda$; for complete generality ϕ should be augmented by an additive constant often called δ, which we have dropped because it would prove irrelevant here. Much of the physics of plane waves is embodied in *the dispersion relation*, namely the function $\omega(\mathbf{k})$ giving the frequency in terms of the wave-vector. For sound waves in air ψ is the excess pressure, and in the rest frame $S^{(0)}$ of the (undisturbed) air the dispersion relation reads

$$\omega^{(0)} = c_s k^{(0)} \qquad \text{(sound waves in still air)}, \tag{2.38}$$

where c_s is a constant (332 m/s at STP) independent of $\mathbf{k}^{(0)}$. Such waves are called nondispersive. But down to Section 2.5.4 the physical nature of the wave is irrelevant, and we are by no means confined to (2.38).

However, for simplicity *we do throughout this book confine ourselves to isotropic media, i.e. media such that in their rest frame the frequency depends only on the magnitude $k^{(0)}$ but not on the direction of* $\mathbf{k}^{(0)}$. Thus we generalize (2.38) to

$$\omega^{(0)} = f^{(0)}(k^{(0)}) \qquad \text{(medium at rest)}, \tag{2.39}$$

with different functions $f^{(0)}$ for different types of waves. (For example, waves on water of depth h have $f^{(0)}(k^{(0)}) = \sqrt{gk^{(0)}\tanh(hk^{(0)})}$, except for very short-wavelength ripples where the surface tension also matters.) The superfix on $f^{(0)}$ reminds one that this function connects frequency to wave-vector in the medium rest frame $S^{(0)}$; we shall see in Section 2.5.3 that in other frames they are connected by other functions $f(\mathbf{k}, \mathbf{w})$, depending not only on $\mathbf{k}$ but also on the medium velocity $\mathbf{w}$.

Whether or not the medium is at rest, and regardless of the dispersion relation, the group and the phase velocities are

$$\mathbf{u}_g = \nabla_{\mathbf{k}}\omega, \qquad \mathbf{u}_p = \hat{\mathbf{k}}\omega/k, \tag{2.40}$$

where $\nabla_{\mathbf{k}} \equiv (\partial/\partial k_1, \partial/\partial k_2, \partial/\partial k_3)$. Recall that u_g is the travel speed of modulations, i.e. of localizable disturbances carried by waves that are nearly but not quite monochromatic; thus u_g is the same as the signal speed (except under some highly unusual conditions). By contrast, u_p is the speed say of the crests,[14] which in strictly monochromatic (unmodulated) waves cannot carry information, though u_p is crucial to the conditions governing standing waves and resonances. Evidently $\hat{\mathbf{k}} = \hat{\mathbf{u}}_p$, i.e. the phase velocity is always parallel to the wave-vector.

In the special case (2.38), phase and group velocities are equal and independent of $k^{(0)}$, so that $\mathbf{u}_g^{(0)} = \mathbf{u}_p^{(0)} = \hat{\mathbf{k}}^{(0)}c_s$. Such waves are called *nondispersive*. The name stems from the passage of light waves through a prism, where the phase speed varies with colour (i.e. with wavelength or equivalently with k or with ω), whence different colours are dispersed in the sense that they are deviated through different angles: only in the idealized case of waves with constant (k-independent) speed would there be no dispersion.

2.5.2 Galilean transformations: ω, $\mathbf{k}$, $\mathbf{u}_g$, and $\mathbf{u}_p$

Given the description of a wave with respect to a frame S, we need to find the description of the same wave with respect to another frame S', i.e. the Galilean transformations between the two frames. This section determines the rules relating $\omega', \mathbf{k}', \mathbf{u}'_g, \mathbf{u}'_p$ to $\omega, \mathbf{k}, \mathbf{u}_g, \mathbf{u}_p$.

The basic fact is that *the phase of a harmonic plane wave, as of any harmonic oscillation, is an invariant.*[15] To see this, notice that the phase *counts* the oscillations: by saying that between two given events ϕ has increased by $2n\pi$, we mean that the disturbance has gone through n cycles, a number independent of our choice of reference frame, and the same for all observers. Similarly, a crest is a crest and a trough a trough for all observers, and all observers must agree on the *number* of crests that have passed any particular observer between any two particular ticks of her own clock. Thus[16]

$$\phi = \omega t - \mathbf{k} \cdot \mathbf{r} = \omega' t' - \mathbf{k}' \cdot \mathbf{r}'. \tag{2.41}$$

In order to determine ω' and $\mathbf{k}'$, we substitute for $t' = t$ and $\mathbf{r}' = \mathbf{r} - \mathbf{v}t$, and rearrange:

$$\omega' t' - \mathbf{k}' \cdot \mathbf{r}' = \omega' t - \mathbf{k}' \cdot (\mathbf{r} - \mathbf{v}t) = (\omega' + \mathbf{k}' \cdot \mathbf{v})t - \mathbf{k}' \cdot \mathbf{r} = \omega t - \mathbf{k} \cdot \mathbf{r}. \tag{2.42}$$

But this is an *identity*, in the sense that it must hold for any arbitrary values of t and $\mathbf{r}$, which is possible only if their coefficients are the same on both sides. Equating the coefficients we obtain

$$\omega = \omega' + \mathbf{k}' \cdot \mathbf{v}, \qquad \mathbf{k} = \mathbf{k}', \tag{2.43}$$

with the inverse

$$\omega' = \omega - \mathbf{k} \cdot \mathbf{v}, \qquad \mathbf{k}' = \mathbf{k}. \tag{2.44}$$

In other words, and most remarkably, the wave-vector is an invariant. Further, $\mathbf{k}' = \mathbf{k}$ entails[17]

$$\nabla_{\mathbf{k}'} = \nabla_{\mathbf{k}}. \tag{2.45}$$

The Galilean transformation for the group velocity follows at once:

$$\mathbf{u}'_g \equiv \nabla_{\mathbf{k}'} \omega' = \nabla_{\mathbf{k}}(\omega - \mathbf{k} \cdot \mathbf{v}) = \nabla_{\mathbf{k}} \omega - \mathbf{v} = \mathbf{u}_g - \mathbf{v}, \qquad \mathbf{u}_g = \mathbf{u}'_g + \mathbf{v}. \tag{2.46}$$

Thus group and particle velocities transform alike. (In particular, both are subject to *aberration*: their directions with respect to S and S' are different.) Since wave groups like particles have identifiable positions, it would have been disconcerting if they transformed differently: a particle and a wave group seen by one observer as travelling together would then be seen by another observer to part company.

The transformation for phase velocities is quite different:

$$\mathbf{u}'_p \equiv \hat{\mathbf{k}}' \frac{\omega'}{\mathbf{k}'} = \hat{\mathbf{k}} \frac{\omega - \mathbf{k} \cdot \mathbf{v}}{k} \;\Rightarrow\; \mathbf{u}'_p = \mathbf{u}_p - \hat{\mathbf{u}}_p (\hat{\mathbf{u}}_p \cdot \mathbf{v}) = \mathbf{u}_p \left(1 - \frac{\mathbf{u}_p \cdot \mathbf{v}}{u_p^2}\right),$$

$$\mathbf{u}_p = \mathbf{u}'_p + \hat{\mathbf{u}}'_p (\hat{\mathbf{u}}'_p \cdot \mathbf{v}) = \mathbf{u}'_p \left(1 + \frac{\mathbf{u}'_p \cdot \mathbf{v}}{u_p'^2}\right),$$

$$\tag{2.47}$$

where hats denote unit vectors. Thus $\mathbf{u}'_p$ remains parallel to $\mathbf{u}_p$: the phase velocity points the same way relative to all inertial frames whose corresponding axes are parallel. In other words phase velocities, unlike group and particle velocities, are free of aberration: this follows directly from the invariance of **k**, which entails $\hat{\mathbf{u}}_p = \hat{\mathbf{k}} = \hat{\mathbf{k}}' = \hat{\mathbf{u}}_p'$.

Nevertheless, by a remarkable coincidence the rule for $\mathbf{u}_p$ does reduce to the rule for $\mathbf{u}_g$ (and thereby to the rule for particle velocities) when **v** and $\mathbf{u}_p$ are parallel or antiparallel. In this special *collinear* case the unit vectors become redundant (except for their signs), and, as claimed, (2.47) yields $u_p' = u_p - v$.

Just as for particle velocities in Section 2.2.2, the transformations (2.46), (2.47) can be regarded as *velocity-combination rules*, in the sense that the unprimed velocities result from combining the primed velocities (relative to S') with the velocity **v** of S' relative to S. Perhaps the most important special cases are those where S' is the rest frame of the medium (usually called $S^{(0)}$), **v** is the velocity of the (undisturbed) medium relative to the laboratory frame S, and $\mathbf{u}_g, \mathbf{u}_p$ become the laboratory group and phase velocities of waves through a moving medium.

Example 2.7. A sound wave in still air (rest frame $S^{(0)}$), of frequency $\nu^{(0)} = 262$ Hz (C_4), propagates in the xy plane at $\theta^{(0)} = 30°$ to the x axis. Determine its frequency relative to a frame $S^{(R)}$ moving at $v = 100$ km/h along the x axis of $S^{(0)}$.

Solution. From (2.38) we have $k^{(0)} = \omega^{(0)}/c_s = 2\pi\nu^{(0)}/c_s$, while $\mathbf{k}^{(0)} = k^{(0)}(\cos\theta^{(0)}, \sin\theta^{(0)}, 0)$. Take $c_s = 332$ m/s, and note that in this case $u_g^{(0)} = u_p^{(0)} = c_s$. Also note $v = 10^5/3600 = 27.8$ m/s, and $\cos\theta^{(0)} = 0.866$.

In (2.44) we identify primed and unprimed with superfixed (R) and (0), divide by 2π, and find

$$\nu^{(R)} = \nu^{(0)}[1 - (v/c_s)\cos\theta^{(0)}] = 262[1 - (27.8/332) \times 0.866]$$

$$= 262[1 - 0.073] = 243 \text{ Hz},$$

just below $B_3 = 247$ Hz. ∎

Example 2.8. A siren is fixed at the origin The wind is blowing at $w = 100$ km/h from the north. Determine (i) the group speed and (ii) the phase speed of sound going north, south, and east.

Solution. Let S be the ground-fixed frame and $S^{(0)}$ the rest frame of the air. The x, y axes point east and north. Then the velocity of $S^{(0)}$ relative to S is $\mathbf{w} = -w\hat{\mathbf{y}}$. Start from the fact that relative to $S^{(0)}$ group and phase speeds are both equal to c_s.

(i) *Group speeds.* By (2.46), $\mathbf{u}_g = \mathbf{u}_g^{(0)} + \mathbf{w} = c_s\hat{\mathbf{u}}_g^{(0)} - w\hat{\mathbf{y}}$. The problem prescribes the direction of $\mathbf{u}_g$. Sound going north means $\mathbf{u}_g = u_g\hat{\mathbf{y}}$, sound and wind are antiparallel, $\mathbf{u}_g^{(0)} = c_s\hat{\mathbf{y}}$, and $u_g = c_s - w$. Similarly, for sound going south, sound and wind are parallel, and $u_g = c_s + w$. This much is obvious. Sound going east means $\mathbf{u}_g = u_g\hat{\mathbf{x}}$; hence $c_s\hat{\mathbf{u}}_g^{(0)} = u_g\hat{\mathbf{x}} + w\hat{\mathbf{y}}$ identifies a right-angled triangle with hypotenuse c_s (draw it!); and by Pythagoras we have $u_g = \sqrt{c_s^2 - w^2} = \sqrt{332^2 - 27.8^2} = 331$ m/s.

(ii) *Phase speeds: correct solution.* For sound going north or south we have the collinear case where phase velocity and group velocities transform alike, and $u_p = c_s \mp w$ exactly as for the group speeds. But sound going east means that the *group* velocity points east, so that we have exactly the same wave as in part (i).

Because in $S^{(0)}$ phase and group velocities are the same, this entails $c_s \hat{\mathbf{u}}_p^{(0)} = \mathbf{u}_g^{(0)} = u_g \hat{\mathbf{x}} + w \hat{\mathbf{y}} = \mathbf{u}_p^{(0)} = c_s \hat{\mathbf{u}}_p^{(0)}$; then the Galilean transformation (2.47) from $S^{(0)}$ to S yields

$$\mathbf{u}_p = \mathbf{u}_p^{(0)}(1 + \mathbf{u}_p^{(0)} \cdot \mathbf{w}/u_p^{(0)2})$$

$$= c_s \hat{\mathbf{u}}_p^{(0)} \left(1 + (u_g \hat{\mathbf{x}} + w \hat{\mathbf{y}}) \cdot (-w \hat{\mathbf{y}})/c_s^2 \right)$$

$$= c_s \hat{\mathbf{u}}_p^{(0)} \left(1 - w^2/c_s^2 \right),$$

whence finally $u_p = c_s \left(1 - w^2/c_s^2 \right) = 332(1 - 27.8^2/332^2) = 330$ m/s. Notice that $\mathbf{u}_p$ does *not* point east.

(iii) *Phase speeds: incorrect solution.* One might (wrongly) have thought that "sound going east" meant $\mathbf{u}_p = u_p \hat{\mathbf{x}}$. Applied to the transformation from S to $S^{(0)}$ the rule (2.47) would then yield

$$\mathbf{u}_p^{(0)} = \mathbf{u}_p - \hat{\mathbf{u}}_p (\hat{\mathbf{u}}_p \cdot \mathbf{w}) = u_p \hat{\mathbf{x}} - \hat{\mathbf{x}}(\hat{\mathbf{x}} \cdot (-w \hat{\mathbf{y}})) = u_p \hat{\mathbf{x}} \qquad \text{(wrong)}.$$

This would make $\mathbf{u}_p^{(0)}$ parallel $\mathbf{u}_p$, and $\mathbf{u}_p^{(0)} = c_s \hat{\mathbf{y}}$ would then entail $u_p = c$. ∎

2.5.3 Galilean Transformations: The Dispersion Relation

In Newtonian physics all waves are vibrations of some material medium, and the equations that govern them are uniquely simpler in the rest frame $S^{(0)}$ of the medium than in any other frame. That is why the dispersion relation (2.39), and its special case (2.38), indicate explicitly that, in the form they are written there, they hold only in $S^{(0)}$.

In order to express the dispersion relation in a form valid relative to any arbitrary inertial frame S, we must first introduce an obviously pertinent parameter, namely the velocity $\mathbf{w}$ of the (undisturbed) medium with respect to S. (Thus the velocity of $S^{(0)}$ relative to S is $\mathbf{v} = \mathbf{w}$.) Typically for relativity theory, the reasoning is simple yet very powerful: we need merely construct an equation that is both (i) equivalent to (2.39) in $S^{(0)}$, and therefore true in $S^{(0)}$ by assumption; and also (ii) form-invariant, and therefore true in all inertial frames by virtue of the relativity principle.

The obvious candidate equation reads

$$\omega - \mathbf{k} \cdot \mathbf{w} = f^{(0)}(k). \qquad (2.48)$$

This satisfies condition (i) because it reduces to (2.39) in $S^{(0)}$ (where $(\omega, \mathbf{k})$ become $(\omega^{(0)}, \mathbf{k}^{(0)})$, and $\mathbf{w}^{(0)} = 0$). It also satisfies condition (ii), in the simplest way possible, because each side is an invariant in its own right (regardless of whether the two sides are equal). The right-hand side is invariant because k is. To confirm that the expression on the left is invariant we need the transformation (2.43), namely $\omega = \omega^{(0)} + \mathbf{k}^{(0)} \cdot \mathbf{v}$, plus the fact that $\mathbf{w}$, being the velocity of the material, transforms according to the standard particle-velocity combination rule (2.9), $\mathbf{w} = \mathbf{w}^{(0)} + \mathbf{v}$. Then we apply the standard test from Section 2.3.3, finding that in

$$\omega - \mathbf{k} \cdot \mathbf{w} = (\omega^{(0)} + \mathbf{k}^{(0)} \cdot \mathbf{v}) - \mathbf{k}^{(0)} \cdot (\mathbf{w}^{(0)} + \mathbf{v}) = \omega^{(0)} - \mathbf{k}^{(0)} \cdot \mathbf{w}^{(0)}$$

the rightmost expression on dropping the superfixes does indeed revert to the original unsuperfixed expression on the left.

Accordingly we adopt (2.48), and rearrange it to obtain the dispersion relation in a form applicable to all inertial frames, superfixed or not:

$$\omega = f(\mathbf{k}, \mathbf{w}) \equiv f^{(0)}(k) + \mathbf{k} \cdot \mathbf{w}. \qquad (2.49)$$

It is left as an exercise for the reader to derive this directly from (2.39), (2.43), (2.44), and then to rederive the group and phase velocity-combination rules directly from (2.49).

To summarize, in (2.49) the function $f^{(0)}$ is determined by the physics of the medium at rest; by contrast, the addend $\mathbf{k} \cdot \mathbf{w}$ follows from the relativity principle alone.

Example 2.9. Waves of wavelength $\lambda = 10$ m move along a river $h = 5$ m deep and flowing at $w = 5$ km/h. What is their frequency[18] ν moving (i) downstream, and (ii) upstream?

Solution. The dispersion relation in the rest frame of the water, quoted just after (2.39), reads $f(k) = \sqrt{gk \tanh(hk)}$, where $k = 2\pi/\lambda$. (i) For motion downstream $\mathbf{k} \cdot \mathbf{w} = kw$ and (2.49) yields

$$\omega = \sqrt{9.81 \frac{2\pi}{10} \tanh\left(5\frac{2\pi}{10}\right)} + \frac{2\pi}{10} \times \frac{5000}{3600} = 2.48 + 0.87 = 3.35 \text{ s}^{-1},$$

whence $\nu = \omega/2\pi = 0.53$ Hz. (ii) For motion upstream $\mathbf{k} \cdot \mathbf{w} = -kw$, whence $\omega = (2.48 - 0.87)/2\pi = 0.26$ Hz. ∎

2.5.4 The Doppler Effect for Sound

The simplest and most important application of the Galilean transformation rules for plane waves is to sound signals through air. Compared with Einstein-relativistic light signals through empty space (Chapter 13), the acoustic case is somewhat simpler by virtue of the Galilean invariance of $\mathbf{k}$, but much more complicated through its dependence on the wind velocity, which has no analogue for light. Even so, Galilean *aberration* is generally straightforward, because the relevant directions are those of the group velocities, transforming by the already familiar rules that apply to particle velocities. Hence this subsection is confined to the Doppler effect. We take sound to be nondispersive in $S^{(0)}$, i.e. subject to (2.38): $\omega^{(0)} = c_s k^{(0)}$, with speed c_s independent of frequency.[19]

The emitter E and the receiver R have velocities $\mathbf{u}_E$, $\mathbf{u}_R$, and the wind blows with velocity $\mathbf{w}$. These velocities are referred to the ground (or laboratory) frame, call it S. The rest frames of E, R are $S^{(E)}$, $S^{(R)}$; the rest frame of the air is $S^{(0)}$. Velocities referred to these frames are identified by the corresponding bracketed superfix. Let $\omega_E^{(E)}$ be the the proper frequency of E, i.e. the frequency at which E oscillates relative to $S^{(E)}$; and let $\omega_R^{(R)}$ be the frequency registered by R. The Doppler effect is the deviation of the ratio $\omega_R^{(R)}/\omega_E^{(E)}$ from unity. It depends on two basic facts: (i) emitters work so that relative to $S^{(E)}$ (but not necessarily to other frames) the frequency $\omega^{(E)}$ of the emitted wave equals $\omega_E^{(E)}$, regardless of $\mathbf{w}^{(E)}$; while (ii) receivers work so that, regardless of $\mathbf{w}^{(R)}$, the frequency $\omega_R^{(R)}$ they register equals the frequency $\omega^{(R)}$ relative to $S^{(R)}$ (but not necessarily to other frames) of the wave being detected. We repeat that *superfixes specify reference frames; frequencies without a suffix are those of the wave under study; while frequencies with suffices E and R refer to emitter and receiver in the way just explained.*

For simplicity we consider only velocities such that u_E, u_R, w, $|\mathbf{u}_E - \mathbf{w}|$, and $|\mathbf{u}_R - \mathbf{w}|$ are all less than c_s.

For the special *collinear* case with E and R both confined to the x axis, and w in the x direction, the result is probably familiar:

$$\frac{\omega_R^{(R)}}{\omega_E^{(E)}} = \frac{c_s \mp (u_R - w)}{c_s \mp (u_E - w)}, \qquad \text{collinear case, } E \text{ to the } \binom{\text{left}}{\text{right}} \text{ of } R, \qquad (2.50)$$

where the upper (lower) signs corresponds to left (right). A typical scenario might have $u_E > u_R > 0$, and E initially to the left so that $\omega_R^{(R)}/\omega_E^{(E)} > 1$; eventually E passes R, upper and lower signs are interchanged, and afterwards $\omega_R^{(R)}/\omega_E^{(E)} < 1$.

The usual elementary but tedious derivation runs as follows. Take E to the left of R, let T_E be the period of the emitter, and calculate the time interval T_R between the arrival at R of two successive wave crests.

The second crest is emitted a time T_E after the first, and at a distance $T_E u_E$ to the right. By then the first crest has progressed through a distance $T_E(c_s + w)$. Thus the spacing between the crests is

$$D = T_E(c_s + w) - T_E u_E = T_E(c + w - u_E);$$

in other words, when the first crest arrives at R, the second is a distance D behind the first. When this second crest is received a time T_R later, it must have travelled a distance $D + T_R u_R$, because meanwhile E will have progressed a distance $T_R u_R$; since the crests move with velocity $c_s + w$, this takes a time

$$T_R = (D + T_R u_R)/(c_s + w).$$

Eliminating D between these two equations we find

$$T_R(c_s + w) = T_E(c_s + w - u_E) + T_R u_R \quad \Rightarrow \quad T_R(c + w - u_R) = T_E(c_s + w - u_E).$$

But $T_E/T_R = \omega_R^{(R)}/\omega_E^{(E)}$, whence the last equation indeed reproduces (2.50).

It is left as an exercise for the reader to construct the analogous argument when E is to the right of R.

Example 2.10. A fast car A moving against a headwind of 10 km/h passes another identical car B moving with the same speed in the opposite direction. As they pass, the driver of A hears the engine note of B fall by an octave (a factor of two). By what factor does the engine note of A fall as heard by B?

Solution. Let the wind speed be v, the speed of each car u, and the engine note emitted by each ω_E. Take A's velocity as positive, so that the wind velocity is $w = -v$.

The driver of A identifies R, E as A, B. Thus $u_E = -u$, $u_R = u$. Before the cars pass, E is to the right, and afterwards to the left, whence

$$\frac{\omega_{A,\text{before}}}{\omega_E} = \frac{c_s + u + v}{c_s - u + v}, \qquad \frac{\omega_{A,\text{after}}}{\omega_E} = \frac{c_s - u - v}{c_s + u - v}.$$

Thus

$$\frac{\omega_{A,\text{after}}}{\omega_{A,\text{before}}} = \left(\frac{c_s - u - v}{c_s + u - v}\right)\left(\frac{c_s - u + v}{c_s + u + v}\right).$$

The driver of B identifies R, E as B, A. Thus $u_E = u, u_R = -u$. Before the cars pass, E is to the left, and afterwards to the right, whence

$$\frac{\omega_{B,\text{before}}}{\omega_E} = \frac{c_s + u - v}{c_s - u - v}, \qquad \frac{\omega_{B,\text{after}}}{\omega_E} = \frac{c_s - u + v}{c_s + u + v}.$$

Thus

$$\frac{\omega_{B,\text{after}}}{\omega_{B,\text{before}}} = \left(\frac{c_s - u + v}{c_s + u + v}\right)\left(\frac{c_s - u - v}{c_s + u - v}\right).$$

But inspection shows that this is the same as $\omega_{A,\text{after}}/\omega_{A,\text{before}}$; therefore B hears the same one-octave drop as A, regardless of the wind. Is this obvious, and could it have been foreseen without calculating each ratio separately? ∎

By contrast to the laborious derivation of (2.50) from first principles just given in small print, we proceed to demonstrate that the ideas introduced in Section 2.5.3 lead to the general result with very little effort, and that special cases like (2.50) then follow almost immediately.

Our tools are the transformation rules (2.43), (2.44), plus the dispersion relation

$$\omega = c_s k + \mathbf{k} \cdot \mathbf{w} = k[c_s + \hat{\mathbf{k}} \cdot \mathbf{w}] \tag{2.51}$$

which is equally valid in any frame, i.e. equally valid with superfixes (E) or (R) on ω and $\mathbf{w}$. The trick is to start by expressing $\omega_R = \omega_R^{(R)}$ in terms of $\omega_E^{(E)} = \omega_E$ for a wave in an arbitrary direction $\hat{\mathbf{k}} = \hat{\mathbf{k}}^{(R)} = \hat{\mathbf{k}}^{(E)}$, leaving this direction to be chosen in the last step, appropriately to the particular situation under study.

(i) In $S^{(E)}$ we have $\mathbf{w}^{(E)} = \mathbf{w} - \mathbf{u}_E$, whence (2.51) yields

$$\omega^{(E)} = k[c_s + \hat{\mathbf{k}} \cdot (\mathbf{w} - \mathbf{u}_E)].$$

(ii) The Galilean transformation from $S^{(E)}$ to $S^{(R)}$, with the relative velocity $\mathbf{v} = \mathbf{u}_R - \mathbf{u}_E$, yields

$$\omega^{(R)} = \omega^{(E)} - \mathbf{k} \cdot \mathbf{v} = \omega^{(E)} - \mathbf{k} \cdot (\mathbf{u}_R - \mathbf{u}_E) = \omega^{(E)} - k[\hat{\mathbf{k}} \cdot (\mathbf{u}_R - \mathbf{u}_E)].$$

Eliminating k between (i) and (ii) one obtains

$$\omega^{(R)} = \omega^{(E)} - \frac{\omega^{(E)}}{[c_s + \hat{\mathbf{k}} \cdot (\mathbf{w} - u_E)]} [\hat{\mathbf{k}} \cdot (\mathbf{u}_R - \mathbf{u}_E)],$$

which rearranges into

$$\omega^{(R)}[c_s + \hat{\mathbf{k}} \cdot (\mathbf{w} - \mathbf{u}_E)] = \omega^{(E)}\{[c_s + \hat{\mathbf{k}} \cdot (\mathbf{w} - \mathbf{u}_E)] - [\hat{\mathbf{k}} \cdot (\mathbf{u}_R - \mathbf{u}_E)]\}$$

$$= \omega^{(E)}\{c_s + \hat{\mathbf{k}} \cdot (\mathbf{w} - \mathbf{u}_R)\}.$$

(iii) Since $\omega^{(R)} = \omega_R^{(R)}$ and $\omega^{(E)} = \omega_E^{(E)}$, we immediately find *the general Doppler formula*

$$\frac{\omega_R^{(R)}}{\omega_E^{(E)}} = \frac{1 - \hat{\mathbf{k}} \cdot (\mathbf{u}_R - \mathbf{w})/c_s}{1 - \hat{\mathbf{k}} \cdot (\mathbf{u}_E - \mathbf{w})/c_s}, \tag{2.52}$$

which is the central result of this section. The expression on the right is Galilean invariant, because $\hat{\mathbf{k}}$ and the velocity differences $\mathbf{u}_{R,E} - \mathbf{w}$ are invariants; this is as it should be, since $\omega_R^{(R)}$ and $\omega_E^{(E)}$ are invariants simply because each is defined by reference to the functioning of a given piece of apparatus in its own rest frame.

Evidently the effect vanishes (the frequencies are the same) if $\mathbf{u}_R = \mathbf{u}_E$, but otherwise it depends on $\mathbf{u}_R$ and $\mathbf{u}_E$ separately, and not only on the relative velocity $(\mathbf{u}_R - \mathbf{u}_E)$. However, if $|\mathbf{u}_R - \mathbf{w}|/c_s$ and $|\mathbf{u}_E - \mathbf{w}|/c_s$ are both much smaller than unity, then a binomial expansion gives

$$\omega_R^{(R)}/\omega_E^{(E)} = 1 - \frac{1}{c_s}\hat{\mathbf{k}}\cdot(\mathbf{u}_R - \mathbf{u}_E) - \frac{1}{c_s^2}(\hat{\mathbf{k}}\cdot(\mathbf{u}_R - \mathbf{u}_E))(\hat{\mathbf{k}}\cdot(\mathbf{u}_E - \mathbf{w})) + \cdots \qquad (2.53)$$

where, remarkably, the relative velocity alone does determine the first term, which is often all one needs.

The general formula (2.52) or the approximation (2.53) can now be applied to special cases, by choosing the direction $\hat{\mathbf{k}}$ so that the wave gets from E to R. One must pay close attention to signs, and may need to remember that it is the group velocity that must point from emission towards reception.

In the *collinear Doppler effect* as described above, the wave from E to R must travel to the right (left) when R is to the right (left); hence we have $\hat{\mathbf{k}} = (\pm 1, 0, 0)$, $\hat{\mathbf{k}}\cdot(\mathbf{u}_{R,E} - \mathbf{w}) = \pm(u_{R,E} - w)$, and (2.52) reduces to (2.50).

In the so-called *transverse cases*, E or R or both move only at right angles to $\mathbf{k}$. For simplicity we consider only $\mathbf{w} = \mathbf{0}$; then $S = S^{(0)}$, the waves are nondispersive, phase and group velocities coincide, and one need not ask which of them determines the pertinent direction $\hat{\mathbf{k}}$. Suppose for instance that R is fixed at the origin, E moves in the xy plane parallel to the x axis, and when it crosses the y axis it sends a signal to R along this axis. Then $\hat{\mathbf{k}}\cdot\mathbf{u}_R = 0 = \hat{\mathbf{k}}\cdot\mathbf{u}_E$, whence $\omega_R = \omega_E$. The same is true if E is fixed and R moves. Thus, in acoustics, there is no transverse Doppler effect.

Example 2.11: Euler's paradox for the Doppler shift.   Accurately to second order in $1/c_s$, compare the Doppler ratios for the three superficially similar cases specified by the first four columns of the table, all with a common value of the relative velocity $\mathbf{u}_R - \mathbf{u}_E$. Explain any differences.

Case	$\mathbf{w}$	$\mathbf{u}_E$	$\mathbf{u}_R$	$\mathbf{u}_R - \mathbf{u}_E$	$\omega_R^{(R)}/\omega_E^{(E)}$ (exact)	$\omega_R^{(R)}/\omega_E^{(E)}$ (to $O(v/c_s)^2$ incl)
1	0	0	$\mathbf{v}$	$\mathbf{v}$	$1 - \hat{\mathbf{k}}\cdot\mathbf{v}/c_s$	$1 - \hat{\mathbf{k}}\cdot\mathbf{v}/c_s$
2	0	$-\mathbf{v}$	0	$\mathbf{v}$	$1/(1 + \hat{\mathbf{k}}\cdot\mathbf{v}/c_s)$	$1 - \hat{\mathbf{k}}\cdot\mathbf{v}/c_s + (\hat{\mathbf{k}}\cdot\mathbf{v}/c_s)^2 + \cdots$
3	$-\mathbf{v}$	$-\mathbf{v}$	0	$\mathbf{v}$	$1 - \hat{\mathbf{k}}\cdot\mathbf{v}/c_s$	$1 - \hat{\mathbf{k}}\cdot\mathbf{v}/c_s$.

Solution. For each case we complete column 6 from (2.52), and then approximate according to (2.53).

Case 3 arises from case 1 through a Galilean transformation with relative velocity $\mathbf{v}$, from the common rest frame of the air and the emitter ($S^{(0)} = S^{(E)}$) to the rest frame of the receiver ($S^{(R)}$). This explains why they have the same value of the Galilean-invariant frequency ratio $\omega_R^{(R)}/\omega_E^{(E)}$. By contrast, case 2 is not Galilean-transformable to cases 1 or 3: for instance, they have different values of the other Galilean-invariant velocity difference $\mathbf{u}_E - \mathbf{w}$. Hence there is no reason why the frequency ratios should be the same, and we see that they are not. However, if the receiver did not know that sound travels through the material medium of the air, then he *would* expect case 2 to be equivalent to case 1; *Euler's paradox*[20] is that in Newtonian physics this expectation, though not intrinsically unreasonable, is nevertheless false. (We shall see later that for light in empty space it turns out to be true.)

On the other hand, for $v/c_s \ll 1$, the leading correction is common to both cases, and the difference is second-order small. Suppose for example that an old-time astronomer (R) looking at a "fixed" star (E) intends to apply these formulae to light instead of sound $(c_s \rightarrow c)$. He might think it plausible to assume case 1 (i.e. that the "fixed" star is fixed with respect to the aether), and implausibe to assume (for instance) case 2 (earth fixed with respect to the aether). In either case v would be the speed of the earth orbiting around the sun. However, since $v/c \simeq 10^{-4}$, he would in practice need to assume only that w/c is not much bigger; that would ensure that the second-order terms (whether specifically from the table above, or more generally from (2.53)), are only of order 10^{-8}, which would have made them unobservably small, allowing him to predict regardless of the missing information. ∎

2.6 Notes

1. To be definite, one might think at first of a system of point particles, or even of just one such particle. (A *point particle* is merely a body whose dimensions are negligible on any length scale relevant to the problem.) In fact the arguments up to Chapter 4, which concern kinematics, apply to all systems, including extended bodies like rods and spheres. But the relativistic dynamics of extended bodies is extremely complicated, and when it comes to dynamics we shall deal only with point particles.
2. The expressions *with respect to* and *relative to* are treated as synonymous.
3. One might imagine a projectile of given mass fired with arbitrarily increasing amounts of propellant; or even a multistage rocket built so that the firing of each stage increases the speed of the remainder with respect to base by the same amount, allowing any final speed to be achieved in sufficiently many stages. Since right now we are concerned only with a physical principle, it does not matter that as engineering designs both propositions are merely comic.
4. The name is borrowed from W. Rindler, *Introduction to special relativity*, Clarendon Press, Oxford, 1983.
5. The symbol **v** is reserved for the velocity of one frame relative to another; particle or wave velocities will be called by some other name, usually **u**.
6. Mathematically speaking, what we have just established is the group property of Galilean transformations regarded as operators $\mathcal{G}(\mathbf{v})$, with the multiplication rule $\mathcal{G}(\mathbf{v}_2)\mathcal{G}(\mathbf{v}_1) = \mathcal{G}(\mathbf{v}_1 + \mathbf{v}_2)$.
7. There is a subtlety here. The orientation of the rod (like its length) is determined by the positions of its two ends *at one and the same time*. In the rest frame S of the rod this proviso happens to be unnecessary, because both ends stay put; but in S', where the rod is moving, one would obviously obtain nonsense if one combined the value of $\mathbf{r}_1$ at one time with the value of $\mathbf{r}_2$ at another time. Hence the simultaneity required for the Galilean invariance of $(\mathbf{r}_2 - \mathbf{r}_1)$. In Chapter 4, simultaneity will prove equally crucial to Lorentz contraction.
8. We give no general proof. A proof would start from the fact that the most general motion of one frame relative to another combines a velocity with a rotation (see rigid-body motion in any good book on mechanics). It is easy to convince oneself that if frame S' has acceleration $-\mathbf{g}$ with respect to an inertial frame S, then a free particle at rest relative to S has acceleration $\mathbf{g}$ relative to S': thus it disobeys Newton's second law relative to S', whence S' is not inertial. Rotations can be ruled out by appeal to equation (2.2).

9. On the right one might perhaps have expected also another correction term $-(f/s)(w_1^2 - w_2^2)$. However, this is forbidden regardless of the relativity principle, because the rule must remain valid if all the velocities are reversed, i.e. if one changes $(u_{1,2}, w_{1,2}) \to (-u_{1,2}, -w_{1,2})$. This operation changes the sign of both sides of (a), so that the equality remains intact; but it leaves the candidate $-(f/s)(w_1^2 - w_2^2)$ unchanged, so that with this term added on the right the equality could not be true both before and after reversing the motion.

10. Do not confuse the suffixed particle label i with suffices indicating vector components, as on the x component p_x of $\mathbf{p}$.

11. The conventional symbol for kinetic energy is T. We avoid it because we want to reserve capitals for totals in a system of particles, while t is already in use for time.

12. We ignore temperature changes and any dissipative effects generating heat.

13. There are others, notably those for the three components of the total angular momentum.

14. This is true only in the direction of $\mathbf{k}$, which will serve our turn. To study the propagation of the phase in other directions one must think more carefully, introducing further concepts associated with rays, and speeds along rays. A start may be made with section 2.4 in W. Rindler, *Introduction to special relativity*, Clarendon Press, Oxford, 1982.

15. To be precise, this need be true only if ψ in (2.37) is observable in the sense of quantum mechanics. The phases of Schrödinger wave functions, which are not directly observable, do change under Galilean transformations: see Problem 12.9.

16. By the same argument the phase of any periodic disturbance is an invariant, whether or not the time-variation is simple-harmonic. A periodic sequence of pulses, say from a radar gun, is a good example. Thus, instead of (2.37), we could have started with $\psi = F(\phi)$, where F is required only to be periodic, in the sense that $F(\phi + 2\pi) = F(\phi)$ for all ϕ.

17. Even at the risk of confusing readers to whom this conclusion is (rightly) obvious, we point out that it depends on the fact that the components of $\mathbf{k}$ are the only independent variables in the problem, while ω is merely a function of $\mathbf{k}$.

18. This example is somewhat artificial. Generally one must infer the frequency of a wave not from its wavelength, but from the frequency of the emitter, as in the next section.

19. The calculations for dispersive waves are very tedious.

20. Actually the paradox is associated with Euler not through the Doppler shift but through aberration: see Section 13.2.3.

3 Einstein's Relativity Principle

3.1 The Existence of an Invariant Speed c

All experimental tests confirm, within their accuracy, that there exists a finite speed c such that anything that moves at this speed relative to one inertial frame moves at the same speed relative to any other inertial frame, no matter how these frames move relative to each other. We say that the speed c is *invariant*. To be more precise, the velocities **u** in question, with $u = c$, can be those of particles, or of wave groups, or of any disturbance capable of carrying information; they can also be the phase velocities of monochromatic waves, though these are velocities only by courtesy, and are not directly related to signal speeds. Moreover, *in vacuo*, that is in empty and field-free space, no true signals (no information) can travel faster than c, and we shall see later that speed c cannot be reached by any combination of lower-than-c speeds, nor exceeded by combining speed c with any other.[1] Section 3.4 makes a start on discussing the evidence for these assertions.

According to Newtonian physics such invariance is impossible. In particular, for collinear motion the Galilean velocity-combination rules of the previous chapter show that any particle, group, or phase velocity measured as u with respect to a frame S becomes $u' = u - v$ with respect to another frame S', where v is the velocity of S' relative to S; thus u' can equal u only if both are infinite.

As far as measurements can tell, the speed of light *in vacuo* is independent of frequency, and equal to the invariant speed[2] c. In principle the distinction is by no means empty: the critical speed c would remain quite unaffected even if (in modern parlance) photons had a nonzero rest mass m; but Section 12.6 shows that light *in vacuo* would then satisfy the dispersion relation

$$\omega = c\sqrt{\kappa^2 + k^2}, \qquad \kappa \equiv mc/\hbar, \tag{3.1}$$

entailing group and phase speeds

$$u_g = \mathrm{d}\omega/\mathrm{d}k = c\sqrt{1-(c\kappa/\omega)^2} < c, \qquad u_p = \omega/k = c/\sqrt{1-(c\kappa/\omega)^2} > c.$$

The value of m is a matter for measurement, without prejudice to the relativity principle. The present upper limit is $m < 4 \times 10^{-16}$ eV/$c^2 \sim 7 \times 10^{-52}$ kg. Since the electron mass is $m_e \sim 5 \times 10^5$ eV/c^2, we see that $m/m_e < 10^{-23}$, small enough to persuade one that $m = 0$, allowing light to be treated as radiation under the standard equations of electromagnetism.

However, on first learning about such waves (for brevity we speak simply of "light") one might reasonably think that, like other more familiar and more tangible waves, they too travel through some medium, call it the aether, and that the speed predicted for them by Maxwell's theory is their speed relative to the rest frame of the aether, call it S_0. If this were so, then with respect to any other frame S moving uniformly relative to S_0, Maxwell's equations and thereby the speed of light would be different, and electromagnetic measurements carried out wholly within S could determine its velocity relative to S_0, contrary to part 1 of the relativity principle as stated in Section 2.3. There is nothing logically absurd in this scenario: if it applied, then electromagnetism would furnish an absolute standard of rest (even without reference to the cosmic background radiation or to the fixed stars). The velocity of the aether relative to an observer would then enter the equations in a way loosely analogous to the way the velocity **w** of the air entered for sound in Section 2.5. Since most of physics is ruled by electromagnetic forces, including all of atomic and condensed-state physics and all of chemistry, frames at rest relative to S_0 would become preferable for most practical purposes; and the relativity principle would lose most of its value, reducing to a curious formal property of purely mechanical problems without significant electromagnetic input.

But in fact there is no aether, in the sense that no electromagnetic experiment ever detects any velocity like **w**: Maxwell's equations apply with respect to all inertial frames, compatibly with part 1 of the relativity principle; in particular, their prediction that (in SI units) the speed of light is $1/\sqrt{\varepsilon_0\mu_0}$ applies in all inertial frames; therefore this speed is indeed invariant; therefore the true invariant speed c featured in part 2 of the principle should be not infinite but finite, and equal to $1/\sqrt{\varepsilon_0\mu_0}$. Indeed the Maxwell theory merely exemplifies true (or at least reasonable) theories generally,[3] all of which we believe must satisfy the principle, and some of which describe other signals propagating at the same invariant speed, like neutrinos (though there have been suggestions that neutrinos do have a small mass, in which case their wave-functions would conform to (3.1)).

Thus there is incompatibility between, on the one hand, (i) the Newtonian/Galilean view of space and time, associated with the invariance of the time-coordinate ($t = t'$), and with the Galilean transformation rules; and on the other hand, (ii) the relativity principle, given the invariance of c. Einstein's theory recognizes this incompatibility; resolves it conformably with the evidence by positing a finite invariant speed in part 2 of the relativity principle; and then determines the transformation rules between inertial frames from the principle thus revised, accepting the results, embodied in the Lorentz transformations, however they might conflict with the everyday-familiar Galilean rules. That is why at first the theory was perceived as a revolutionary reorganization of basic concepts. But it would be just as reasonable to call it a thoroughly conservative reassertion of the relativity principle, firmly rooted in Newtonian physics, but threatened a century ago by preconceptions about an aether, under the tacit but incapacitating misapprehension that the invariant speed is infinite rather than finite.

3.2 The Relativity Principle

The facts just outlined make part 2 of the Galilean relativity principle untenable, except as an approximation at low speeds. The true relativity principle can be formulated as follows:

1. *(a) All true equations in physics (i.e. all "laws of nature", and not only Newton's first law) assume the same mathematical form relative to all inertial frames. Equivalently, (b) no experiment performed wholly within one inertial frame can detect its motion relative to any other inertial frame.*

2. *There exists an invariant speed c that is* finite, *in the sense that anything that moves at this speed relative to any inertial frame moves at the same speed relative to any other.*

- We call this Einstein's relativity principle, though only for contrast: strictly speaking the attribution is unnecessary, since nature observes only one relativity principle, namely this one. Part 1 is the same as it was in Section 2.3; part 2 is different.
- Only the *speed c* is invariant: the directions of signals having this speed, i.e. their *velocities*, are generally different with respect to different frames.
- The transformation rules between inertial frames must be determined compatibly with part 2. They are called Lorentz transformations, are necessarily different from the Galilean, and will occupy much of the rest of this book.
- Except that c is finite instead of infinite, *everything* in Section 2.3 applies exactly as before, and the reader should now go through that section again very carefully. It could and perhaps should have been reprinted at this point, with only two changes: (i) in equation (2.14) the relation between $\mathbf{A}', \mathbf{B}'$ and $\mathbf{A}, \mathbf{B}$ must at this stage be left open; and (ii) Examples 2.2, 2.3, and 2.4 apply the principle in the right way but in the mistaken belief that c is infinite, or only under conditions where it may be approximated as if it were. We leave it as an exercise to make the appropriate revisions once the Lorentz transformations become available in Chapter 4.
  - The most characteristic and the most radical step of the Einsteinian revision is that it adopts the invariance of c as axiomatic for all of physics, i.e. as experimental input into a general theory that can then make many and very varied predictions with comparative ease. This contrasts sharply with earlier attempts to derive it as output from some specific theory of light complying with Newtonian relativity (and therefore necessarily different from Maxwell's equations), by methods invariably obscure and unpersuasive even when logically self-consistent. The crucial point, whose importance remains undiminished, is that the invariance of c is a simple and general fact, far better incorporated into our basic assumptions than left to be recovered as if from scratch by different methods in different contexts. Physics gains because everyday working methods are simplified, a simplification so enormous and in practice so indispensable that it has wider implications: namely that concepts, however venerable through long service (like the Newtonian views of space and time), are not unalterable dogmas, but are subject to revision in the light of experience. Major revisions

have always required uncomfortable readjustments of intuition and common sense, which is what identifies them as major. That intuition and so-called common sense are not generally reliable guides to stubborn facts is of course the hardest to learn and the most beneficial lesson of science.

3.3 Curtain-Raiser: Light Clocks

To exploit Einsteinian relativity efficiently one needs the Lorentz transformations. However, even before introducing them in the next chapter, we can establish two important consequences directly from the invariance of c. They are derived by analysing the rates of peculiar time-keeping devices called light clocks, hypothetical but with nothing against them in theory. The reason why perfectly general conclusions can be drawn from so special a case lies in the relativity principle itself: if a given constant velocity with respect to a frame S had different effects on differently constructed clocks all at rest with respect to another frame S', then an observer in S' could use the differences between their readings to determine the velocity of S' relative to S, without comparing S' and S directly. But the relativity principle forbids this; hence all the rates must be the same.

3.3.1 Transverse Light Clock: Time Dilation

Consider two parallel mirrors mounted rigidly on a bench fixed relative to an inertial frame S', and separated by a distance l_0 in the y' direction. The suffix 0 is used here instead of a prime merely by common habit, and l_0 is called the *rest-length* of the apparatus. A light signal bounces normally between the mirrors, as shown in Figure 3.1(a) (which does not show the bench). The period T' relative to S' is given by (speed × time = distance covered), i.e. by

$$cT' = 2l_0. \tag{3.2}$$

Let S' be in standard configuration with S (see Section 2.2.2), so that S' moves with velocity v along the x axis of S. With respect to the y direction there is complete symmetry between the two frames, whence the lateral dimension has the same value l_0 with respect to S and S'. (We shall argue this point further when we derive the Lorentz transformations in Section 4.3.) Relative to S the path of the signal is the zig-zag shown in Figure 3.1(b), still travelled at the same speed c. From the figure, the period T relative to S is given by Pythagoras as

$$cT = 2\sqrt{l_0^2 + (vT/2)^2} = \sqrt{(2l_0)^2 + (vT)^2}. \tag{3.3}$$

To relate T and T' we eliminate l_0 between (3.2) and (3.3): $c^2T^2 = c^2T'^2 + v^2T^2$ yields

$$T' = T\sqrt{1 - v^2/c^2}. \tag{3.4}$$

Thus the time lapse between successive bounces on the same mirror is less with respect to S' than with respect to S; in other words the bounces take less time in the rest frame S' of the apparatus than in any other frame. This is the only frame where the successive bounces are events happening at the same place. Time as measured in S' is called the *proper time* of the clock. Conversely, regarding the system as a clock at rest in S' but moving in S, with the bounces for ticks, we say that *moving clocks run slow*, by a

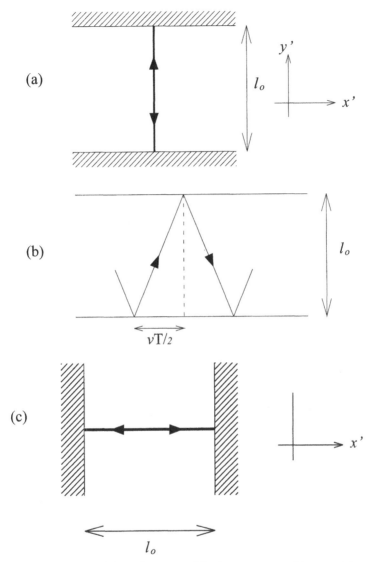

Figure 3.1 Light clocks, consisting of two parallel mirrors at rest with respect to an inertial frame S'. Relative to S, the frame S' moves with speed v in the positive x direction. (a) The path with respect to S' of a light signal bouncing between the mirrors. (b) The same path as in (a) but shown with respect to S. In this case the motion of S' relative to S is *transverse* to the light path in S'. (c) The clock has been turned through $90°$. The trajectory is shown with respect to S'; but now the motion of S' relative to S is *parallel* to the light path in S'.

factor $\sqrt{1 - v^2/c^2}$, where v is the speed of the clock with respect to the rest frame of the observer.

Of course all mechanisms, and not only periodic ones like clocks, are slower by the same factor in their rest frames than in other frames relative to which they are moving at speed v. Radioactive decays are one example of such nonperiodic processes.

Example 3.1.   Some cosmic rays hitting the upper atmosphere produce very high-speed muons. These particles decay spontaneously (into an electron and two neutrinos), so that given $N(0)$ muons, the average number surviving after a time lapse t' relative to their rest frame is $N(0) \exp(-t'/T_0)$, where $T_0 \approx 2.2 \times 10^{-6}$ s. Let F_1 be the mean flux, vertically downward into a detector, of muons in some small velocity range around v_1, measured at a height of $h_1 = 6265$ feet above sea level up a mountain. Muons are slowed down by collisions with the air: those that had speed v_1 at h_1 would have a lower mean speed v_2 if detected vertically or almost vertically below, at a height of $h_2 = 10$ feet. Let F_2 be the mean flux of such muons (in the corresponding velocity range) into the same detector at h_2. An experiment[4] measured $F_1 = 563/\text{hour}$, with the appropriate average of v_1 and v_2 equal to v such that $\gamma(v) \equiv \sqrt{1 - v^2/c^2} = 8.4$. Estimate F_2 and compare it with the measured value of $408/\text{hour}$.

Solution. For a rough estimate we pretend that the muons always move at their average speed v. But $v = c\sqrt{1 - 1/\gamma^2} \approx 0.99c$, and the descent is through $h = 6255$ feet ≈ 1907 m; hence the travel time (with respect to ground) is $t = h/v \approx 6.4 \times 10^{-6}$ s. If we did not know about time dilation we would predict $F_2 = F_1 \exp(-6.4/2.2) = 563 \times 0.055 \approx 31/\text{hour}$. But the time lapse in the muons' rest frame is less (since it is in this frame that the decays all happen at the same place), namely $t' = t/\gamma \approx 0.76 \times 10^{-6}$ s; therefore the correct estimate reads $F_2 = F_1 \exp(-0.76/2.2) = 563 \times 563 \approx 399/\text{hour}$. ∎

3.3.2 Longitudinal Light Clock: Lorentz Contraction

Consider the same apparatus but with the mirrors separated in the x' direction, as shown in Figure 3.1(c). The period with respect to the rest frame S' is still given by (3.2). Let S' and S be in standard configuration as before, so that now they move relative to each other parallel to the light. We want to determine the mirror separation l with respect to S, and do this via the period T, whose relation to T' we already know from time dilation.

In S, let a forward traverse take time T_1. To make this traverse the light travels a distance cT_1 equal to the separation l plus the further distance vT_1 through which the target mirror recedes before the light catches up with it:

$$cT_1 = l + vT_1, \qquad (c - v)T_1 = l.$$

Similarly the backward traverse time T_2 is governed by

$$(c + v)T_2 = l.$$

Thus the period is

$$T = T_1 + T_2 = \frac{l}{c - v} + \frac{l}{c + v} = \frac{2cl}{c^2 - v^2}. \tag{3.5}$$

But substituting (3.2) and (3.5) into (3.4) we find

$$T' = 2l_0/c = T\sqrt{1 - v^2/c^2} = \frac{2cl}{c^2 - v^2}\sqrt{1 - v^2/c^2} = \frac{2l/c}{\sqrt{1 - v^2/c^2}}.$$

Hence finally

$$l = l_0\sqrt{1 - v^2/c^2}; \tag{3.6}$$

with respect to S, relative to which the apparatus is moving, its length in the direction of motion is less than its rest length l_0.

Evidently, neither conclusion above is compatible with Galilean transformations, which yield the same value in all inertial frames for the time interval between events (Section 2.2.1), and for the length of any object (Section 2.2.2; see especially note 7 in Chapter 2). But at low speeds the differences are only of relative order $(v/c)^2$, which is often negligible within the accuracies required. Just how small $(v/c)^2$ is in many cases of interest is shown by some examples quoted at the end of Section 4.1 below.

Example 3.2. A particle traverses our galaxy at speed $u = c/3$. How long does this take (a) in the rest frame S of the galaxy? and (b) in the rest frame S' of the particle? (Take the galactic diameter as $D = 10^5$ light years. A light year is the distance that light travels in a year.)

Solution. (a) This is trivial, and no "relativity" is needed: $T = D/u = (10^5 \text{years.c})/(c/3) = 3 \times 10^5$ years.

(b) We want the time interval between two events, namely the coincidence of the particle first with one end of a galactic diameter and then with the other. These events happen at the same place relative to the rest frame of the particle; hence the time lapsel in this frame is the lesser, and by time-dilation $T' = \sqrt{1 - u^2/c^2}\,T = \sqrt{8/9}\,T = 2.83 \times 10^5$ years.

Alternatively, relative to S' the galaxy is moving at speed u, so that it is Lorentz-contracted, with $D' = \sqrt{1 - u^2/c^2}D$. Hence the time it takes to pass the stationary particle is $D'/u = \sqrt{1 - u^2/c^2}D/u = \sqrt{1 - u^2/c^2}\,T$, the conclusion already drawn directly from time-dilation. ∎

3.4 Evidence

3.4.1 Preamble

Before going on to explore more systematically the implications of the relativity principle as formulated in Section 3.2, a logician might think it necessary to quote conclusive evidence for part 2, namely for the invariance of the speed c, in practice for the invariance of the speed of light. The assumptions of parts 1 and 2 are the *inputs* to the theory, whence by mathematical means it derives the predictions forming its *output*. Other experiments should then test the predictions one by one, partly in order to find the limits of applicability of the theory, beyond which its failure or inadequacy indicates new physics.

Actually of course physical theories are neither developed nor assessed in this sequence. How far we trust them depends on how well they tally with experiments overall and on how efficiently they predict, regardless of fine distinctions between input and output as long as they are linked without manifest logical and mathematical errors: useful theories are invariably up and running long before their axioms and the precise status of their internal connections have been clarified to the satisfaction of those who both understand and care about such things. Here we recall also the comments in Chapter 1 on experimental evidence generally.

At present the best evidence for relativity theory comes from the phenomenal accuracy of its predictions about the motion of very fast charged particles in and around modern accelerators. This and other similar evidence is overwhelming: if future experiments find light speeds different from c, one would try to identify causes from beyond the already known and uncontroversial limits of the theory (see Chapter 1 and Section 2.3.2), long before contemplating any revisions of the theory within those limits. Meanwhile, the assumed invariance of c leads to a theory so closely integrated into physics at large that at present the circumstantial evidence would be unarguable even without direct confirmation.

A few spectacular pieces of direct evidence are described briefly below. All tests used to be very difficult because all sources were far slower than c, as the motions of laboratory receivers still are. Fortunately, modern accelerators provide very fast sources, and modern lasers allow previously undreamt-of accuracy in measurements of frequencies and wavelengths.

3.4.2 Testing Ballistic Models: Does Light Speed Depend on the Source Velocity? (Low Velocity)[5]

Ballistic models envisage light as a stream of particles. Then it is natural to expect that the light speed measured in collinear motion would be $c' = c + u$, where c is the light velocity relative to the source, and u the velocity of the source relative to the receiver; just as in Newtonian physics we reckon the speed of a bullet fired at us to be the sum of the muzzle velocity and of the approach velocity of the gun. In fact, a so-called *emission-theory* variant has been proposed even for electromagnetism,[6] with the built-in property that $c' = c + u$. It modifies Maxwell's equations in a way which has been claimed to be self-consistent, though it has not been explored thoroughly enough to guarantee this.

All such models contrast sharply with wave motion of a medium (like sound, or electromagnetic waves as pictured on the aether theory), where the speed of the waves is constant relative to the medium, and quite independent of the velocity of the source. What we believe today is that $c' = c$. The test we describe supposes that

$$c' = c + ku, \tag{3.7}$$

and looks for an upper bound on the constant k. On the ballistic model $k = 1$, but the invariance of c would be disproved by any value of k differing from zero by more than the experimental error. Conversely, invariance would be confirmed if and to the extent that the upper bound is much smaller than unity. Analyses are simplified by assuming $k \ll 1$ from the start, in accordance with already known fact. The test is remarkable in being completely clean: all the velocities are measured as velocities, and the analysis requires no results that the theory can produce only at a later stage.

Consider a small X-ray star closely circling a large companion as shown in Figure 3.2, so that from earth the source is eclipsed between points A and B of its orbit. The orbital period is T, the orbital speed $u \ll c$, and the distance to earth D. The light reaching us from A travels with speed $c - ku$, with a time of flight

$$\frac{D}{(c - ku)} = \frac{D}{c(1 - ku/c)} \approx \frac{D}{c}(1 + ku/c),$$

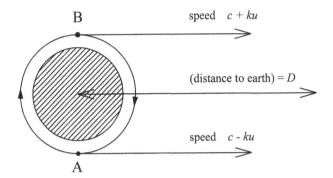

Figure 3.2 Apparatus to check whether light from a source with velocity **v** might move with velocity **c** + k**v**. A star (shown shaded) has a small X-ray source circling it in a tight orbit. An observer on earth sees the source disappearing at A and reappearing at B. If the coefficient k were large enough, then the source would reappear before it disappears.

where the approximation relies on $ku \ll c$. Similarly, light from B, emitted a time $T/2$ later, travels at speed $c + ku$, and has a time of flight $(D/c)(1 - ku/c)$. If the light from B arrives before the light from A, then we see a double image. Conversely, if we never see a double image, as in fact we do not, then

$$\frac{T}{2} + \frac{D}{c}\left(1 - \frac{ku}{c}\right) > \frac{D}{c}\left(1 + \frac{ku}{c}\right) \Rightarrow \frac{T}{2} > \frac{2Dku}{c^2},$$

$$k < (c/u)(cT/4D) = c^2T/4Du. \tag{3.8}$$

Even if c/u is very large, the right-hand expression can become small for large enough D/cT.

The binary X-ray source SMCX-1 has $u = 3 \times 10^5$ m/s, measured through its Doppler shift (see Sections 13.3–13.5: the argument here is not circular, because u/c is small enough for the shift to be calculated without having to rely on relativity theory). Further, $T = 3.9$ days $= 3.4 \times 10^5$ s, which is directly visible; and $D \geq 60$ kpc $= 1.9 \times 10^{21}$ m, inferred from the location of the system in the small Magellanic cloud. Thus one finds

$$k < \frac{(9 \times 10^{16})(3.4 \times 10^5)}{4(1.9 \times 10^{21})(3 \times 10^5)} = 1.3 \times 10^{-5}. \tag{3.9}$$

(i) If X-rays are pictured as waves, then the c' in these formulae is the *group velocity*, because it measures the actual speed with which signals reach us from the source.

(ii) Since the Doppler shift observed from earth, call it $\Delta\omega(t)$, depends on the velocity of the source along the line of sight, it oscillates from a maximum at B to a minimum at A, and if the radius r of the source orbit were truly negligible, then for $k = 0$ the oscillations would be strictly sinusoidal. But for nonzero k they are not; in particular, $\Delta\omega$ passes through zero when the phase of the oscillations differs by a small angle, call it Φ, from the exact halfway point between maximum and minimum. It turns out that the measured upper limit $\Phi < 0.05$ can improve the bound (3.9) by several orders of magnitude. A full analysis must allow also for the varying distance from source to earth due to the circulation of the source, i.e. for the finite value of r/D, and is rather tedious. But it rewards one with the spectacularly low limit

$$k < \frac{cT}{D} \cdot \frac{\Phi}{2\pi} < 4 \times 10^{-10}. \tag{3.10}$$

(iii) As tests of Ritz's emission theory such measurements are valid only if one can disregard the effects of the intergalactic or interstellar medium, which is reasonable for X-rays but not for visible light.

Although emission theories have not been applied in detail to propagation through a medium, a plausible argument runs as follows. A distance x down the beam, the refractive index n of the medium has induced a phase shift $\Delta\phi = 2\pi x(n-1)/\lambda$ relative to a beam in empty space. This is merely another way of saying that the medium changes the phase velocity from c to c/n, and the group velocity correspondingly. The phase shift is brought about by interference between light arriving directly from the star, and light that has been scattered and rescattered in the forward direction by the particles that constitute the medium. It seems reasonable to infer that the amplitude of the scattered component has become comparable to that of the direct component when $\Delta\phi$ has risen to be of order 2π, or in other words at a distance of the order of[7] $L \sim \lambda/|n-1|$. On the other hand it is clear, at least for a tenuous medium with $|n-1| \ll 1$, that between scattering events (i.e. for an overwhelmingly preponderant fraction of the time) all components of the light travel through empty space. But scattering might reasonably be viewed as absorption followed by re-emission. If so, then even according to emission theories any scattered light would propagate with speed c rather than $c + ku$, because the scatterers in the medium know nothing of the motion of the star. The efficacy of the test would gain nothing from travel over distances larger than L, where almost all the light has been scattered at least once. Therefore the formulae above apply only if $D < L$; otherwise L replaces D.

The refractive index of intergalactic (as of interstellar) space is essentially that of an electron plasma, with $|n-1| \approx 4\pi Ne^2/4\pi\varepsilon_0 m\omega^2$, where N is the electron number density and m the electron mass. The small Magellanic cloud is a galaxy different from ours, so one should use the intergalactic density $N \sim 10/\text{m}^3$. The measurement detected X-rays of energy $\hbar\omega = 70$ keV. It is left as an exercise to verify that the corresponding frequency does satisfy the criterion $D < L$, but that optical frequencies, corresponding to about 5 eV, would not. (Nor would the electron densities inside the galaxy, which are higher by a factor of order 10^5.)

3.4.3 Testing Ballistic Models: Does Light Speed Depend on the Source Velocity? (High Velocity)[8]

This is an accelerator-based test of (3.7). Purists might object that the analysis assumes the relativistic connection between particle energy and speed, even though this will be derived only in Chapter 9.

The set-up is sketched in Figure 3.3. The then proton accelerator at CERN in Geneva, operating in short pulses, delivered 19.2 GeV protons, which collide with a target T and produce, amongst other debris, very fast neutral pions, particles with a rest energy of $mc^2 \approx 135$ MeV, decaying into two γ-rays (photons) with a mean life of 8.4×10^{-17} s in their rest frame. Hence practically all such pions decay within a few microns from the target, even allowing for time dilation at speeds as close to c as we shall estimate presently. Some of the resulting photons pass through the lead collimator C, emerging in a narrow beam freed from the potentially troublesome background of charged particles by the sweeping magnets M.

Some of the photons of each pulse are detected in the first counter, and some others from the same pulse in the second counter, a distance $D \approx 30$ m downstream. Modern electronics can measure the time lag t between them directly, giving their speed

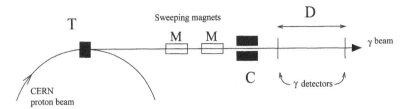

Figure 3.3 Apparatus to measure the speed of γ-rays from the decay of neutral pions moving very nearly as fast as light. Description in the text.

$c' = D/t = (2.9977 \pm 0.0004) \times 10^8$ m/s. The counters are biased to respond only to photons so energetic that they must have been produced by pions with an energy ε of at least 6 GeV = 6000 MeV $\approx 45mc^2$. (The kinematics ensures also that the parent pions were travelling near enough in the direction of the eventually collimated gamma rays.) As we shall see later from Chapter 9, the pion speed $u = \beta c$ is connected to ε by $\beta = 1/\sqrt{1 - (mc^2/\varepsilon)^2}$, so that $\beta > \sqrt{1 - 1/45^2} = 0.99975$. Since $u \approx c$, we have $c' = c + ku \approx c(1 + k)$; given $c = 2.9979 \times 10^8$ m/s, the conclusion regarding (3.7) is that

$$k = \left(c'/c\right) - 1 = (-3 \pm 13) \times 10^{-5}. \tag{3.11}$$

This is comparable with (3.9) but far less stringent than (3.10). One can check that the effects of the medium (air) are negligible here too.

3.4.4 Testing Aether Theories: Does Light Speed Depend on the Velocity of the Apparatus Relative to the Fixed Stars?[9]

This experiment contrasts sharply with those just described, in two respects. (i) Both source and receiver are fixed in the laboratory and thereby stationary relative to each other, so that ballistic models would make $c' = c$ and predict no effect at all. But according to the aether theory light waves travel at speed c relative to the aether: if the laboratory has velocity u with respect to the aether, then relative to the laboratory the light speed is once again given by (3.7) (with $k = 1$). The experiment looks for the effects of the velocity of the laboratory due to the rotation of the earth. (ii) What is measured is not the group but the phase velocity of the light, since it is the phase velocity that governs the effectively standing wave which is crucial to interpreting the results.

As sketched in Figure 3.4, the light beams from two identically constructed lasers of very nearly equal frequencies ν_1, ν_2 are brought together on a photodetector, whose output current fluctuates with the beat frequency $\Delta\nu \equiv \nu_2 - \nu_1$. The point is that the frequency $\Delta\nu$ is low and easily measured, while the optical frequencies ν_1, ν_2 are not. Beam 1 passes through a cell containing iodine vapour, and its frequency is locked to a nearby natural vibration frequency of the iodine molecules. There is no reason to expect molecular frequencies to be affected by u, so that ν_1 is regarded as remaining fixed.

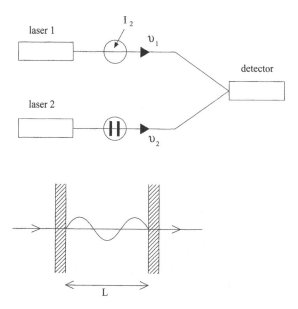

Figure 3.4 Apparatus to test whether the phase velocity of light depends on the receiver velocity. The insert magnifies the two partially-silvered mirrors, and illustrates how they constrain the wavelength through $L = n\lambda/2$, showing the case $n = 3$. Full description in the text.

Beam 2 passes normally through two highly-silvered parallel mirrors a fixed distance L apart, setting up an (almost) standing wave between them, and its frequency ν_2 is governed, through its wavelength $\lambda_2 = c'/\nu_2$, by the resonance condition that a whole number of half waves must fit between the mirrors, as shown in the figure. Thus $L = N\lambda_2/2$, and small continuous variations of c' cannot change the integer N, because integers cannot change continuously. Hence $\lambda_2 = 2L/N$ remains fixed throughout, and any small variations of c' modulate only ν_2, and thereby the beat frequency. From $\nu_2 = c'/\lambda_2$ with fixed λ_2 we see that a small shift $\delta c' = ku$ in c' produces a small shift in ν_2 given by $\delta\nu_2/\nu_2 = \delta c'/c = ku/c$. In the denominator we replace ν_2 by ν, which can be thought of indiscriminately as either ν_1 or ν_2 or their average, because here the difference between them does not matter. Finally, since ν_1 is fixed we have $\delta\Delta\nu = \delta\nu_2$, so that $\delta\Delta\nu/\nu = ku/c$, giving the upper limit

$$k < \frac{|\delta\Delta\nu|_{max}}{\nu} \cdot \frac{c}{u}. \tag{3.12}$$

The experiment records $\Delta\nu(t)$ as a function of time, with results exemplified by Figure 3.5. In this jungle one looks for a Fourier component with a one-day period, of the form

$$\delta\Delta\nu = A\sin\left(\frac{2\pi t}{(1\text{ day})} + \chi\right),$$

where A is just the $|\delta\Delta\nu|_{max}$ that we want, and χ is an irrelevant phase angle. In fact the (rather sophisticated) data analyis finds nothing but noise, and sets an upper limit $A < 100$ Hz at a 90% confidence level. (This is quite plausible even at first sight: simple inspection of the figure suggests that there can be no one-day periodicity with

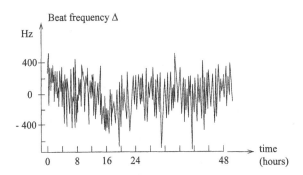

Figure 3.5 Typical results obtained with the apparatus sketched in Figure 3.4 (from Hils & Hall, referenced in note 9). Vertical axis: the beat frequency $\Delta\nu$, in Hz. Horizontal axis: time in hours. One is looking for a Fourier component of $\Delta\nu(t)$ with a 24 h period, having an amplitude of A Hz, say. But all one observes is noise, yielding (after analysis) an upper limit $A < 100\,\text{Hz}$. A limit of this order seems reasonable to the naked eye, noting that the oscillations of $\Delta\nu(t)$ as a whole (including all Fourier components) hardly ever exceed 400 Hz. Reproduced by permission of the American Physical Society.

amplitude A exceeding a few hundred hertz). Given the (rough) values $\nu \sim 5 \times 10^{14}$ Hz and $u \sim 500$ m/s one obtains

$$k < \frac{100}{5 \times 10^{14}} \cdot \frac{3 \times 10^8}{500} \sim 10^{-7}. \tag{3.13}$$

It is remarkable how such a low upper bound on k emerges in spite of the large second factor in (3.12), which is vastly overcompensated by the first factor, small to the tune of the ratio of the minute upper limit on A to the optically high frequency ν.

3.5 Notes

1. The reader may wonder why we subject this speed limit to so many conditions, instead of saying simply that *the speed c is invariant and cannot be exceeded*, or words to that effect. The reason is that the simple statement can fail under perfectly normal and familiar conditions; these then require special pleading, which may obscure the issue retrospectively. In fact all the stated conditions are essential. Their application to particles is usually obvious, to waves less so. For instance, the closing speed of two particles carries no information, and is not bounded by c (see Section 4.2.4). By the same dispensation, phase velocities too may and often do exceed c, both *in vacuo* and in material structures like waveguides. Moreover, as a consequence of quantum effects even light signals travelling through empty space can go more slowly than c in strong magnetic fields, and faster than c in certain strong gravitational fields. (Admittedly the latter are beyond the remit of relativity theory.)

2. By an unfortunate quirk, the current internationally agreed choice of units takes this equality for granted, so that numerically $c = 1/\sqrt{\varepsilon_0\mu_0} = 2.997792458$ m/s is exact by definition. Under this convention (and if one does not question the standard of time), any measured departures from the relativity principle would show up not as variations in the speed of light but as variations in the length of the metre. What is agreed by committees does not always illuminate fundamental physics.

3. Of course the connection between electromagnetism and relativity theory is central to the history of both: see the histories cited in Chapter 1. The connection is also vital experimentally, because light waves remain by far the most accessible signals with speeds at or near c.

4. This is a famous demonstration experiment by D. H. Frisch and J. H. Smith, "Measurement of the relativistic time dilation using μ-mesons", beautifully described and analysed in the *American Journal of Physics* (1963), volume **31**, page 342. The account is instructive also about several other aspects of relativity.

5. This account is adapted from K. Brecher, "Is the speed of light independent of the velocity of the source?", *Physical Review Letters* (1977), volume **39**, page 1051.

6. W. Ritz, *Oeuvres*, Gauthier-Villars, Paris, 1911. The paper in question is "Recherches critiques sur l'électrodynamique générale", originally published in 1908.

7. This critical distance L happens to coincide with the so-called extinction length introduced by some discussions of the extinction theorem in perfectly conventional electromagnetic theory. The relevance of L to our problem is unaffected by the fact that most comments on the extinction length are misconceived (as to why, see H. Fearn, D. F. V. James, and P. W. Milonni, "Microscopic approach to reflection, transmission, and the Ewald-Oseen extinction theorem", *American Journal of Physics* (1996), volume **64**, page 986).

8. T. Alväger *et al.*, "Test of the second postulate of special relativity in the GeV region", *Physics Letters* (1964), volume **12**, page 260.

9. D. Hils & J. L. Hall, "Improved Kennedy-Thorndike experiment to test special relativity", *Physical Review Letters* (1990), volume **64**, page 1697. The experiment was done at the Joint Institute for Laboratory Astrophysics in Boulder, Colorado. Our symbols are different from those of the paper, and we exploit its results in a simpler but less searching manner.

Part II
Kinematics

4 Lorentz Transformations

4.1 Statement of the Transformations

Replacing the Galilean relativity principle by Einstein's amounts to replacing the Galilean transformations of event coordinates by the Lorentz transformations. Both are established in Section 4.3, by reasoning and from physical input common to both until the inputs diverge right at the end. But in order to become familiar with the Lorentz transformations we start by simply stating them, and deriving their simplest consequences,[1] namely time dilation and the Lorentz contraction (already anticipated in Chapter 3), and the rule for transforming particle velocities.

This book considers only transformations between two frames S and S' in *standard configuration*[2] (see Section 2.2.2): Figures 4.1 remind us of the relation between them. The figures on the left and right are labelled from the points of view of S and of S' respectively.

The direct and the inverse transformations read

$$t' = \frac{t - vx/c^2}{\sqrt{1 - v^2/c^2}}, \quad x' = \frac{x - vt}{\sqrt{1 - v^2/c^2}}, \quad y' = y, \quad z' = z; \tag{4.1}$$

$$t = \frac{t' + vx'/c^2}{\sqrt{1 - v^2/c^2}}, \quad x = \frac{x' + vt'}{\sqrt{1 - v^2/c^2}}, \quad y = y', \quad z = z'. \tag{4.2}$$

Often it proves convenient to use the dimensionless variables

$$\beta(v) \equiv \frac{v}{c}, \quad \gamma(v) \equiv \frac{1}{\sqrt{1 - v^2/c^2}} = \frac{1}{\sqrt{1 - \beta^2}}, \tag{4.3}$$

$$\beta^2 + 1/\gamma^2 = 1, \quad 0 \leq |\beta| < 1, \quad 1 \leq \gamma. \tag{4.4}$$

They turn (4.1), (4.2) into

$$t' = \gamma[t - \beta x/c], \quad x' = \gamma[x - \beta ct], \quad y' = y, \quad z' = z; \tag{4.5}$$

$$t = \gamma[t' + \beta x'/c], \quad x = \gamma[x' + \beta ct'], \quad y = y', \quad z = z'. \tag{4.6}$$

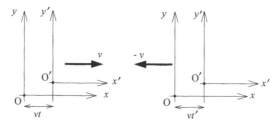

Figure 4.1   Two inertial frames S and S' in standard configuration. The x' and x axes are shown offset for visibility, though in fact they coincide. Left-hand figure from the point of view of S; right-hand figure from the point of view of S'. Relative to S, the frame S' moves to the right with speed v; relative to S', the frame S moves to the left with speed v. The origins O and O' coincide at time $t = t' = 0$.

We shall also write $\boldsymbol{\beta}(\mathbf{u}) \equiv \mathbf{u}/c$ and $\gamma(u) \equiv \sqrt{1 - u^2/c^2}$, with particle velocities $\mathbf{u}$ as their argument.

- The inverses (4.2), (4.6) follow from the direct transformations (4.1), (4.5) just as in the Galilean case, by the rule *to interchange primed and unprimed and reverse the sign of v.* Of course one gets the same result by solving (4.1) algebraically, as if they were equations for the unprimed variables (regarded as unknowns) in terms of the primed (regarded as known).
- Mathematically, the limit $v/c \to 0$ reduces the Lorentz to the Galilean transformations (2.6), (2.7). In reality of course one cannot vary c, but the limit reminds one that in many cases where $v^2/c^2 \ll 1$ the two kinds of transformations differ only negligibly.
- Often one wants to consider low-velocity regimes where all pertinent speeds, particle speeds u relative to inertial frames as well as speeds v of inertial frames relative to each other, are far below c, so that one has both $u^2/c^2 \ll 1$ and $v^2/c^2 \ll 1$. This might sensibly be called the Galilean or perhaps the Galilean-relativistic limit, but in fact it is almost universally referred to as "*the nonrelativistic limit*", a deplorable choice of words in view of the discussion near the end of Section 3.1, but past remedy. Sometimes one speaks of "the limit $c \to \infty$", somewhat improperly since c is a dimensional quantity whose value depends on what units one uses. By contrast, $u/c \sim 1$ is the norm in high-energy particle physics, in the more precise sense that $\gamma(u) \gg 1$. One calls this *the extreme-relativistic limit*.
- It is worth illustrating just how small $(v/c)^2$ is in many important cases. (i) The speed of sound in air: $v \approx 300$ m/s, $(v/c)^2 \approx 10^{-12}$. (ii) The speed of a point on the equator due to the rotation of the earth: $v \approx 460$ m/s. (iii) The speed of the earth along its orbit: $v \approx 30$ km/s, $(v/c)^2 \approx 10^{-8}$. (v) Our speed relative to the cosmic background radiation (see Example E.1 in Appendix E): $v \approx 370$ km/s, $(v/c)^2 \approx 1.5 \times 10^{-6}$. (vi) The speed of the electron in the ground state of the Bohr hydrogen atom (typical of electrons in atoms and molecules generally): $v/c = e^2/4\pi\varepsilon_0\hbar c \approx 1/137$ (the "fine-structure constant"), $(v/c)^2 \approx 5 \times 10^{-5}$. (Note that this is independent of the electron mass.) (vii) The mean-square speeds of nucleons (protons and neutrons) in heavy nuclei can be estimated very roughly from their mean kinetic energy $mv^2/2 \sim 30$ MeV, where $m \sim 938$ MeV/c^2 is the nucleon mass. Thus $(v/c)^2 \sim 60/938 = 0.06$.

4.2 Some Simple Consequences

4.2.1 Coordinate Differences (Intervals)

Consider two events having coordinates $(t_1, \mathbf{r}_1)$, $(t_2, \mathbf{r}_2)$ relative to S and $(t_1', \mathbf{r}_1')$, $(t_2', \mathbf{r}_2')$ relative to S'. Let $\Delta t \equiv t_2 - t_1$, $\Delta \mathbf{r} \equiv \mathbf{r}_2 - \mathbf{r}_1$ be the intervals between them.[3] Evidently $(\Delta t', \Delta \mathbf{r}')$ are related to $(\Delta t, \Delta \mathbf{r})$ by the same formulae that relate the individual coordinates themselves. Thus (4.1), (4.2) entail

$$\Delta t' = \gamma(\Delta t - v\Delta x/c^2), \quad \Delta x' = \gamma(\Delta x - v\Delta t), \quad \Delta y' = \Delta y, \quad \Delta z' = \Delta z, \quad (4.7)$$

$$\Delta t = \gamma(\Delta t' + v\Delta x'/c^2), \quad \Delta x = \gamma(\Delta x' + v\Delta t'), \quad \Delta y = \Delta y', \quad \Delta z = \Delta z'. \quad (4.8)$$

In strict logic this hardly needs saying, because "the coordinates themselves" merely measure the interval between the event in question and another event chosen as the origin.

What is generally perceived as the most striking innovation in Einsteinian physics is that $t' \neq t$, and the consequent "relativity of simultaneity", i.e. the fact that $\Delta t = 0$ does not imply $\Delta t' = 0$: two events simultaneous with respect to one inertial frame are not simultaneous with respect to another.

4.2.2 Time Dilation

Consider the special case of two events happening at the same place with respect to S' (like the ticks of a clock at rest in S'), i.e. with separation $\Delta \mathbf{r}' = \mathbf{0}$. Then by (4.8)

$$\Delta t \mid_{\Delta \mathbf{r}'=0} = \gamma \Delta t', \quad (4.9)$$

as anticipated in Section 3.2.1. The notation on the left reminds one that Δt here separates two events of a very special kind; and one must always remember that the time lapse is least in the rest frame of the clock.

4.2.3 Lorentz Contraction

Consider an object, say a rod, fixed relative to S', where its length is l_0, and the length of its projection onto the x' axis is $l_{0,x}$. (Tradition dictates the suffix 0 instead of a prime: l_0 is called *the rest length* of the rod.) If the rod lies along the axis, then $l_{0,x} = l_0$. In any case $l_{0,x}$ is the difference between the x' coordinates of the ends of the rod:

$$\Delta x' \equiv x_2' - x_1' \equiv l_{0,x}.$$

Since the rod is at rest, it does not matter whether x_2' and x_1' are determined at the same time or at different times relative to S'. But relative to S the rod is moving in the x direction with velocity v. We ask: what is the length l_x of the projection of the rod onto the x axis, measured with respect to S? Here one must bear in mind that with respect to S the end-point coordinates x_2, x_1 vary with the time.

Evidently $l_x = \Delta x = x_2 - x_1$, where this coordinate difference must be taken *at a given time t with respect to S*, i.e. with $\Delta t = 0$, and emphatically not with $\Delta t' = 0$. More explicitly, let the end-point positions be measured, and call these measurements events 1 and 2; then $l_x = \Delta x$ if and only if $\Delta t = 0$. Remarkably, the quickest way to Δx is not through the inverse transformations (4.8) which seem tailormade to give the unprimed in terms of the primed quantities, but through the forward transformations (4.7) giving the primed in terms of the unprimed. From the second of equations (4.7) we find immediately that

$$\Delta t = 0 \quad \Rightarrow \quad \Delta x = \Delta x'/\gamma = \sqrt{1 - v^2/c^2}\,\Delta x', \tag{4.10}$$

or in other words that

$$l_x = l_{0,x}\sqrt{1 - v^2/c^2} < l_{0,x}. \tag{4.11}$$

When the rod is moving, it is shorter in the direction of motion than it is at rest, as anticipated in Section 3.3.2. This change in length is called[4] the Lorentz contraction.

There is no Lorentz contraction of lengths measured perpendicularly to **v**, because coordinates in these directions are unaffected by Lorentz transformations:

$$l_y = l_{0,y}, \qquad l_z = l_{0,z}. \tag{4.12}$$

Example 4.1.   A rod of rest length l_0 is fixed in S' at an angle θ_0 to the x' axis. What is its length l and its orientation relative to S?

Solution. Choose axes so that the rod lies in the xy and $x'y'$ plane. Then we can forget all about the z direction, and have

$$l_{0,x} = l_0 \cos\theta_0, \quad l_{0,y} = l_0 \sin\theta_0; \qquad l_x = l\cos\theta, \quad l_y = l\sin\theta.$$

Transforming these components in turn using (4.11) and (4.12) we find

$$l_x = l\cos\theta = \sqrt{1 - v^2/c^2}\,l_0 \cos\theta_0, \quad l_y = l\sin\theta = l_0 \sin\theta_0. \tag{4.13}$$

The angle θ between the rod and the x axis is given by

$$\tan\theta = \frac{l_y}{l_x} = \frac{\sin\theta_0}{\sqrt{1 - v^2/c^2}\cos\theta_0} = \gamma\tan\theta_0, \tag{4.14}$$

up to the ambiguity of the inverse tangent,[5] which must be resolved by fitting the sign of $\sin\theta$ or of $\cos\theta$. Equation (4.13) shows that the signs of $\sin\theta$ and $\sin\theta_0$ ($\cos\theta$ and $\cos\theta_0$) are the same; hence θ lies in the same quadrant as θ_0. The higher v, the more the rod leans away from the direction of motion, and $v \to c$ entails $|\tan\theta| \to \infty$, $|\theta| \to \pi/2$. Compare this with the Galilean result $\theta = \theta_0$ from Example 2.1.

Pythagoras gives the length relative to S:

$$l = \sqrt{l_x^2 + l_y^2} = l_0\sqrt{1 - (v/c)^2 \cos^2\theta_0}. \quad \blacksquare$$

Some comments are in order.

(i) The Lorentz contraction requires reasoning notably more subtle than does time dilation. It is too easily forgotten that one must consider the positions of the two ends of the rod *simultaneously with respect to the laboratory frame S*; unless this restriction is incorporated into the argument carefully and explicitly, one is apt to conclude that the rod expands. It is easy to see why: consider two events separated by $\Delta x'$ with respect to S', but simultaneous with respect to S' and not (as in the argument leading to the Lorentz contraction) with respect to S. Then $\Delta t' = 0$, and (4.8) yields

$\Delta x = \gamma \Delta x' > \Delta x'$, a perfectly correct conclusion but irrelevant to the length of the rod.

(ii) The Lorentz contraction is not merely apparent but real. Words like "the rod appears to be contracted" mislead and should be avoided. By contrast, it is a challenging problem to determine *the visual appearance* of a moving object, as registered for instance on a photographic plate: the plate responds to light which has left different points of the object at different times but all of which arrives at the plate at one and the same time. The full solution is very intricate and well beyond our scope.[6] For example, snapshots show small and distant objects not contracted but turned through an angle; while snapshots of spheres always have a circular outline, but generally distort the surface. Illustrations in some older popular books showing visible contraction should be ignored.

(iii) One is certainly entitled to ask *why* a uniformly moving body is shorter than it is at rest. The reason has nothing to do with the stresses that it must inevitably have experienced under acceleration: in the first place these can be made as small as one likes by making the acceleration smaller and letting it last longer; and in any case, unless they exceed the elastic limits of the material, the strains so induced[7] disappear once the body reverts to uniform motion.

The true reason can depend only on the internal forces responsible for the structure of the body. In ordinary matter these are electromagnetic forces; and given two charges both moving at the same velocity **v**, the electromagnetic force between them certainly does depend on **v**. There is nothing esoteric about this: stationary charges give rise and respond only to electric fields, while moving charges give rise and respond to magnetic fields as well. A sufficiently hard-working theorist could predict the Lorentz contraction from Maxwell's equations with never a mention of Lorentz transformations: one is confident that he can, because the equations respect Einstein's relativity principle whether he notices this or not. However, if he does know that the equations respect the principle, then he can derive the Lorentz contraction as above, that is to say with perfect generality and in a way wholly independent of any details of electromagnetism and of the quantum theory of matter.

Exactly the same applies to other bodies, like atomic nuclei, whose structure is governed wholly or partly by forces other than electromagnetic: we believe that all forces, even if our knowledge of them is incomplete, conform to Einstein's relativity principle, which is all that the argument requires.

(iv) Similarly one is entitled to ask *why* the rate of a clock depends on its velocity. The answer is the same, except that the mechanisms responsible now include the dynamic as well as the equilibrium properties of its structure.

4.2.4 Particle Velocity

One reasons as in deriving the Galilean rules (2.9). Let a particle have velocity[8] **u**′ relative to S', so that over a small time interval $\delta t'$ its displacement is $\delta \mathbf{r}' = \mathbf{u}'\delta t'$. Over the same interval referred to S, where the velocity is **u**, one has $\delta \mathbf{r} = \mathbf{u}\delta t$. We use the Lorentz transformations (4.7) to express the primed in terms of the unprimed coordinate differences:

$$\delta x' = u_x'\delta t' \;\Rightarrow\; \gamma(\delta x - v\delta t) = u_x'\gamma\left(\delta t - v\delta x/c^2\right);$$

with $\delta x = u_x \delta t$ this yields

$$\delta t(u_x - v) = u'_x \delta t(1 - vu_x/c^2) \;\Rightarrow\; u'_x = \frac{u_x - v}{1 - vu_x/c^2}.$$

Perpendicularly to **v** one finds

$$\delta y' = u'_y \delta t' \;\Rightarrow\; \delta y = u'_y \gamma(\delta t - v\delta x/c^2),$$

$$\delta t u_y = u'_y \gamma \delta t(1 - vu_x/c^2) \;\Rightarrow\; u'_y = \frac{u_y}{\gamma(1 - v\delta x/c^2)},$$

and similarly for the z component. Collecting the results and their inverses we display the Lorentz-transformation rules for particle velocities:[9]

$$u'_x = \frac{u_x - v}{(1 - vu_x/c^2)}, \quad u'_{y,z} = \frac{u_{y,z}}{\gamma(1 - vu_x/c^2)} = \frac{u_{y,z}\sqrt{1 - v^2/c^2}}{(1 - vu_x/c^2)}, \qquad (4.15)$$

$$u_x = \frac{u'_x + v}{(1 + vu'_x/c^2)}, \quad u_{y,z} = \frac{u'_{y,z}}{\gamma(1 + vu'_x/c^2)} = \frac{u'_{y,z}\sqrt{1 - v^2/c^2}}{(1 + vu'_x/c^2)}. \qquad (4.16)$$

Clearly these rules are more complicated than the Lorentz transformations for the coordinates. In the nonrelativistic limit $v/c \to 0$ they reduce as they should to the Galilean rules (2.9). The denominators in (4.16) say could be written as $(1 + \mathbf{v}\cdot\mathbf{u}'/c^2)$, but right now this is not much of a simplification.

Evidently the velocities $\mathbf{u}'$ and $\mathbf{u}$ on the one hand and $\mathbf{v}$ on the other hand have entered the argument quite differently, $\mathbf{u}'$ and $\mathbf{u}$ being particle velocities, while $\mathbf{v}$ is the velocity of one frame relative to another. Nevertheless, in an obvious sense the velocity $\mathbf{u}$ results from combining (is the resultant of) the velocities $\mathbf{u}'$ and $\mathbf{v}$, and it can be helpful to speak of it in these terms, and to call (4.16) the *velocity-combination rule for particles*. (Problem 4.9 is one example where this way point of view proves useful.) For instance, we see that the *collinear* (or *unidirectional*) formula

$$u_x = \frac{u'_x + v}{1 + vu'_x/c^2} \qquad (4.17)$$

is symmetric in u'_x and v. Also, in view of

$$\lim_{v \to c} u_x = \frac{u'_x + c}{1 + u'_x/c} = c$$

it shows that if v or u'_x or both tend to c, then u_x also tends to c: in other words, by combining achievable collinear speeds one can approach but cannot exceed c. In fact this is just a special case of a general result applying to all particle velocities, collinear or not. With $u^2 \equiv u_x^2 + u_y^2 + u_z^2$, some straightforward though tedious algebra[10] leads from (4.16) to

$$1 - u^2/c^2 = \frac{(1 - u'^2/c^2)(1 - v^2/c^2)}{(1 + \mathbf{v}\cdot\mathbf{u}'/c^2)^2}, \qquad (4.18)$$

symmetric in $\mathbf{u}'$ and $\mathbf{v}$. The right-hand side vanishes if u' or v or both are equal to c, regardless of directions; then the left-hand side must vanish too, entailing $u = c$. The inverse of (4.18) reads

$$1 - u'^2/c^2 = \frac{\left(1 - u^2/c^2\right)\left(1 - v^2/c^2\right)}{(1 - \mathbf{v} \cdot \mathbf{u}/c^2)^2},\tag{4.19}$$

symmetric in $\mathbf{u}$ and $\mathbf{v}$. Rearrangement then yields the transformation rules for the speed:

$$u^2 = \frac{(\mathbf{u}' - \mathbf{v})^2 - (\mathbf{u}' \times \mathbf{v})^2}{(1 + \mathbf{v} \cdot \mathbf{u}'/c^2)^2}, \qquad u'^2 = \frac{(\mathbf{u} + \mathbf{v})^2 - (\mathbf{u} \times \mathbf{v})^2}{(1 - \mathbf{v} \cdot \mathbf{u}'/c^2)^2}.\tag{4.20}$$

Example 4.2. Relative to S', a particle moves with speed u' in the $x'y'$ plane, at an angle ϕ' to the x' axis. What are (a) its direction and (b) its speed relative to S? For definiteness, take $0 < \phi' < \pi/2$, so that $u'_{x,y}$ are both positive.

Solution. (a) To transform the angle, express it in terms of the components of the velocity. Thus

$$\tan\phi \equiv \frac{u_y}{u_x} = \frac{u'_y/\gamma(v)\left(1 + vu'_x/c^2\right)}{(u'_x + v)/(1 + vu'_x/c^2)} = \frac{u'\sin\phi'}{\gamma(v)(u'\cos\phi' + v)} = \frac{\sqrt{1 - v^2/c^2}\,u'\sin\phi'}{u'\cos\phi' + v},\tag{4.21}$$

differing by the square root from the Galilean Example 2.2. As in Example 4.1, the ambiguity of the inverse tangent is resolved by fitting the signs of the cosine or the sine:

$$u\cos\phi = u_x = \frac{u'_x + v}{(1 + vu'_x/c^2)}, \qquad u\sin\phi = u_y = \frac{u'_y}{\gamma(1 + vu'_x/c^2)}.\tag{4.22}$$

Note first that the denominator $\left(1 + vu'_x/c^2\right)$ is aways positive (even if $vu'_x < 0$), because $|v|$ and $|u'_x|$ are both less than c. In our case u'_y and thereby $\sin\phi$ are positive; u'_x is positive, but $u'_x + v$ and thereby $\cos\phi$ can have either sign, depending on the sign and magnitude of v. If $v > -u'_x$, then ϕ, like ϕ', lies in the first quadrant. But if $v < -u'_x$, then $\tan\phi$ and $\cos\phi$ are both negative, and ϕ lies in the second quadrant. Thus, $v \to +c$ entails $\phi \to 0$, while $v \to -c$ entails $\phi \to \pi$: with respect to S all velocities veer to $\mathbf{v}$, whatever their direction with respect to S'. Contrast this with the transformation of the angle θ in Example 4.1, and see Problem 4.12.

(b) It would be possible but silly to tackle this question from scratch, since the answer is given directly by the general formula (4.18), with $\mathbf{v} \cdot \mathbf{u}' = vu'\cos\phi'$. ∎

Finally some comments.

- *At light speed*, with $u' = c = u$, only the directions need transforming, and the rules (4.21), (4.22) simplify to

$$\cos\phi' = \frac{(\cos\phi - \beta)}{(1 - \beta\cos\phi)}, \qquad \sin\phi' = \frac{\sin\phi}{\gamma(1 - \beta\cos\phi)}, \qquad \tan\phi' = \frac{\sin\phi}{\gamma(\cos\phi - \beta)}.\tag{4.23}$$

Simpler still is the consequent formula

$$\tan(\phi'/2) = \sqrt{\frac{1+\beta}{1-\beta}} \cdot \tan(\phi/2), \qquad (u' = u = c). \tag{4.24}$$

The proof is easiest by hindsight:

$$\tan\frac{\phi'}{2} = \frac{\sin\phi'}{1+\cos\phi'} = \frac{1}{\gamma(1-\beta)} \cdot \frac{\sin\phi}{(1+\cos\phi)} = \sqrt{\frac{1+\beta}{1-\beta}} \cdot \tan\frac{\phi}{2}, \tag{4.25}$$

where the first step is a standard identity; the second follows on using (4.23) and then simplifying; and the last merely reverses the first, but in terms of ϕ' instead of ϕ.

- Consider two successive collinear Lorentz transformations, both in standard configuration: the first from S to S' having velocity v_1 relative to S, and the second from S' to S'' having velocity v_2 relative to S'. The result is equivalent to a single transformation directly from S to S'' having some velocity, call it v_{12}, relative to S (see Section 2.2.2 on the combination of Galilean transformations). Unsurprisingly, the resultant v_{12} is given by the inverse velocity-combination rule (4.16), if one identifies $(v_1, v_2, v_{12}) \rightarrow (u', v, u)$. This should be checked as an exercise, using (4.17) and (4.18) as appropriate. (By contrast, Appendix B, Section B.4.2, shows that if the two relative velocities are not parallel, then in general the axes of S'' are rotated relative to those of S, so that the initial and the final frames are not in standard configuration with each other.)

- It may need stressing that the speed limit applies only to information-carrying signals, or equivalently to the velocity of one thing relative to another; and that there are physically relevant speeds that can exceed c because they are not of this kind. Two examples make the point. (i) You swivel a torch (or a well-focused laser beam) so that the light spot moves across a screen. By making the distance from torch to screen large enough you can in principle make the spot move as fast as you like. No paradox is involved, because the spot cannot carry a signal directly from one point of the screen to another (though it can of course carry information from you to the screen). (ii) As observed from the laboratory, two particles move in opposite directions along the x axis, each with speed $3c/4$. Call their coordinates x_1, x_2, so that the distance between them is $s \equiv x_1 - x_2$. The rate of change of s is called the *closing speed* (as observed in the laboratory). Evidently $|ds/dt| = 3c/2$. No paradox is involved, because s is not the distance between the particles in the rest frame of either. By contrast, Problem 4.9 shows that the speed of one particle measured by an observer sitting on the other certainly is bounded by c.

4.2.5   Particle Acceleration

We reason as we did for velocities, considering now changes of velocity rather than of position over a small time interval.

The *longitudinal acceleration* a_x changes the velocity u_x by

$$a_x\delta t = \delta u_x = \delta\left[\frac{u'_x + v}{1 + u'_x v/c^2}\right] = \frac{\delta u'_x}{(1 + u'_x v/c^2)} - \frac{(u'_x + v)v\delta u'_x/c^2}{(1 + u'_x v/c^2)^2} = \frac{\delta u'_x(1 - v^2/c^2)}{(1 + u'_x v/c^2)^2}.$$

Substituting $\delta u'_x = a'_x \delta t'$ we find

$$a_x \delta t = \frac{a'_x \delta t'}{\gamma^2 (1 + u'_x v/c^2)^2}.$$

Finally δt is expressed in terms of primed variables by the inverse Lorentz transformation

$$\delta t = \delta [\gamma(t' + vx'/c^2)] = \gamma(\delta t' + v u'_x \delta t'/c^2) = \delta t' \gamma(1 + v u'_x/c^2);$$

substitution and rearrangement then yield

$$a_x = \frac{a'_x}{\gamma^3 (1 + v u'_x/c^2)^3}, \qquad a'_x = \frac{a_x}{\gamma^3 (1 - v u_x/c^2)^3}. \qquad (4.26)$$

The second relation is the inverse of the first, given by the rule to interchange primed and unprimed and to reverse the sign of v. (It is instructive to check the inverse by substituting from (4.15) for u'_x in the first denominator, and rearranging.)

The *transverse acceleration* follows similarly from the changes in $u_{y,z}$:

$$a_{y,z} \delta t = \delta u_{y,z} = \delta \left[\frac{u'_{y,z}}{\gamma(1 + v u'_x/c^2)} \right] = \frac{\delta u'_{y,z}}{\gamma(1 + v u'_x/c^2)} - \frac{u'_{y,z} v \delta u'_x/c^2}{\gamma(1 + v u'_x/c^2)^2},$$

$$a_{y,z} \delta t' \gamma(1 + v u'_x/c^2) = \frac{a'_{y,z} \delta t'}{\gamma(1 + v u'_x/c^2)} - \frac{u'_{y,z} v a'_x \delta t'/c^2}{\gamma(1 + v u'_x/c^2)^2},$$

$$a_{y,z} = \frac{1}{\gamma^2 (1 + u'_x v/c^2)^2} a'_{y,z} - \frac{v u'_{y,z}/c^2}{\gamma^2 (1 + u'_x v/c^2)^3} a'_x, \qquad (4.27)$$

$$a'_{y,z} = \frac{1}{\gamma^2 (1 - u_x v/c^2)^2} a_{y,z} + \frac{v u_{y,z}/c^2}{\gamma^2 (1 - u_x v/c^2)^3} a_x. \qquad (4.28)$$

Chapter 8 will return to these transformations with a better toolkit.

Note that if the transverse velocity $u_{y,z}$ is nonzero, then $a'_{y,z}$ can be nonzero even when $a_{y,z}$ vanishes. This reflects a geometric complication in transforming accelerations that is absent for velocities. The transform $\mathbf{u}$ depends on only two vectors, $\mathbf{u}'$ and $\mathbf{v}$, and lies in the plane they define. But the transform $\mathbf{a}$ depends on three vectors, namely on $\mathbf{a}'$, $\mathbf{u}'$, and $\mathbf{v}$, which need not be coplanar; and in the general case where they are not, $\mathbf{a}$ is not confined to the plane spanned by $\mathbf{a}'$ and $\mathbf{v}$.

Example 4.3. Relative to S', a particle oscillates along the y' axis:

$$x' = 0 = z', \qquad y' = A' \sin(\omega' t'), \qquad \text{(i)}$$

where A', ω' are constants. (a) Find its acceleration relative to S (in standard configuration with S'), as a function of time t. (b) Confirm your result by first determining the trajectory $\mathbf{r}(t)$ of the particle relative to S.

Solution. (a) Differentiating (i) twice with respect to t' we find

$$u'_{x,z} = 0 = a'_{x,z}, \qquad u'_y = \omega' A' \cos(\omega' t'), \qquad a'_y = -\omega'^2 A' \sin(\omega' t').$$

Substitution into (4.26), (4.28) yields

$$a_{x,z} = 0, \qquad a_y = a'_y/\gamma^2 = -\left(\omega'^2 A'/\gamma^2\right)\sin(\omega' t'). \tag{ii}$$

This expression for a_y is true but unhelpful to the observer in S, whose clocks indicate not t' but t. However, the Lorentz transformation $t = \gamma(t' + vx'/c^2) = \gamma t'$ entails $t' = t/\gamma$, whence

$$a_y = -A\omega^2 \sin(\omega t), \qquad \omega \equiv \omega'/\gamma, \qquad A \equiv A'. \tag{iii}$$

(b) The trajectory relative to S is found by Lorentz-transforming the coordinates and then eliminating t' in favour of t. We have

$$x = \gamma(x' + vt') = \gamma vt', \qquad y = y' = A'\sin(\omega' t'), \qquad z = z' = 0$$

and $t' = t/\gamma$ as above, whence

$$x = vt, \qquad y = A\sin(\omega t), \qquad z = 0. \tag{iv}$$

with ω and A as in (iii). From (iv) we calculate the acceleration by differentiating twice with respect to t. Evidently this recovers $a_{x,z} = 0$, $a_y = -A\omega^2 \sin(\omega t)$, confirming (ii) and (iii). ∎

4.3 Derivation of the Transformations

Having had some practice in applying them, we now derive the Lorentz-transformation rules (4.1) for event coordinates. *Derivation* here means giving mathematical expression to several symmetries of space and time so familiar that it can seem pedantic to spell them out, and then marrying them to Newton's first law and to the relativity principle. The same reasoning with just one change right at the end yields the Galilean transformations instead.

We are concerned with the coordinates $(t, \mathbf{r})$ and $(t', \mathbf{r}')$ of a given event relative to two frames in standard configuration, and keep an eye on Figure 4.1.

(i) From symmetry alone one can conclude that

$$y' = y, \qquad z' = z,$$

because the contrary would be absurd. The point is that S' and S are distinguished only by the sign of the velocity ($\pm v$) of one relative to the other. If we choose to reverse the direction of the x and x' axes, i.e. if we choose to take the axes as positive to the left rather than to the right, then the sign of v is reversed; this means that in any mathematical expression S' and S exchange roles. On the other hand, because this choice is purely arbitrary, it can have no physical consequences. Now suppose we had assumed $y' > y$; then interchanging the roles of S' and S would entail a change to $y > y'$, a change that does have physical consequences: in other words the relative magnitudes of y and y' would then distinguish absolutely between left and right, which in fact is impossible. Hence the inequality $y' > y$ is absurd, the opposite inequality is absurd for the same reasons, and one must indeed have $y' = y$. Similarly we must have $z' = z$. The underlying *left-right symmetry* is often called reflection invariance.

(ii) Furthermore, t', x' cannot depend on y, z, whose values could be changed simply by choosing the origin of S at a different point of the yz plane, which has no effect on physics. Formally, we are appealing to the *homogeneity* of space. The upshot is that in standard configuration we can forget all about y, y' and z, z', and need relate only (t', x') to (t, x).

(iii) Because Newton's first law holds in both frames (is invariant), motion such that $x = ut$ with constant u must entail $x' = u't'$ with some constant u'. This is guaranteed if the transformation is *linear*:[11]

$$x' = \gamma(v)[x - vt], \qquad t' = v\eta(v)x + \alpha(v)t, \tag{4.29}$$

where the as yet arbitrary coefficients γ, η, α can depend only on v (but not on the event, i.e. not on the coordinates). We have written $v\eta(v)$ by hindsight in order to save some algebra later: it is just a fancy way of expressing an arbitrary function of v, which could just as easily have been represented by a single symbol rather than as a product. The right-hand side of the first relation is proportional to $(x - vt)$ simply because in standard configuration the origin O' of S' (i.e. the point $x' = 0$) is, with respect S, at $x = vt$. Similarly, the second relation has no additive constant on the right, because in standard configuration $x = 0$ and $t = 0$ entail $t' = 0$.

(iv) By virtue again of *left-right symmetry*, we are free to reverse the direction of the x axis, so that (4.29) must continue to hold on replacing $x \to -x$, $x' \to -x'$, and $v \to -v$. Hence

$$-x' = \gamma(-v)[-x + vt] \;\Rightarrow\; x' = \gamma(-v)[x - vt],$$

$$t' = -v\eta(-v)(-x) + \alpha(-v)t \;\Rightarrow\; t' = v\eta(-v) + \alpha(-v)t.$$

Comparison with (4.29) shows that

$$\gamma(-v) = \gamma(v), \quad \eta(-v) = \eta(v), \quad \alpha(-v) = \alpha(v); \tag{4.30}$$

in other words the coefficients are even functions of v, depending only on the magnitude of v but not on its sign. Thus we can safely set $\gamma(-v) = \gamma(v) = \gamma$.

(v) Now write down *the inverse transformation*, following first one and then the other of our familiar prescriptions, and compare the results. Solving (4.29) algebraically for x, t in terms of x', t' yields

$$x = \frac{\alpha x' + \gamma vt'}{\gamma[\alpha + v^2\eta]}, \qquad t = \frac{-v\eta x' + \gamma t'}{\gamma[\alpha + v^2\eta]}.$$

On the other hand, reversing the sign of v (which by (4.30) has no effect on the coefficients) and interchanging primed and unprimed yields

$$x = \gamma[x' + vt'], \qquad t = -v\eta x' + \alpha t'.$$

But the two expressions for x must be equal for all values of x' and of t', which is possible only if the coefficients of x' and of t' are the same in both. Equating the coefficients we find

$$\frac{\alpha}{\gamma[\alpha + v^2\eta]} = \gamma \Rightarrow \frac{1}{\alpha + v^2\eta} = \frac{\gamma^2}{\alpha},$$

$$\frac{\gamma v}{\gamma[\alpha + v^2\eta]} = \gamma v \Rightarrow \frac{1}{\alpha + v^2\eta} = \gamma.$$

Together these equations entail

$$\alpha = \gamma, \quad \eta = (1/\gamma - \gamma)/v^2, \tag{4.31}$$

leaving γ as the only unknown. Substituting into (4.29) we find

$$x' = \gamma[x - vt], \qquad t' = \frac{1/\gamma - \gamma}{v} x + \gamma t. \tag{4.32}$$

(vi) Only now do we invoke the existence of an invariant speed c, whether finite or infinite. Let a signal travelling at this speed in the x (hence also x') direction be emitted at $t = 0 = t'$ from the then coincident origins O, O'. Subsequently its position satisfies $x/t = c = x'/t'$. With these special values (4.32) yields

$$c = \frac{x'}{t'} = \frac{\gamma[x - vt]}{(1/\gamma - \gamma)x/v + \gamma t} = \frac{\gamma[c - v]}{(1/\gamma - \gamma)c/v + \gamma}. \tag{4.33}$$

If c is finite, then (4.33) rearranges straightforwardly to $(1 - v^2/c^2)\gamma = 1/\gamma$, whence

$$\gamma = 1/\sqrt{1 - v^2/c^2}, \qquad (1/\gamma - \gamma)/v = -\gamma v/c^2. \tag{4.34}$$

Substitution into (4.32) then reproduces as promised the Lorentz transformations (4.1).

If on the other hand c is infinite, then (4.33) leads to

$$\lim_{c \to \infty} \frac{\gamma[c - v]}{(1/\gamma - \gamma)c/v + \gamma} = \frac{\gamma}{(1/\gamma - \gamma)/v} = \frac{\gamma v}{1/\gamma - \gamma} = c \to \infty$$

$$\Rightarrow \; 1/\gamma - \gamma = 0 \Rightarrow \gamma = 1, \tag{4.35}$$

which reproduces the Galilean transformations directly. Of course we already know that they can also be reproduced indirectly, by first deriving the Lorentz transformations as if for finite c, and taking the limit $c \to \infty$ of (4.34) afterwards.

4.4 Evidence: Time Dilation

Evidence for the assumptions made by the theory, i.e. for its input, was outlined in Chapter 3. At this point one is interested in evidence testing its output: specifically its first three simple predictions, about time dilation, Lorentz contraction, and the particle-velocity combination rule. We have stressed that theories are assessed through the totality of their consequences: in logic there is no need, and in practice it is obviously impossible to confirm every conceivable implication. Nevertheless, it would be pleasing to be able to cite evidence for each of our three predictions independently of the other two. Or one might settle for two out of three, on the grounds that kinematics deals (only) with time and space: correspondingly there are just two essentials in which Lorentz and Galilean transformations have scope to differ, and do differ, namely in their effect on the time and on the space coordinates; whence two tests should suffice even for propaganda. Therefore it is disappointing, if only superficially, that at this level of simplicity one can report only one test, for time dilation.

The reason why the Lorentz contraction has not been observed is obvious: particles and nuclei can be made to move fast but their linear dimensions cannot be measured directly (though sometimes they can be inferred from other observations); and macroscopic bodies whose lengths we can measure cannot be made to move fast enough. By contrast, the velocity combination rule would be easy though perhaps expensive to test. One possible scenario might use the decay $\Sigma^+ \to \Lambda^0 + \pi^+$ of charged Σ^+ into neutral Λ^0 hyperon plus charged pion. The speed of the pion relative to the parent Σ^+ is known (Chapter 9 shows how it can be inferred from the masses); the laboratory speed of the parent could be inferred from its momentum, or measured; and the laboratory speed of the daughter π^+ would have to be measured directly, say as in the (different) experiment sketched below. But if any such test has been reported, references to it have become effectively untraceable: whereas reported failures would certainly be well remembered.

Thus we must settle for a well-authenticated test just of time dilation.[12] Charged pions (particles of rest mass $m \approx 140$ MeV/c^2) decay into a mu-meson plus a neutrino ($\pi^\pm \to \mu^\pm + \nu$), with a mean life in their own rest frame of $T_0 = 2.60 \times 10^{-8}$ s, the same for both signs of the charge. This means that of an initial number $N(0)$ of pions *at rest*, the average number surviving after time t is $N(t) = N(0) \exp(-t/T_0)$. The experiment compares T_0 with the mean life in flight at speed $u = \beta c$, call it T_β. In Example 3.1 we have seen that time dilation predicts $(T_\beta)_\text{theory} = T_0/\sqrt{1 - \beta^2}$, so that of an initial $N(0)$ of mesons *in flight*, the average number predicted to survive after time t is $N(0) \exp(-t/T_\beta) = N(0) \exp(-t\sqrt{1 - \beta^2}/T_0)$.

Pions are produced in bunches by letting bunches of fast protons from the Berkeley cyclotron hit a target. They are selected for momentum (and thereby for β) by a magnetic spectrometer; but their velocity is checked in the course of the experiment by direct time-of-flight measurements, so that one knows β independently of the relativistic dynamics of the selection process (to which we shall come only in Appendix C, Section C.2). The pions then coast down a straight evacuated drift tube, traversing six Cherenkov counters filled with liquid deuterium, approximately 2 m apart, which indicate the passage of each bunch without absorbing or slowing the pions appreciably. Such counters work by emitting flashes of light when traversed by charges moving faster than the speed of light in the medium; the light is then registered by photodetectors connected to each counter. As a given batch of pions moves down the line, the time interval between flashes from successive counters determines the pion speed. The intensity of the flashes is proportional to the number of pions, so that the intensity ratios determine the relative decrease of pion number with distance downstream; since the speed is known, this translates directly into a decrease of $N(t)/N(0)$ with time t. It is one of the beauties of the experiment that one need not determine the absolute number of pions.

Measurements were made with $\beta = 0.912$, entailing a dilation factor $\gamma = 1/\sqrt{1 - \beta^2} = 2.44$. They were analysed by choosing $(T_\beta)_\text{exp}$ so as to get the best fit between $\exp(-t/(T_\beta)_{\exp})$ and the measured $N(t)/N(0)$; the results are reported as

$$(T_\beta)_{\exp} = (T_\beta)_{\text{theory}} \quad \text{within} \pm 0.4\%, \tag{4.36}$$

verifying the prediction within the indicated experimental error.

One merit of this design is that the particles move uniformly, so that the theory applies without any special pleading. The next chapter notes that in fact the theory applies equally to particles under any practically achievable acceleration; once this is accepted, muons circulating in a storage ring provide a check of comparable accuracy.

4.5 Notes

1. In principle one could invert the argument, and derive the Lorentz transformation directly from time dilation plus Lorentz contraction (see, for example, W. Rindler, "Einstein's priority in recognizing time dilation physically, *American Journal of Physics* (1970), volume **38**, page 1111). The fact that speed c is found to transform into speed c might then be regarded as a self-consistency check. We follow a different route so as to avoid obscuring the message that it is the Lorentz transformations that are the key both to the principles of the theory and to its efficient application.

2. For completeness, Appendix A does give the Lorentz transformation between two frames in arbitrary configuration, whose relative velocity need not point along the x axis.

3. These Δ should not be confused with the same symbol used with a different meaning in conservation laws, as in Section 2.4.

4. It is also called the FitzGerald or the Lorentz-FitzGerald contraction, though FitzGerald's name seems, unfairly, to be fading. But see J. S. Bell "George Francis FitzGerald", *Physics World*, September 1992, page 31.

5. We could have used the sine or the cosine instead of the tangent, but they are less convenient because they would force one to find the length ratio l/l_0 before the angle.

6. Comparatively accessible discussions include A Peres, "Relativistic telemetry", *American Journal of Physics* (1987), volume **55**, page 516; G. D. Scott and H. J. van Driel, "Geometrical appearance at relativistic speeds", *ibid.* (1970), volume **38**, page 971; and the original account by J. Terrell, "Invisibility of the Lorentz contraction", *Physics Review* (1959), volume **116**, page 1041.

7. For examples, see Section 6.3.

8. The Cartesian components of $\mathbf{u}'$ will be written as u'_x and so on. In standard configuration (where the corresponding primed and unprimed axes are parallel by definition) the literally correct $u'_{x'}$ is unnecessary, and would be intolerably pedantic.

9. The Lorents transformations of wave velocities will be discussed in Chapter 12. Group velocities turn out to transform exactly like particle velocities, which one would have expected for the same reasons as in the Galilean case (see Section 2.5.2). Phase velocities transform quite differently, unless they are parallel to $\mathbf{v}$, in which case they too happen to obey the rules (4.15), (4.16) for particles.

10. See also Problem 4.9. The invariants introduced in Chapter 7 will lead to (4.18) with far less effort. But (4.18) should be derived at this point, tedium and all, in order to appreciate just how much one does gain by using invariants.

11. Linearity means that x, t occur on the right at most to the first power, to the exclusion for instance of x^2, t^2, xt, $\sqrt{xt^3}$, and so on. In fact linearity is necessary as well as sufficient to guarantee the invariance of Newton's first law, but we do not prove this here.

12. A. J. Greenberg *et al.*, "Charged-pion lifetime and a limit on a fundamental length", *Physical Review Letters* (1969), volume **23**, page 1267. The experiment was done at the Lawrence Radiation Laboratory in Berkeley, California.

5 Invariant Intervals and Space–Time Diagrams

5.1 Intervals Timelike, Spacelike, and Lightlike

Given the time and space intervals $(\Delta t, \Delta \mathbf{r})$ between two events, it proves useful to define the combination

$$\Delta s^2 \equiv c^2 (\Delta t)^2 - (\Delta \mathbf{r})^2 = c^2 (\Delta t)^2 - (\Delta x)^2 - (\Delta y)^2 - (\Delta z)^2, \qquad (5.1)$$

where Δs^2 is to be regarded as a single symbol, and not as the square of some quantity Δs. In particular, Δs^2 can be positive, negative, or zero.

It is a crucially important fact, to be proved presently, that Lorentz transformations leave Δs^2 unchanged. What this means is the following: transformation changes Δt into $\Delta t' \neq \Delta t$ and $\Delta \mathbf{r}$ into $\Delta \mathbf{r}' \neq \Delta \mathbf{r}$, but nevertheless

$$c^2 (\Delta t')^2 - (\Delta x')^2 - (\Delta y')^2 - (\Delta z')^2 = c^2 (\Delta t)^2 - (\Delta x)^2 - (\Delta y)^2 - (\Delta z)^2. \qquad (5.2)$$

We say that Δs^2 is *invariant*, or *an invariant*. In exploiting the relativity principle, invariants generally will tend to take centre-stage from here on. Though less prominent they are useful also in Newtonian physics: Section 2.5.3 for instance exploited the Galilean invariance of $\omega - \mathbf{k} \cdot \mathbf{w}$ in discussing harmonic plane waves. (As a rule it is of course different combinations that are left invariant by Galilean and Lorentz transformations respectively.)

To prove (5.2) it is convenient to work in standard configuration, with ct instead of t, and to write the Lorentz transformations (4.7), (4.8) in terms of $\beta \equiv v/c$, $\gamma \equiv 1/\sqrt{1 - \beta^2}$:

$$c\Delta t' = \gamma(c\Delta t - \beta \Delta x), \quad \Delta x' = \gamma(\Delta x - \beta c\Delta t), \quad \Delta y' = \Delta y, \quad \Delta z' = \Delta z, \qquad (5.3)$$

$$c\Delta t = \gamma(c\Delta t' + \beta \Delta x'), \quad \Delta x = \gamma(\Delta x' + \beta c\Delta t'), \quad \Delta y = \Delta y', \quad \Delta z = \Delta z'. \qquad (5.4)$$

Thus we need show only that

$$c^2 \Delta t^2 - \Delta x^2 = c^2 \Delta t'^2 - \Delta x'^2, \qquad (5.5)$$

in spite of the fact that $\Delta t \neq \Delta t'$, $\Delta x \neq \Delta x'$. The proof is by explicit algebra from (5.3):

$$
\begin{aligned}
c^2 \Delta t'^2 - \Delta x'^2 &= \gamma^2 (c\Delta t - \beta \Delta x)^2 - \gamma^2 (\Delta x - \beta c \Delta t)^2 \\
&= \gamma^2 \{ (c^2 \Delta t^2 - 2\beta c \Delta t \Delta x + \beta^2 \Delta x^2) - (\Delta x^2 - 2\beta c \Delta t \Delta x + \beta^2 c^2 \Delta t^2) \} \\
&= \gamma^2 \{ c^2 \Delta t^2 (1 - \beta^2) - \Delta x^2 (1 - \beta^2) \} = c^2 \Delta t^2 - \Delta x^2. \quad \blacksquare
\end{aligned}
\tag{5.6}
$$

By the same token the combination $s^2 \equiv c^2 t^2 - \mathbf{r}^2$ is invariant, because the coordinates $(t, \mathbf{r})$ of an event are just the intervals between it and the event at the origin $(0, \mathbf{0})$ (see Section 4.2.1). Hence the relations above apply equally on dropping the Δ's.

As we shall see, the defining property of invariants is also the property that makes them useful in problem solving: (i) their value can be determined in whatever inertial frame proves convenient; (ii) one can equate expressions for them evaluated in different frames.

Evidently one can classify intervals according to their value of Δs^2: the classification is invariant because Δs^2 is invariant, and it turns out to be very useful.

- If Δs^2 is negative, we call the interval *spacelike*, because its space part $-\Delta \mathbf{r}^2$ then dominates the time part $c^2 \Delta t^2$. Given a spacelike interval, one can always find a frame S' where the events are simultaneous, i.e. such that

$$
c\Delta t' = \gamma (c\Delta t - \beta \Delta x) = 0;
$$

one need only choose $\beta = c\Delta t / \Delta x$, which is compatible with $|\beta| < 1$ because $c^2 \Delta t^2 < \Delta x^2$ by assumption. Expressing Δs^2 in S' we see that

$$
\Delta s^2 = -\Delta \mathbf{r}'^2 |_{\Delta t' = 0};
\tag{5.7}
$$

in other words $-\Delta s^2$ is the squared distance between the events in the frame where they are simultaneous, and $\sqrt{|\Delta s^2|}$ is called the *proper distance* between spacelike-separated events.

- If Δs^2 is positive we call the interval *timelike*. Given a timelike interval, one can always find a frame S' where the events happen at the same place:

$$
\Delta x' = \gamma (\Delta x - \beta c \Delta t) = 0
$$

is achieved by choosing $\beta = \Delta x / c\Delta t$, compatibly again with $|\beta| < 1$. In S' we have

$$
\Delta s^2 = c^2 \Delta t'^2 |_{\Delta r' = 0},
\tag{5.8}
$$

and $\sqrt{\Delta s^2 / c^2}$ is called the *proper time interval* between timelike-separated events.

- If Δs^2 is zero we call the interval *lightlike*. Then $|\Delta \mathbf{r}| = c\Delta t$, so that the events can be linked by a light signal. Evidently this is a limiting case dividing spacelike from timelike, though we shall see presently that it has more in common with the latter.

Besides the sign of Δs^2, the sign of Δt also deserves attention.

- For timelike or lightlike intervals ($\Delta s^2 \geq 0$), the sign of $\Delta t'$ is the same as the sign of Δt. This is evident from (5.3). If $c\Delta t > 0$, then $c\Delta t' < 0$ would require $\beta > c\Delta t / \Delta x$, which is impossible since $|\beta| < 1$ always, and $|c\Delta t / \Delta x| \geq 1$ by assumption. Thus *for timelike and lightlike separations the sign of t is invariant: all observers agree as to which event happened first.*

• For spacelike intervals $(\Delta s^2 < 0)$, the time interval can have different signs with respect to different observers, who therefore need not agree as to which happened first. Again this is clear from (5.3). Suppose $\Delta t > 0$; then we can make $\Delta t' < 0$ by choosing β so that $1 > \beta > |c\Delta t/\Delta x|$, which is perfectly possible because now the rightmost expression is less than 1.

Example 5.1. Re-derive time dilation from the invariance properties of Δs^2 and of Δt.

Solution. Consider two events that happen at the same place with respect to S', so that $\Delta \mathbf{r}' = 0$. For definiteness we suppose that both events happen at the origin O'. With respect to S one has $\Delta \mathbf{r} = \mathbf{v}\Delta t$, where $\mathbf{v}$ is the velocity of S' wrt S. Hence

$$\Delta s^2 = c^2 \Delta t'^2 = c^2 \Delta t^2 - \Delta r^2 = \Delta t^2 (c^2 - v^2) \Rightarrow \Delta t' = \pm \Delta t \sqrt{1 - v^2/c^2}.$$

But Δs^2 is positive, so that Δt and $\Delta t'$ must have the same sign, and the positive root is wanted. ∎

5.2 Space–Time Diagrams and the Light Cone

The different types of interval can be illustrated on so-called space–time diagrams. Every such diagram is drawn with respect to some specific inertial frame S. One axis indicates the space coordinate x. The other is effectively the time axis, but indicates ct rather than t, so that quantities with the same physical dimensions are plotted along both axes. The second and third space-coordinate axes must be imagined. This is fairly easy for the second, say for the y axis, which can be pictured as coming out of the paper at right angles to ct and to x; but not so easy for the third.

Points on a space–time diagram will be labelled with CAPITAL SANS SERIF SYMBOLS; *the origin for instance, i.e. the* event *that happens at the point* $\mathbf{r} = \mathbf{0}$ *at time* $t = 0$, *is called* O, *if only because the symbol O has been preempted for the geometrical point* $\mathbf{r} = \mathbf{0}$.

To guide the eye, we draw the two 45° lines, which are trajectories of light signals through O (i.e. of signals that pass through the point $\mathbf{r} = \mathbf{0}$ at time $t = 0$). If we do picture the y axis too, then the set of all such trajectories forms a right-circular double cone around the time axis, with apex at O, called *the light cone through* O. However, generally we shall just ignore y and z, reducing the light cone to the two sloping lines Figure 5.1.

Any event can be indicated by a point on the diagram. Timelike events (shorthand for "events timelike separated from O") lie inside the light cone, like A and B. Because the sign of t is invariant, one distinguishes sharply between A with $t_A > 0$, said to be *inside the future light cone*, and B with $t_B < 0$, said to be *inside the past light cone*.

Lightlike events lie on the light cone, and for them too one distinguishes between future and past; for instance C lies *on the future light cone*.

Spacelike events like D lie *outside the light cone*. Because the sign of their time coordinate depends on one's choice of the frame S, it is usually irrelevant whether spacelike points lie above or below the x axis.

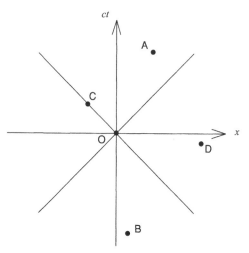

Figure 5.1 Space–time diagram drawn with respect to an inertial frame S, *showing the light cone through the origin* O. The event A is in the forward light cone; B is in the past light cone; C is on the future light cone; D is outside the light cone. For the implications see the text.

The trajectories of particles and of other localizable signals like wave groups are lines on the diagram, called *world lines*.[1] They are restricted by the speed limit c. For example, motion in the x direction at constant velocity u is represented by a straight line at an angle ϕ to the time axis, where $|\tan \phi| = |u|/c < 1$, so that $|\phi| \leq 45°$. If the velocity $u(t)$ of a particle changes, then its world line is curved, with an angle $\phi(t) = \tan^{-1}(u(t)/c)$ that varies but always remains between $\pm 45°$.

Thus signals *from* O can reach any event in or on the future light cone (like A or C); this means that the point x_A where such an event might happen can be reached by a signal arriving no later than the time t_A when the event is scheduled. Likewise signals can be received *at* O from any event in or on the past light cone. World lines passing through O have always been and will always remain trapped within the light cone through O: they cannot cross it because entry or exit would require $|\phi| > 45°$, i.e. speed greater than c, which is impossible. Hence no signal can bridge the gap between O and any event outside the light cone. Some possible and impossible world lines are shown in Figure 5.2.

The constraints just described are said to represent *causality*: signals from O can cause or forestall events in or on the future light cone, and signals from events in or on the past light cone can cause or forestall events at O. But nothing that is done or happens at O can cause or be caused by anything that happens outside the light cone; that is why there is nothing paradoxical about inverting the temporal order of such events relative to O, i.e. changing the sign of their time coordinate. As we saw, this can be done merely by choosing a different reference frame, a choice void of physical consequences.

It is worth stressing that one and the same diagram may usefully show light cones drawn through several different events. The future cones through spacelike-separated events eventually intersect, and their overlap is open to influence from either apex.

A space–time diagram drawn with respect to a different inertial frame S' looks exactly like Figure 5.1, except that its axes are labelled by ct' and x'. Any set of events

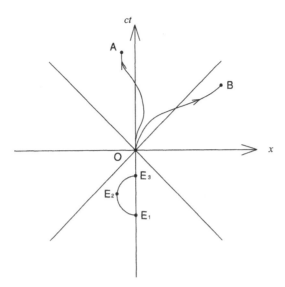

Figure 5.2 Space–time diagram with the same origin, axes, and light cone as in Figure 5.1, plus some would-be trajectories through O. The trajectory OA is possible, because its inclination to the time axis is always less than 45°, showing that the speed along the trajectory is always less than c. The trajectory OB is impossible, as is immediately obvious from the fact that it crosses the light cone. The trajectory $E_1E_2E_3O$ is equally impossible, even though it does not cross the light cone, because near E_1 and E_2 its slope indicates speeds greater than c (in fact infinite speeds at these points).

can be entered on either diagram, with different coordinates (so that a given event appears at different points relative to the two different sets of axes), but with the same invariant separations Δs^2 between any pair of events; in particular, events inside the future light cone of either diagram are inside the future light cone of the other. (Note that, given two frames S and S' in standard configuration, their respective space–time diagrams have a common origin: thus $O = O'$, whereas their ordinary (geometric) origins O and O' coincide only at time $t = 0 = t'$.)

Example 5.2. Draw space–time diagrams showing the trips in Problem 4.4, from the points of view (a) of the earth, and (b) of the probe.

Solution. (a) Figure 5.3(a) is drawn in the earth-fixed frame S. The world line of the earth is the time axis; the world line of the star is parallel to this axis at a distance of $L = 20$ light years. The two legs of the probe's trip are the two sloping lines: events O, A, B are the probe's departure, its landing and immediate takeoff at the star, and its return to earth. Each leg takes (earth) time $T = L/u$.

(b) Let S', S'' be the rest frames of the probe on its way out and back. Since these are different frames, a separate diagram is needed for each leg. The transfer at the star from S' to S'', unavoidable if the probe is to get back, is called *frame hopping*. Figure 5.3(b) shows the outward leg. Notice that the distance between earth and star is Lorentz-contracted: $L' = L\sqrt{1 - u^2/c^2}$. The journey time is $T' = L'/u$. The return leg is shown in Figure 5.3(c). ■

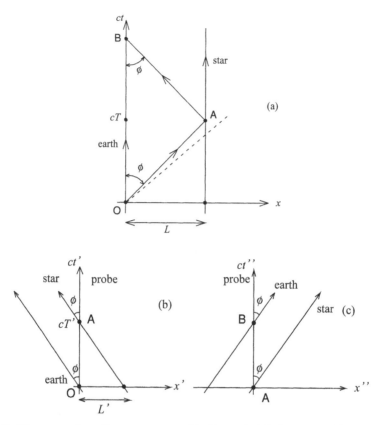

Figure 5.3 The space–time diagrams requested in Example 5.2, drawn roughly to scale. (a) In the rest frame of earth and star. The broken line is part of the light cone through O. Earth trajectory: OB along the time axis; star trajectory: the vertical line $x = L = 20$ light years. Probe trajectory: OA outward, AB return, inclined to the time axis at angles $\phi = \tan^{-1} u/c$ (with the value of u given by the eventual solution to the problem). Note that O $=$ O$'$. (b) In the rest frame of the probe on the outward leg. The trajectories of earth and star are the lines leaning to the left, at the same angle $\phi = \tan^{-1} u/c$. The distance between them is Lorentz-contracted to $L' = L\sqrt{1 - u^2/c^2}$. (c) In the rest frame of the probe on the return leg.

In addition to comparing the different space–time diagrams drawn in different frames S and S', it is instructive to indicate the primed axes on the unprimed diagram, as in Figure 5.4. In standard configuration the primed and unprimed origins are the same: O $=$ O$'$. The ct' axis is the straight line whose equation is $x' = 0 \Rightarrow x = \beta ct$, inclined to the ct axis at an angle of $\theta = \tan^{-1}\beta < 45°$. It is just the trajectory of the (ordinary) origin O$'$ of frame S'; any line parallel to it is the trajectory of some other point with some other fixed value of x'. Similarly the x' axis has the equation $ct' = 0 \Rightarrow ct = \beta x$, inclined by the same angle θ but to the x axis. It is the locus of points that are simultaneous with O$'$ relative to S'; other lines parallel to it correspond to other constant values of t'.

Figure 5.4 demonstrates how a Lorentz transformation can reverse the sign of the time coordinate of an event E spacelike-separated from O $=$ O$'$: points between the x axis and the light cone all have positive t, but those below the x' axis have negative t'.

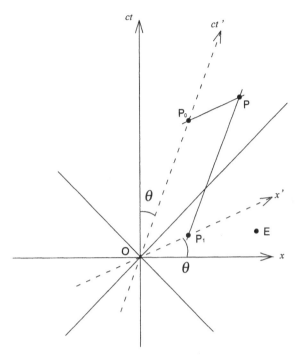

Figure 5.4 Space–time diagram with the same origin, axes, and light cone as in Figure 5.1. Broken lines: the x' and ct' axes of another inertial frame S' in standard configuration with S. They are inclined to the corresponding unprimed axes by angles $\theta = \tan^{-1}\beta$. The lines through P are parallel to the primed axes. The x coordinate of the event P is indicated by the foot of the perpendicular (not shown) dropped from P onto the x axis, *not because this is a perpendicular, but because it is the line through P parallel to the t axis*; and similarly for the t coordinate. The (ct', x') coordinates of P are indicated by P_0, P_1 respectively. However, the scale along the primed axes is greater than the scale along the unprimed by a factor $\sqrt{(1+\beta^2)/(1-\beta^2)}$. For details see the text, and Problem 5.5.

Oblique axes can cause confusion unless one realizes that the projection P_1 of an event P onto the x' axis is not the foot of the perpendicular from P onto the axis, but the intercept of the line through P parallel to the ct' axis; and similarly for the projection P_0 onto the time axis. It follows that the primed and unprimed axes have different scales: if intervals $\Delta x = 1$ m are mapped into intervals of length say b along the x axis of Figure 5.4, then intervals $\Delta x' = 1$ m are mapped into intervals of length $b\sqrt{(1+\beta^2)/(1-\beta^2)} > b$ along the x' axis (see Problem 5.5).

5.3 Note

1. We treat *trajectory* and *world line* as synonymous.

6 Proper Time and Nonuniform Motion

6.1 Preamble: Nonuniform Motion

The relativity principle connects observations referred to different inertial frames, all moving uniformly with respect to each other; but it says nothing about the experiences of observers who are accelerated. In particular, the theory entails, as we have seen, that a uniformly moving clock runs slow by a factor $\sqrt{1 - \beta^2}$ compared with an identically constructed clock at rest; but it is clear that no such blanket statements can be made about an accelerated clock, whose rate may be affected by the acceleration stresses, and can be foreseen only if we know how the clock mechanism will respond to them.[1] For example, the biological clock of a pilot will stop if the acceleration stresses kill him (somewhere above $10g$). But often the effects of the stresses are negligible: the quartz watch of the dead pilot will probably continue to run normally, though it too may be stopped by the deceleration when the plane crashes. Obviously, such effects operate in Newtonian as much as in Einsteinian physics. Timekeeping mechanisms not measurably affected by any acceleration stresses in question will be called *robust*: Section 6.3 outlines some important examples. In the limiting case of total indifference to any such stresses one speaks of an *ideal clock*.

However, even though relativity theory alone cannot determine it, we do quite often need to know the time lapse measured by an accelerated clock, for instance to predict the decay rate of radioactive particles in nonuniform motion. To answer such questions, one adopts the following so-called *clock hypothesis: at any instant, an accelerating* ideal *clock advances at the same rate as would an inertial clock having, momentarily, the same velocity.* In other words, to an *ideal* clock acceleration as such is irrelevant: it matters only indirectly, insofar as it affects the speed. The time registered by such a clock is called its *proper time τ.*

The merit of this hypothesis is that experiment confirms it (see Section 6.4); but we stress again that logically speaking it is not a consequence of relativity theory, but an additional assumption. On the other hand it is a highly plausible assumption even ahead of the evidence, on at least two grounds. First, it is commended by the simplicity and the sheer elegance of the formal definition of the proper time, as should become apparent from the next section. Second, any other assumption would open the door to unlimited complications: if the rates even of ideal clocks were to depend not only on the speed (as dictated by relativity), but on other characteristics of the motion besides, then there is no reason to focus just on the acceleration. Rather, all the time derivatives of the velocity would become equally plausible determinants; every such derivative would have to be admitted into the expression for the rate, with some appropriately dimensional coefficient needing to be found empirically; and any candidate theory requiring so much undetermined input would be too short of predictive power to earn its keep.

In all essentials, what we say about the effects of accelerations on measured times applies also to their effects on lengths; indeed we shall see that in imperfectly robust clocks it is often stress-induced length changes that govern the change in their rates. Much the same remarks apply to rotating systems.

Finally, some possible confusion might be forestalled by recalling from Section 2.3.4 the distinction between active and passive transformations. The prime problem in relativity theory is to relate the *descriptions of a given physical system* referred to different inertial frames, all of which move uniformly with respect to each other. This is the problem solved by the passive transformations. That relations of the same form can be interpreted as active transformations is not obvious, though it may seem so, and though it is true. But *the systems to be described* naturally include those undergoing acceleration or rotation. How to describe such a system relative to an inertial frame is merely a question of kinematics, to be resolved by inspection; but its physical response to acceleration or rotation is a question of its mechanics. For example, a uniformly moving rod is subject to Lorentz contraction, but relative to its rest frame it reverts to its rest length; by contrast, the expansion or contraction of the same rod under acceleration (see Section 6.3) is observed even in its momentary rest frame. To bring such differences into sharper focus, it may help to think very briefly about transformations between inertial and accelerated frames, which are not considered anywhere else in this book, and which we shall not try to explicate at all. Whether such a transformation once written down is to be understood as passive or active is no longer a matter of free choice, because the description of an accelerated system referred to the inertial frame does not have the same form as the description of the unaccelerated system referred to the noninertial frame. This is obvious from the example at the start of this section: whether the pilot is alive or dead is not a question of one's point of view.

6.2 The Proper Time τ

In order to define the proper time of an arbitrarily moving particle in the most useful way, it is best to start by looking for an invariant measure of the advance of the particle along its trajectory. A typical trajectory is shown in Figure 6.1. This space–time diagram is drawn in some frame S, where the trajectory is described by the position $\mathbf{r}(t)$ as a function of time. Consider now the event P_1 at $(ct, \mathbf{r}(t))$ on the trajectory, where the velocity is $\mathbf{u}(t) = d\mathbf{r}/dt$; consider also the small element $P_1 P_2$ along the trajectory next after P_1. The coordinate increments are $c\,dt$ and $d\mathbf{r} = \mathbf{u}\,dt$. The squared invariant interval between P_1 and P_2 is

$$ds^2 = c^2(dt)^2 - (d\mathbf{r})^2 = c^2(dt)^2 - u^2(dt)^2 = (c\,dt)^2(1 - u^2(t)/c^2).$$

It is timelike (because the particle speed u is less than c), and as in Section 5.1 we define the corresponding *increment of proper time* through $d\tau \equiv \sqrt{ds^2/c^2}$, so that

$$d\tau \equiv dt\sqrt{1 - u^2(t)/c^2}. \tag{6.1}$$

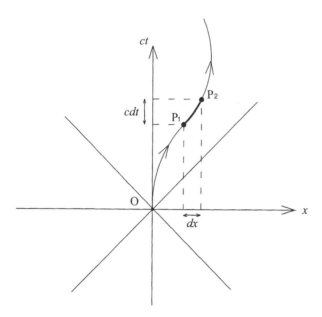

Figure 6.1 Space–time diagram with respect to an inertial frame S, and the light cone through the origin. The world line, being curved, indicates a trajectory described with varying velocity $u(t)$. The text considers an infinitesimal portion $P_1 P_2$ of the trajectory, with increments dt and $dx = u(t)dt$, in order to define the corresponding invariant proper-time increment $d\tau \equiv dt\sqrt{1 - u^2(t)/c^2}$.

From its definition $d\tau$ is an *invariant*: describing the same element of the same trajectory with respect to another frame S' one would find $d\tau \equiv dt'\sqrt{1 - u'^2(t')/c^2}$, even though $dt' \neq dt$ and $u'^2 \neq u^2$.

Any event P on the trajectory can now be assigned a unique value of τ_P by adding up successive increments $d\tau$:

$$\tau_P - \tau_O = \int_O^P d\tau = \int_{t_O}^{t_P} dt\sqrt{1 - u^2(t)/c^2}. \tag{6.2}$$

We might for instance choose to measure τ from O, so that $\tau_O = 0$. The line integral $\int_O^P d\tau$ is understood to be evaluated along the trajectory in question: along a different trajectory between the same two events O and P the increase in proper time would generally be different (see Problem 6.1). By contrast, the rightmost integral in (6.2) is wholly self-contained, and the function $u(t)$ in its integrand contains enough information to evaluate $\tau_P - \tau_O$ explicitly. Of course the same trajectory could equally well be specified with respect to another frame S' by prescribing $\mathbf{r}'(t')$, with $\mathbf{u}'(t') = d\mathbf{r}'(t')/dt'$; then one one would have to evaluate

$$\tau_P - \tau_O = \int_{t'_O}^{t'_P} dt'\sqrt{1 - u'^2(t')/c^2},$$

a calculation possibly quite different from (6.2), but with the same numerical result.

More interesting is the option of specifying the trajectory not through $\mathbf{r}$ as a function of t, but by giving both t and $\mathbf{r}$ as functions of τ. Evidently τ is a kinematic analogue of arc length measured along a given curve in geometry: both are intrinsic properties of a curve, independent of what coordinate frame one may have chosen.

To visualize the physical significance of the small increment $d\tau$ at P, we define the inertial frame S_P that has velocity $\mathbf{u}(t)$ relative to S for all time (and not only at time t). A possible alternative name[2] is $S'(t)$, appropriate if the trajectory is to be described specifically relative to S. We call this frame *the instantaneous rest frame of the particle at* P (*or equivalently at t*), or its *instantaneously* (*or momentarily*) *co-moving frame there*. While $\mathbf{u}$ is changing, a different instantaneous rest frame corresponds to each point of the trajectory; that is why $S'(t)$ has been written with argument t. For example, relative to S, the velocity difference between $S'(t + dt)$ and $S'(t)$ is $\mathbf{a}(t)\,dt$, where $\mathbf{a}(t) \equiv d\mathbf{u}/dt$ is the acceleration at time t.

Accordingly, $d\tau$ can be viewed as the time interval, measured in $S'(t)$, between two events on the trajectory happening at times t and $t + dt$ with respect to S: an *inertial* clock with the velocity $\mathbf{u}(t)$ would advance by $d\tau$ while a clock fixed in S advances by dt.

Example 6.1.   Our speed relative to the cosmic background radiation is approximately 370 km/s (see Appendix E, Section E.1). By how many seconds a year does our proper time drop behind that of the cosmos?

Solution.   In self-explanatory notation, with $T \equiv (1 \text{ year})$,

$$\tau = \int_0^T dt\sqrt{1 - (u/c)^2} \approx \int_0^T dt(1 - u^2/2c^2) = T - T(u/c)^2/2,$$

$$\Delta t \equiv T - \tau \approx T(u/c)^2/2,$$

$$\Delta t = \frac{1}{2} \cdot (365 \times 24 \times 60 \times 60) \cdot \left(\frac{3.7 \times 10^5}{3 \times 10^8}\right)^2 = 24 \text{ s}.$$

(Since in this case u is constant, the question could have been asked and answered directly in terms of time dilation, without explicit mention of proper time.)   ∎

Example 6.2.   By how many seconds a year does proper time in Singapore drop behind a hypothetical reference clock fixed relative to but far away from the sun, (a) due to the orbital motion of the earth; (b) due to its orbital motion jointly with its rotation? (The orbital speed is $u_1 = 3 \times 10^4$ m/s, and the speed of rotation at the equator is $u_2 = 460$ m/s. Pretend that the axis of rotation is perpendicular to the orbital plane.)

Solution.[3]   (a) Since the orbital speed is (near enough) constant, we have, exactly as in Example 6.1, that

$$\Delta t \approx \frac{1}{2}T\left(\frac{u_1}{c}\right)^2 = \frac{1}{2} \cdot (365 \times 24 \times 60 \times 60) \cdot \left(\frac{3 \times 10^4}{3 \times 10^8}\right)^2 = 0.16 \text{ s}.$$

(b) Because both speeds are so much smaller than c, they can be combined according to the Galilean rules. Then, with $\omega \equiv 2\pi/(1 \text{ day})$, the velocities $u_\parallel$ and $u_\perp$ of

Singapore along and at right angles to the orbit are $u_{\parallel} = u_1 + u_2 \cos \omega t$ and $u_{\perp} = u_2 \sin \omega t$, whence $u^2 = u_{\parallel}^2 + u_{\perp}^2 = u_1^2 + 2u_1 u_2 \cos \omega t + u_2^2$. We require

$$\tau = \int_0^T dt \sqrt{1 - u^2(t)/c^2} \approx T - \frac{1}{2} \int_0^T dt\, u^2/c^2.$$

Now subdivide the integral into days; over a day $\cos \omega t$ averages to zero, whence u^2 averages to $u_1^2 + u_2^2$. Therefore Δt is found from the result of part (a) simply by replacing u_1^2 with $(u_1^2 + u_2^2)$, i.e. by multiplying with $(1 + u_2^2/u_1^2) \approx (1.00024)$. Thus the difference due to the rotation of the earth is only of second order in u_2/u_1, and negligible at the accuracy to which we are working. ∎

6.3 Effects of Acceleration Stresses

6.3.1 Dimensional Estimates

In Section 6.1 we conjectured that a robust enough clock indicates its own proper time τ: figuratively speaking, each of its ticks takes precisely as long as the corresponding tick of a comparison clock fixed in the instantaneously co-moving inertial frame. Before citing evidence in the next section, we estimate how high an acceleration supposedly robust clocks might be expected to tolerate before their rate is affected after all. What one finds is that nuclei and elementary particles make ideal clocks for all conceivable purposes; right now atoms do too, though this may change if they are probed just beyond the limits of current technology. But ordinary solids do not.

A purely dimensional argument helps to set the scene, and to see the clock hypothesis in context. We note that Nature supplies a universal scale of speed, namely c, allowing the rate of a clock to depend on its speed u through the dimensionless ratio u/c, the dependence being negligible as long as $u/c \ll 1$. By contrast, she admits no universal scale a_0 of acceleration. Since acceleration has the dimensions $[L][T^{-2}]$, such a scale factor could be written as $a_0 \sim c/T_0$ or $a_0 \sim c^2/L_0$, requiring either some fundamental time T_0 or some fundamental length L_0. Current physics however supplies neither. It might if it featured some truly fundamental mass M_0, which would then admit the estimates $T_0 \sim \hbar/M_0 c^2$ or $L_0 \sim \hbar/M_0 c$ (where $\hbar$ is Planck's constant); but the particles known to high-energy physics are too numerous and too disparate, and their masses too various, for the mass of any particular one of them to be even remotely credible as a candidate for such an M_0. Thus deprived of a universal a_0, we cannot so much as contemplate the possibility that accelerations a might affect clock rates through some universal function of a/a_0; rather one needs to estimate, separately for different kinds of clocks, typical values of a_0 such that the clocks can be expected to be effectively robust as long as their accelerations are well below a_0. Mostly this exercise requires only Newtonian physics. In some cases it helps to remember that the speed of longitudinal (compressional) sound waves in a solid is $c_s = \sqrt{Y/\rho}$, where Y is Young's modulus and ρ the density.

To assess the robustness of timekeeping in particular systems, one must of course know or guess something about them. The obvious approach writes $a_0 \sim u_0^2/r_0$, and then tries to identify a plausibly pertinent speed u_0 and length r_0. Often r_0 is just the system size.

- Typically of *laboratory apparatus*, consider a vibrating mechanical system with linear dimensions $r_0 \sim 0.01$ m, made of aluminium. Then one would estimate $u_0 \sim c_s = 6400$ m/s, so that $a_0 \sim 10^9$ m/s^2.
- As regards the motions inside *atoms*, one expects $u_0^2 \sim (c/137)^2$ as explained near the end of Section 4.1, and $r_0 \sim$ (Bohr radius) $\sim 5 \times 10^{-11}$ m, so that $a_0 \sim 10^{23}$ m/s^2: as yet, no man-made accelerations have had any measurable effect on atomic frequencies or spectral lines. But one day they might. Ultracentrifuges can produce $a \sim 10^6$ m/s^2, which might well entail relative frequency shifts of order $a/a_0 \sim 10^{-17}$; and accuracies of one part in 10^{17} are contemplated for atomic clocks now being designed. Admittedly one can see little chance of the two kinds of apparatus being combined.
- For the motion of protons and neutrons inside *nuclei* we have estimated $u_0^2/c^2 \sim 0.06$, while nuclear radii are typically of order $r_0 \sim 10^{-14}$ m. Thus $a_0 \sim 10^{29}$ m/s^2: nuclei are more far robust than atoms, because they are smaller and the speeds inside them are higher.
- Estimates regarding *elementary particles* are bound to be far more tentative than for atoms and nuclei, because we know far less about their structure. But it seems not unreasonable to expect any internal speeds u to be comparable to c, and linear dimensions to be comparable to but somewhat smaller those of nuclei, say $r_0 \sim 10^{-15}$ m/s. On this basis one would expect $a_0 \sim c^2/r_0 \sim 10^{32}$ m/s^2.

An alternative dimensional estimate based on $c, \hbar$, and the particle mass m would read $a_0 \sim mc^3/\hbar$. For instance, a classic test described in the next section features mu-mesons, which have $m \sim 100$ MeV/c^2, whence $a_0 \sim 10^{48}$ m/s^2. Or one might suspect that electromagnetic forces stemming from the charge e of the muon could also be relevant. Because the fine-structure constant $\alpha \equiv e^2/4\pi\varepsilon_0\hbar c \approx 1/137$ is a pure number, purely dimensional arguments would then fail, admitting as they do $a_0 \sim \alpha^n(mc^3/\hbar)$ with arbitrary n. However, the bracketed factor is so large that a/a_0 remains negligible for any plausible value of n. For example, one might suppose that $ma_0 \sim force \sim e^2/4\pi\varepsilon_0$ (distance)^2, while (distance) $\sim \hbar/mc$, whence $a_0 \sim \alpha mc^3/\hbar$, suggesting $n = 1$ and $a_0 \sim 10^{46}$ m/s^2.

6.3.2 Three Models

For simplicity the accelerations are taken as constant, and small enough for speeds relative to the initial rest frame S to remain far below c. Then Newtonian/Galilean physics suffices to describe the clocks with respect to S.

Model 1.   Consider a uniform straight rod of length L, density ρ and Young's modulus Y, clamped at one end and free at the other, and subject to lengthwise accelerations a, which must of course be induced by forces exerted through the clamp. Count a as positive if the rod is pushed. Then a changes the length by ΔL, where $\Delta L/L = -a\rho L/2Y = -aL/2c_s^2$.

For longitudinal vibrations of the unaccelerated and unstressed rod, the fundamental wavelength and frequency are given by $\lambda/4 = L$, $f = c_s/\lambda = c_s/4L$. To first order in a the speed of sound is unchanged, so that f and the period $T = 1/f$ are shifted by Δf and ΔT, where $\Delta T/T = -\Delta f/f = \Delta L/L = -aL/2c_s^2$. A clock that measures time by counting these periods would therefore run fast by a factor $(1 + aL/2c_s^2)$. Notice how this conclusion tallies with our naive dimensional estimate $a_0 \sim c_s^2/L$, and fractional corrections of relative order a/a_0.

Alternatively, consider a parallel-plate capacitor with capacitance $C = \varepsilon_0 \varepsilon_r A/l$, where A, l are the area and separation of the plates. The plates are square, and made of the same material as the rod envisaged above. The capacitor is part of an $\mathcal{LC}$ circuit of frequency $f = 1/2\pi\sqrt{\mathcal{LC}}$. If the plates are clamped along one edge, and the system is accelerated in the plane of the plates perpendicularly to this edge, then A changes by the same factor as did the length of the rod: $\Delta A/A = -a\sqrt{A}/2c_s^2$; hence $\Delta f/f = -\Delta C/2C = a\sqrt{A}/4c_s^2$, with evident consequences for a clock counting the oscillations.

Model 2. Now envisage a thin, hollow, right circular cylinder, radius R. Rotation around the axis with angular velocity ω requires tension which induces a change $\Delta R/R = \omega^2 R^2/c_s^2 = u^2/c_s^2$, where $u = \omega R$ is the tangential velocity of the rim. Suppose the cylinder has length $l \gg R$, and that it is wound with N turns of wire, so that its self-inductance is $\mathcal{L} = \mu_0 \mu_r N^2 \pi R^2/l$. Then the rotation induces $\Delta\mathcal{L}/\mathcal{L} = 2\Delta R/R = 2u^2/c_s^2$, again with corresponding changes in the timekeeping of an $\mathcal{LC}$ circuit using the inductance.

It is noteworthy that a uniform rod of length R rotating around a perpendicular axis through one end expands differently, namely with $\Delta R/R = u^2/3c_s^2$, where u is the speed of the moving end.

Model 3. Consider, finally, the longitudinal light clock from Section 3.3.2, which relies on the traverse times of light rather than on mechanical vibrations. Suppose that, with respect to its instantaneous rest frame, the clock is made to accelerate longitudinally, in such a way that the separation of the mirrors remains constant. With the clock at rest, the traverse times of the light pulse are l/c in both directions, and the period is $T = 2l/c$. We determine corrections only to leading order in $1/c$, and only for a small enough that the speed gained in one traverse is much smaller than c: thus $a(l/c) \ll c$, or $al/c^2 \ll 1$.

The positions of the trailing and the leading mirror at time t are $at^2/2$ and $(l + at^2/2)$. Let the light pulse leave the trailing mirror at $t = 0$. An easy calculation[4] shows that it reaches the leading mirror at time $t_1 = (l/c)[1 + (al/c^2)/2 + \cdots]$, and that its passage back to the trailing mirror takes a further time $t_2 = (l/c)[1 - 3(al/c^2)/2 + \cdots]$. Thus the period of the clock is

$$T + \Delta T = t_1 + t_2 = (2l/c)[1 - (al/c^2)/2 + \cdots] \Rightarrow \Delta T/T = -(al/c^2)/2.$$

This too tallies with the naive dimensional estimate of fractional corrections of the order a/a_0 with $a_0 \sim u_0^2/r_0$, but now of course with $u_0 = c$ and $r_0 = l$.

Evidently the light clock is more robust than the others, by the enormous factor $(c/c_s)^2$. But its advantage depends on the strict constancy of the distance l between the mirrors: if they were mounted so as to make l vulnerable to acceleration stresses in their support (e.g. in an optical bench), then our first example shows that the light clock too would suffer fractional corrections of order (al/c_s^2) instead of only (al/c^2).

One obvious way to try and make real laboratory clocks more robust is to build them of stiffer materials, with the largest possible Y, the smallest possible ρ, and thus with the fastest possible c_s. It is amusing though of no practical importance to observe that there is a theoretical limit to such improvements, because c_s cannot exceed c. Theorists have in fact made a start on exploring the elastic properties of maximally stiff materials that are assumed to realize this limit. Interest in the problem can be triggered by enquiring into the Lorentz contraction of the rim and of the spokes of a rotating wheel (or even into the contraction of a solid disk): the material of the rim moves along its own length and is subject to Lorentz contraction, while the material of the spoke moves laterally, and is not. However, the apparent paradox is illusory. For real materials the strains due to the rotation stresses dominate the Lorentz contraction by the enormous factor $(c/c_s)^2$, and the question reduces to a well-defined though nontrivial problem in ordinary nonrelativistic elasticity. For idealized maximally stiff materials with $c_s = c$ the question likewise reduces to a problem in elasticity,[5] albeit of a novel relativistic kind, which seems not to have been fully worked out as yet. By contrast, the paradox imagined in earlier days stemmed from regarding the material as totally immune from stress-induced strains, as if Y were infinite, a scenario inadmissible under the Einsteinian assumption that sound signals, like all other signals, are incapable of moving faster than c.

6.4 Evidence

We cite two tests confirming that robust enough clocks indicate their proper time regardless of acceleration.

The classic experiment[6] uses very fast negative mu-mesons (muons) in a storage ring. Their mass is $m = 105.66$ MeV$/c^2$, and at rest they decay (into an electron plus two neutrinos) with a mean life of $T_0 = (2.198 \pm 0.001)$ μs. The experiment measures their mean life T in fast flight around a circle.

The muons are produced by letting many successive bunches of protons from a high-energy accelerator hit a fixed target; they are then channelled into the storage ring where they are made to move in a circle by a strong B field. There they decay slowly enough to build up an appreciable circulating current, which is essential if its decay is to be followed for long enough to allow reasonable accuracy. The measurement starts when there are enough muons in the ring, say at time $t = 0$. The intensity $I(t)$ of the muon current is then monitored by detecting the electrons from the decays; all one needs are the relative values $I(t)/I(0) = \exp(-t/T)$, where T is determined by a best fit to the observations. The design parameters envisage an orbit radius close to $r = 2.5$ m and a circulation period $2\pi/\omega$ close to 53×10^{-3} s. However, these values are far too rough to determine the speed u. Instead, u is inferred from the muon momentum $p = 1.284$ GeV$/c$, which is in turn inferred from the dynamics of the muons in the accurately known fields. Chapter 9 will show that $\gamma \equiv 1/\sqrt{1 - u^2/c^2} = \sqrt{p^2/m^2c^2 + 1}$, whence $\gamma = 12.14$.

Because the speed is constant, the proper time τ is related to the laboratory time t by $\tau = t/\gamma$. Our conjecture then predicts $I(t)/I(0) = \exp(-\tau/T_{\text{theory}})$, with $T_{\text{theory}} = \gamma T_0 = (26.68 \pm 0.01)$ μs. The experimental result is $T_{\text{exp}} = (26.37 \pm 0.065\,$μs; the main sources of error are muon losses from the beam other than through decay. Thus the conjecture is confirmed within the experimental error of roughly 1%.

As regards robustness, note that $u = c\sqrt{1 - 1/\gamma^2} \approx c$ entails $a \sim c^2/r \sim 4 \times 10^{16}$ m/s^2. But even the lowest estimate in Section 6.3.1 gave $a_0 \sim 10^{32}$ m/s^2; hence $a/a_0 < 10^{-16}$, so that in this experiment the internal clock governing the decays is quite certainly immune to the acceleration stresses. That the acceleration is irrelevant is accepted so universally (if tacitly) that the result has always been taken as eminently satisfactory confirmation of time dilation as such.

This scenario has the peculiarity that the speed is constant, and the acceleration always at right angles to the velocity. Another complementary experiment[7] studies the mean life of positive and negative sigma hyperons $\Sigma^\pm$ (mass 1189 MeV/c^2) while they are decelerated through successive random collisions in a liquid-hydrogen target. The initial speeds vary between $u/c = 0.17$ and 0.48. Within the experimental error of a few percent, the measured survival rates are reported to agree with the theoretical prediction $\exp(-\tau/T_0)$, where $T_0 = 8.0 \times 10^{-11}$ s is the mean life at rest, and τ the proper time calculated from the time-varying laboratory velocities. To this accuracy therefore the experiment confirms that the internal timekeeping of the Σ's, too, is governed wholly by time dilation; it remains unaffected by the decelerations, whose mean values range up to 10^{16} m/s^2.

6.5 Notes

1. We exclude accelerations due to gravity. A body falling freely in a constant homogeneous gravitational field experiences no acceleration stresses, and its rest frame is effectively inertial. This is the observation from which the general theory of gravitation has envolved. Its implications are beyond the remit of this book.

2. The prime on $S'(t)$ is just the usual remainder that this is a frame different from S.

3. The question is carefully worded. In fact the proper time is additionally affected by the gravitational potential Φ due to the sun: even if u_1 and u_2 were zero, one would have $\tau = t\sqrt{1 + 2\Phi/c^2} \approx t(1 + \Phi/c^2)$, with t the time measured by the reference clock (where Φ is negligible). But $\Phi = -GM_S/R$ while $GM_S M_E/R^2 = M_E u_1^2/R$, so that $\Phi/c^2 = -u_1^2/c^2$. Thus the gravitational effect has the same sign as the time dilation due to the orbital motion, and twice the magnitude.

4. Time dilation and Lorentz contraction are second-order small, and can be ignored. For instance, the velocity increment in one traverse is $u \sim a(l/c)$, entailing a contraction $\Delta l \sim l u^2/c^2$, and a consequent correction to the traverse time of the order of $\Delta l/c \sim (l/c)(al/c^2)^2 \ll \Delta T \sim (l/c)(al/c^2)$.

5. J. E. Hogarth and W. H. McCrea. "The relativistically rigid rod", *Proceedings of the Cambridge Philosophical Society* (1952), volume **48**, page 616; and W. H. McCrea, "Rotating relativistic ring", *Nature* (1971), volume **234**, page 399. To follow these papers in detail the reader will first have to study Chapter 9 and Appendix C.

6. It is reviewed by J. Bailey and E. Picasso, "Measurement of the relativistic time-dilation using μ mesons", *Progress in Nuclear Physics* (1970), volume **12**, page 43. The measurements were made at CERN in Geneva. We have already met decaying mu mesons in Example 3.1.

7. C. E. Ross *et al.* "$\Sigma^\pm$ lifetimes and longitudinal acceleration", *Nature* (1980), volume **286**, page 244.

7 Four-Vectors

Sections 2.3 and 3.2 explained that the relativity principle requires any true equation in physics to have the same mathematical form with respect to all inertial frames; in other words it must retain its form (must be form-invariant) under Lorentz transformations. To satisfy and to exploit this requirement efficiently, one must organize physical variables into standard sets whose transformation rules are easy to recognize and easy to handle. The simplest such sets, and the only ones used in this book, are the *four-vectors* that we are about to define.

It is four-vectors that will supply the key, in Chapter 9, to exploiting relativistic *kinematics* in order to revise particle *dynamics* compatibly with the relativity principle. Likewise they will prove essential, in Chapter 12, to a relativistic understanding of plane waves. Allied to these fundamental virtues, they have several very practical advantages. (i) They provide the clearest possible view of Lorentz transformations in general, and of the transformations of particle velocity and acceleration more especially. (ii) They streamline many kinematic calculations of kinds that we have done already (especially some of the harder ones, like the proofs of (4.18) and in Problem 4.9. (iii) In particle dynamics, they alone can avoid intolerable labour in any nontrivial application of energy and momentum conservation. (iv) Similarly, they can expedite many calculations on aberration and on the Doppler effect for waves.

Section 7.1 defines four-vectors, and the invariants formed as four-scalar products of such vectors. These are crucial to everything that follows. Section 7.2 introduces the four-vector associated with particle velocity. Section 7.3 is a brief aside on some simplicities peculiar to one-dimensional problems, where four-vectors effectively reduce to two-vectors. Finally, Section 7.4 outlines how four-vectors make it easy to keep equations conformable to the relativity principle, and uses four-velocities to illustrate this, if somewhat primitively. (The really important applications follow in Chapter 9 and beyond.) Four-accelerations are deferred to Chapter 8.

7.1 Definitions

The Lorentz invariance of $s^2 \equiv c^2 t^2 - \mathbf{r}^2$, which was central to Chapters 5 and 6, makes it attractive to specify an event in terms of the variables $(ct, x, y, z) = (ct, \mathbf{r})$.

Hence we define a new quantity $\vec{X}$, called the *coordinate four-vector* of the event, and specified with respect to an inertial frame S by its four components

$$\vec{X} = (X_0, X_1, X_2, X_3) \equiv (ct, x, y, z). \tag{7.1}$$

The components will be labelled by Greek suffices, and written typically as X_μ, with $\mu = 0, 1, 2, 3$. One may abbreviate (7.1) to

$$\vec{X} = (X_0, \mathbf{X}), \qquad \mathbf{X} \equiv (X_1, X_2, X_3) \equiv (x, y, z). \tag{7.2}$$

The second equality in (7.1) is an ordinary definition, introducing X_0 as just another name for ct, and so on. But the first equality must be understood somewhat differently, namely in the sense in which one might, for an ordinary three-dimensional position vector, write $\mathbf{r} = (x, y, z)$. Here, $\mathbf{r}$ on the left is a physical quantity in its own right, and the coordinates (x, y, z) on the right merely identify or represent $\mathbf{r}$ through its components with respect to whatever coordinate frame S one happens to have adopted. It is a familiar fact that with respect to a different frame S' *the same vector* $\mathbf{r}$ is represented by *a different set of components* (x', y', z'). Therefore, speaking strictly, one ought to say, not that (x, y, z) *is equal* to $\mathbf{r}$, but that *it represents* $\mathbf{r}$ relative to a particular set of coordinate axes. Likewise, the four-vector $\vec{X}$ is to be regarded as a physical quantity in its own right, *represented* with respect to our inertial frame S by its four components X_μ. The mathematically proper and wholly unambiguous definitions of ordinary and of four-vectors, with emphasis on the transformation rules for their components (between primed and unprimed) is sketched in Appendix B, which you should read now unless you have found the comments in this paragraph unnecessary or self-evident. From here on, however, we shall generally follow the physicist's habit of ignoring the difference between "equal to" and "represented by", and shall write $\vec{X} = (X_0, X_1, X_2, X_3)$ and so on without apology, except for an odd caution about ambiguities in the notation (which never mislead in practice, and which one soon ceases to notice).

Lorentz transformations relate the components X_μ of $\vec{X}$ to the components X'_μ of the same four-vector relative to another inertial frame S'; in standard configuration, with the coefficients taken from (say) equations (5.3), (5.4) we have

$$X'_0 = \gamma(X_0 - \beta X_1), \qquad X_1 = \gamma(X_1 - \beta X_0), \qquad X_{2,3} = X'_{2,3}; \tag{7.3}$$

$$X_0 = \gamma(X'_0 + \beta X'_1), \qquad X_1 = \gamma(X'_1 + \beta X'_0), \qquad X'_{2,3} = X_{2,3}. \tag{7.4}$$

Pursuing the analogy with ordinary vectors, *we now define as a four-vector, say* $\vec{A} = (A_0, A_1, A_2, A_3)$, *any set of four physical quantities that under Lorentz transformations change according to the same rules as do the* X_μ, i.e. such that

$$A'_0 = \gamma(A_0 - \beta A_1), \qquad A'_1 = \gamma(A_1 - \beta A_0), \qquad A'_{2,3} = A_{2,3}, \tag{7.5}$$

with the inverse obtained similarly from (7.4) by replacing X's with A's. How any given quantities do transform must be discovered from their physics: sets of four variables qualified to be the components of a four-vector must either be discovered by inspection, or they must be constructed by appropriate definitions, as will be done presently.

One calls A_0 the *time component of the four-vector* $\vec{A}$, and (A_1, A_2, A_3) its *space components*. Ordinary vectors like $\mathbf{A} = (A_1, A_2, A_3)$ are re-named *three-vectors*; thus a four-vector can be said to consist of, or to contain, a time component and a three-vector.

In the light of the comments just below equation (7.2), we stress that one must be prepared for peculiar-looking statements like[1]

$$\vec{A} = (A_0, A_1, A_2, A_3) = (A_0', A_1', A_2', A_3');$$ (7.6)

this merely identifies the components of $\vec{A}$ with respect both to S and to S', so that there is no contradiction even though $A_\mu' \neq A_\mu$.

Given two four-vectors $\vec{A}$ and $\vec{B}$, we define their *four-scalar product* by

$$\vec{A} \cdot \vec{B} \equiv A_0 B_0 - \mathbf{A} \cdot \mathbf{B} = A_0 B_0 - A_1 B_1 - A_2 B_2 - A_3 B_3.$$ (7.7)

The important special case where $\vec{B} = \vec{A}$ reads

$$\vec{A} \cdot \vec{A} = A_0 A_0 - \mathbf{A} \cdot \mathbf{A} = A_0 A_0 - A_1 A_1 - A_2 A_2 - A_3 A_3.$$ (7.8)

To avoid confusion with $\vec{A} \cdot \vec{A}$, it is evidently safest to abandon the symbol A^2 for the three-scalar product $\mathbf{A} \cdot \mathbf{A}$.

Just as *three-scalar products* $\mathbf{A} \cdot \mathbf{B}$ are invariant under a change to another coordinate system which changes the values of the components of $\mathbf{A}$ and of $\mathbf{B}$, so four-scalar products are invariant[2] under Lorentz transformations:

$$A_0 B_0 - A_1 B_1 - A_2 B_2 - A_3 B_3 = A_0' B_0' - A_1' B_1' - A_2' B_2' - A_3' B_3'.$$ (7.9)

To prove (7.9) one substitutes for the A_μ' from (7.5), for the B_μ' from the corresponding equation got by replacing A's with B's, and evaluates all the products. The calculation proceeds exactly like the proof (5.5) of the special case $\vec{B} = \vec{A} = \vec{X}$, and has in effect been called for already in Problem 5.3.

Four-vectors with $\vec{A} \cdot \vec{A}$ positive, negative, and zero are called timelike, spacelike, and lightlike; by the same argument as in Section 5.1, the time components of timelike and of lightlike four-vectors retain their sign under Lorentz transformations, while those of spacelike four-vectors need not.

Example 7.1. Consider a set of four physical variables that provisionally we call $(B_0, \mathbf{B})$, without knowing as yet how they Lorentz-transform (i.e. without knowing anything about the B_μ'); but suppose we do know that the combination $A_0 B_0 - A_1 B_1 - A_2 B_2 - A_3 B_3$ is invariant for *any* four-vector $\vec{A}$. Now prove *the converse*[3] of (7.9), namely that $(B_0, \mathbf{B})$ too must transform like a four-vector, i.e. according to (7.5). This will be crucial when we come to plane waves in Chapter 12, and to wave equations in Appendix D.

Solution. The proof is straightforward. The invariance we assume means that

$$A_0 B_0 - A_1 B_1 - A_2 B_2 - A_3 B_3 = A_0' B_0' - A_1' B_1' - A_2' B_2' - A_3' B_3'$$
$$= \gamma(A_0 - \beta A_1) B_0' - \gamma(A_1 - \beta A_0) B_1' - A_2' B_2' - A_3' B_3'$$
$$= A_0 \gamma(B_0' + \beta B_1') - A_1 \gamma(B_1' + \beta B_0') - A_2 B_2' - A_3 B_3',$$

where the second step uses the known transformation rules for the A_μ, and the third step is merely a rearrangement. The crucial point is that the leftmost and rightmost expressions are equal for *any* values of the A_μ, which is possible only if each A_μ has the same coefficient on both sides. But equating the coefficients of A_0 we obtain

$B_0 = \gamma(B_0' + \beta B_1')$, which is indeed the (inverse) Lorentz transform of the time component of a four vector. Equating the coefficients of the other A_μ we recover the appropriate rules for the other B_μ. ■

Most arguments and calculations involving four-vectors can be compressed a little further, into matrix form, which stresses both the similarities and the differences between four-vector and three-vector algebra. We outline this approach in Appendix B, but do not use it elsewhere.

7.2 Particle Four-Velocity

Our only four-vectors so far are the $\vec{X}$, the coordinate four-vectors of events, whose transformation rules serve to define four-vectors in general. The point is that in relativity these rules are the simplest one ever finds (apart from the trivial rule for four-scalars, which do not change at all). By contrast, in Section 4.2.4 the transformation rules for particle velocities $\mathbf{u} \equiv \mathrm{d}\mathbf{r}/\mathrm{d}t$ turned out to be much more complicated, precisely because $\mathbf{u}$ is one component $\mathrm{d}\mathbf{r}$ of a four-vector divided by another component $\mathrm{d}t$: any simplicity in the rules for numerator and denominator separately can hardly be expected to survive for their ratio.

To avoid these complications in calculating with velocities, one looks for a set of four variables that do constitute a four-vector, call it $\vec{U}$, but are nevertheless related to $\mathbf{u}$ closely enough to allow easy passage between them. The key is the invariance (i.e. the four-scalar nature) of the small increments $\mathrm{d}\tau = \mathrm{d}t\sqrt{1 - u^2/c^2}$ in the proper time of the particle: we observe that the four-vector $\mathrm{d}\vec{X}$ remains a four-vector on division by $\mathrm{d}\tau$, just as in Galilean physics the three-vector $\mathrm{d}\mathbf{r}$ remains a three-vector on division by the Galilean scalar $\mathrm{d}t$ to give $\mathbf{u}$. Hence we define

$$\vec{U} \equiv \frac{\mathrm{d}\vec{X}}{\mathrm{d}\tau} = \frac{1}{\sqrt{1 - u^2/c^2}} \cdot \frac{\mathrm{d}}{\mathrm{d}t}(ct, \mathbf{r}) = \frac{1}{\sqrt{1 - u^2/c^2}} \left(c, \frac{\mathrm{d}\mathbf{r}}{\mathrm{d}t} \right) = \frac{1}{\sqrt{1 - u^2/c^2}}(c, \mathbf{u}),$$

$$(7.10)$$

or equivalently

$$\vec{U} \equiv (U_0, \mathbf{U}), \qquad U_0 \equiv \frac{c}{\sqrt{1 - u^2/c^2}}, \qquad \mathbf{U} \equiv \frac{\mathbf{u}}{\sqrt{1 - u^2/c^2}}. \qquad (7.11)$$

One must be careful never to confuse capital $\mathbf{U}$ with lower-case $\mathbf{u}$.

Remarkably, (7.8) with $A \to U$ gives the invariant

$$\vec{U} \cdot \vec{U} = U_0^2 - \mathbf{U} \cdot \mathbf{U} = \frac{c^2 - \mathbf{u} \cdot \mathbf{u}}{(\sqrt{1 - u^2/c^2})^2} = c^2, \qquad (7.12)$$

which has the same value c^2 for all particles, irrespective of their three-velocity $\mathbf{u}$. Note that $\vec{U}$ is positive timelike.

Given $\vec{U}$, equation (7.11) immediately yields the three-velocity as

$$\mathbf{u}/c = \mathbf{U}/U_0, \qquad (7.13)$$

a relation crucial in problem solving. Both sides are dimensionless: to obtain **u** itself one multiplies through by c.

Example 7.2. Derive the velocity-combination rules (4.15) by Lorentz-transforming the U_μ.

Solution. The strategy is typical of many such problems. (i) The variable one wants (in this case **u**$'$) is expressed in terms of the components of suitable four-vectors with respect to the frame S' (here in terms of the components U'_μ of the four-vector $\vec{U}$); (ii) the transformation rules (7.5) are used to express the primed in terms of the unprimed components (here the U'_μ in terms of the U_μ); (iii) the unprimed components are expressed in terms of the quantities that were to be transformed (here in terms of **u**). Thus the potentially difficult part of the problem, namely the passage from S to S', reduces to a routine application of the standard four-vector transformation rules in step (ii).

(i) Equation (7.13) expresses our target as **u**$'/c =$ **U**$'/U'_0$.
(ii) Equations (7.5) yield

$$U'_0 = \gamma(U_0 - \beta U_1), \qquad U'_1 = \gamma(U_1 - \beta U_0), \qquad U'_{2,3} = U_{2,3}.$$

Thus

$$u'_x/c = \frac{(U_1 - \beta U_0)}{(U_0 - \beta U_1)}, \qquad u'_{y,z}/c = \frac{U_{y,z}}{\gamma(U_0 - \beta U_1)}.$$

(iii) Finally we must express all the U_μ on the right in terms of **u**. In view again of (7.13) this is done by dividing top and bottom by U_0:

$$u'_x/c = \frac{(U_1/U_0 - \beta)}{(1 - \beta U_1/U_0)} = \frac{(u_1/c - \beta)}{(1 - \beta u_1/c)}, \qquad u'_{y,z}/c = \frac{U_{y,z}/U_0}{\gamma(1 - \beta U_1/U_0)} = \frac{u_{y,z}/c}{\gamma(1 - \beta u_1/c)},$$

which tallies with (4.15). ∎

So far the symbols[4] $\beta = v/c$ and $\gamma = 1/\sqrt{1 - v^2/c^2}$ have been reserved for functions of the relative velocity **v** in the Lorentz transformation from S to S'. But the expressions for the U_μ make it convenient to introduce the same abbreviations for the same functions of the particle velocity **u**. Hence we define, for *any* velocity **u**,

$$\beta(u) \equiv u/c, \qquad \gamma(u) \equiv 1/\sqrt{1 - u^2/c^2} = 1/\sqrt{1 - \beta^2(u)}, \qquad \beta^2(\mathbf{u}) = 1 - 1/\gamma^2(u).$$
$$(7.14)$$

In this notation one has

$$U_0 = \gamma(u)c, \qquad \mathbf{U} = \gamma(u)\mathbf{u} = \gamma(u)c\beta(\mathbf{u}). \qquad (7.15)$$

Similarly, increments in the proper time of the particle may now be written as

$$d\tau = dt/\gamma(u). \qquad (7.16)$$

Conversely

$$dt = \gamma(u)d\tau = (U_0(\tau)/c)d\tau. \qquad (7.17)$$

The arguments of β and γ may be omitted if they are obvious from the context. In other cases, as in the examples below, one must distinguish very carefully between $\gamma(u)$ and $\gamma(v)$.

Example 7.3.   Derive the Lorentz-transformation rule for $\gamma(u) \equiv 1/\sqrt{1 - u^2/c^2}$.

Solution. The problem is to express $1/\sqrt{1 - u'^2/c^2}$ in terms of $\mathbf{u}$ and $\mathbf{v}$. The straightforward but laborious method finds the three components of $\mathbf{u}'$ from the velocity-combination rule, combines them into u'^2, and then constructs $\gamma(u')$. The result for $1/\gamma^2(u')$ was quoted in (4.19). The easy method follows: to appreciate its virtues the two calculations should be looked at side by side.

The strategy is to express $\gamma(u')$ in terms of U'_0, and then use the standard Lorentz transformation of the U_μ:

$$\gamma(u')c = U'_0 = \gamma(v)\{U_0 - \beta(v)U_1\} = \gamma(v)\{\gamma(u)c - \beta(v)\gamma(u)u_1\}.$$

Divide through by c, and note that $\beta(v)u_1/c = (v/c)u_1/c = (\mathbf{v}\cdot\mathbf{u})/c^2$, where the last step follows because in standard configuration v points along the x axis. Hence

$$\gamma(u') = \gamma(v)\gamma(u)(1 - \mathbf{v}\cdot\mathbf{u}/c^2), \qquad \gamma(u) = \gamma(v)\gamma(u')(1 + \mathbf{v}\cdot\mathbf{u}'/c^2). \tag{7.18}$$

The second relation is the inverse of the first, given by interchanging primed and unprimed and reversing the sign of $\mathbf{v}$. Squaring and taking reciprocals one recovers (4.18). ∎

Example 7.4: relative speed.   With respect to S, particles A, B have three-velocities $\mathbf{a}$, $\mathbf{b}$. Their four-velocities are $\vec{A}$ and $\vec{B}$. The three-velocity $\mathbf{u}(A, B)$ of B relative to A is defined as the velocity of B in the rest frame of A. Derive $u(A, B)$ from the invariant $\vec{A} \cdot \vec{B}$. (Problem 4.9 asked the same question: again the new calculation and the old should be looked at side by side. In fact it will appear presently that we are solving the problem from Example 7.3 by a different method.)

Solution. Again the strategy is typical. Since $\vec{A} \cdot \vec{B}$ is an invariant, it can be evaluated in any frame. We evaluate it (i) in the laboratory frame S, where the input information is supplied; (ii) evaluate it again in the rest frame S' of A, where the answer is wanted; and (iii) find the answer by equating the two expressions.

(i) In S one has $\vec{A} = (A_0, \mathbf{A}) = \gamma(a)(c, \mathbf{a})$, with a similar expression for $\vec{B}$; hence

$$\vec{A} \cdot \vec{B} = A_0 B_0 - \mathbf{A}\cdot\mathbf{B} = \gamma(a)\gamma(b)(c^2 - \mathbf{a}\cdot\mathbf{b}).$$

(ii) In S' one has $\mathbf{a}' = \mathbf{0}$ and $\mathbf{b}' \equiv \mathbf{u}(A, B)$. Thus $\vec{A} = (A'_0, \mathbf{A}') = (c, \mathbf{0})$, and

$$\vec{B} = (B'_0, \mathbf{B}') = \left(B'_0, \mathbf{u}(A, B)\right) = \gamma(u(A, B))(c, \mathbf{u}(A, B)).$$

Hence

$$\vec{A} \cdot \vec{B} = A'_0 B'_0 - \mathbf{A}'\cdot\mathbf{B}' = A'_0 B'_0 = c^2\gamma(u(A, B)). \tag{7.19}$$

(iii) Equating the two expressions for $\vec{A} \cdot \vec{B}$ and dividing by c^2 we obtain

$$\gamma(u(A, B)) = \gamma(a)\gamma(b)(1 - \mathbf{a}\cdot\mathbf{b}/c^2). \tag{7.20}$$

Now square, take reciprocals, and rearrange:

$$1 - u^2(A, B)/c^2 = \frac{(1 - a^2/c^2)(1 - b^2/c^2)}{(1 - \mathbf{a} \cdot \mathbf{b}/c^2)^2}, \tag{7.21}$$

$$u^2(A, B) = c^2 \frac{(1 - \mathbf{a} \cdot \mathbf{b}/c^2)^2 - (1 - a^2/c^2)(1 - b^2/c^2)}{(1 - \mathbf{a} \cdot \mathbf{b}/c^2)^2}$$

$$= \frac{(\mathbf{a} - \mathbf{b})^2 - a^2 b^2/c^2 + (\mathbf{a} \cdot \mathbf{b})^2/c^2}{(1 - \mathbf{a} \cdot \mathbf{b}/c^2)^2} = \frac{(\mathbf{a} - \mathbf{b})^2 - (\mathbf{a} \times \mathbf{b})^2/c^2}{(1 - \mathbf{a} \cdot \mathbf{b}/c^2)^2}, \tag{7.22}$$

tallying with Problem 4.9. In the nonrelativistic limit ($a/c \to 0$, $b/c \to 0$) the right-hand side reduces correctly to the Galilean result $(\mathbf{a} - \mathbf{b})^2$.

There is an unmistakable ressemblance between (7.20) and (7.18) from Example 7.3. To make the connection, we identify $\mathbf{a} \to -\mathbf{v}$ (the velocity of the frame S relative to S'), and $\mathbf{b} \to \mathbf{u}'$ (the velocity of some particle relative to S'). Then $\mathbf{u}(A, B)$ is the velocity of the particle relative to S, whence we identify also $\mathbf{u}(A, B) \to \mathbf{u}$. Since $\gamma(-\mathbf{v}) = \gamma(\mathbf{v})$, these substitutions translate (7.20) precisely into the second of equations (7.18) ∎.

Remarkably, no useful velocity four-vector exists for the special but important case of light-speed particles, with $u = c$. This is obvious from the start, because $\vec{U}$ is constructed using the particle's proper time, and equation (6.1) already shows that proper time cannot be defined at speed c. It is also obvious from the fact that the vanishing denominators cause the definitions (7.11) to fail if $u = c$. Fortunately this lack never impedes actual calculations with velocities, because (i) there is no need to transform magnitudes, since $u = c$ entails $u' = c$; and (ii) the transformation of directions involves only the ratios of the $u_{x,y,z}$ to each other or to u, and from Section 4.2.4 we can already transform these ratios regardless of the speed. (Equation (7.13) shows that they are just the same for the space components U of $\vec{U}$ as for the corresponding components of $\mathbf{u}$.) Indeed the advantage lies with light speed, which alone validates the remarkable half-angle transformation formulae (4.23)–(4.25). Lastly, in dynamics (Chapter 9), where $\vec{U}$ enters the definition of the four-momentum $\vec{p}$, it turns out (somewhat miraculously) that $\vec{p}$ unlike $\vec{U}$ does make sense even at light speed.

7.3 Lorentz Transformations in 1 + 1 Dimensions

The Lorentz transformation, say in the form (7.5), depends on $\mathbf{v}$, the velocity of frame S' relative to frame S, and we can define the associated four-velocity $\vec{V}$. In standard configuration one has $(v_x, v_y, v_z) = (v, 0, 0)$, and a notation patterned on (7.14), (7.15) yields

$$(V_0, V_1) = \gamma(v)c(1, \beta(v)). \tag{7.23}$$

In terms of $\vec{V}$, the transformation rules (7.5) take the remarkably simple form

$$A'_0 = \frac{1}{c}(V_0 A_0 - V_1 A_1), \qquad A'_1 = \frac{1}{c}(V_0 A_1 - V_1 A_0), \qquad A'_{2,3} = A_{2,3}, \tag{7.24}$$

$$A_0 = \frac{1}{c}(V_0 A'_0 + V_1 A'_1), \qquad A_1 = \frac{1}{c}(V_0 A'_1 + V_1 A'_0), \qquad A_{2,3} = A'_{2,3}. \tag{7.25}$$

In a world with 1 + 1 dimensions (one space, one time) these rules (discarding all y and z components) would be truly fascinating. In the real (1 + 3 dimensional) world they are just a mild curiosity, though occasionally useful when all the vectors in a problem have their space components pointing along **v**, so that with our standard choice of axes their y and z components vanish. Perhaps the most important example is linear acceleration, considered in Chapter 8.

7.4 Covariance

The relativity principle asserts that all true equations of physics assume the same mathematical form with respect to all inertial frames. What this means was explained at length in Sections 2.3 and 3.2: true equations (and candidate equations worth considering) must be form-invariant under Lorentz transformations. Alternatively but equivalently, form-invariant equations are described as *covariant*. Covariance is guaranteed and obvious for equations expressed as[5] (four-scalar) = (four-scalar) or (four-vector) = (four-vector); often one chooses the standard form (four-scalar) = 0 or (four-vector) = $\vec{0}$, though the arrow on $\vec{0}$ is generally omitted.

For example, given a four-vector $\vec{A} = (A_0, \mathbf{A})$ and a four-scalar Q, a candidate equation like

$$A_0 = Q \tag{i}$$

is inadmissible. It might be true by coincidence in some particular inertial frame S, but would fail in other frames S'. To see that (i) fails the test for covariance spelled out in Section 2.3, assume standard configuration, and note that the Lorentz transformation from S' to S yields $A_0 = \gamma(A_0' + \beta A_1')$ but $Q = Q'$. Hence (i) entails

$$\gamma(A_0' + \beta A_1') = Q', \tag{ii}$$

which on dropping the primes would contradict (i).

On the other hand, given another four-vector $\vec{B}$, the candidate equation

$$\vec{A} = \vec{B} \Rightarrow A_\mu = B_\mu, \quad (\mu = 0, \dots, 3). \tag{7.26}$$

passes muster. For instance, $A_0 = B_0$ re-expressed in terms of the primed components reads

$$\gamma(A_0' + \beta A_1') = \gamma(B_0' + \beta B_1'); \tag{7.27}$$

on dropping the primes this gives $\gamma(A_0 + \beta A_1) = \gamma(B_0 + \beta B_1)$, which is true because the original (7.26) entails both $A_0 = B_0$ and $A_1 = B_1$.

Conversely, suppose we have a true equation, say $a = b$; and that on examining the Lorentz-transformation properties of a and b we discover that they are the time components of four-vectors $\vec{A}, \vec{B}$, i.e. that $a = A_0$ and $b = B_0$. Then the four-vectors themselves must be equal: in other words one must have $\vec{A} = \vec{B}$, meaning that $A_\mu = B_\mu$ for $\mu = 1, 2, 3$ as well as for $\mu = 0$. Thus we have automatically secured three new equations, namely $\mathbf{A} = \mathbf{B}$. To see all this, we need only re-express $A_0 = B_0$ in terms of the primed components, which yields (7.27). But this must be true for all values of β, which is possible only if the coefficients of β are equal, i.e. only if $A_1' = B_1'$ also. The other components are shown to be equal in pairs by considering relative velocities **v** in other directions. The same conclusions follow (see Problem 7.2) from the equality $A_\mu = B_\mu$ between any pair of components, i.e. equally from $\mu = 0$ or 1 or 2 or 3.

Coordinate and velocity four-vectors alone cannot illustrate convincingly the enormous advantages of requiring covariance; they will become apparent only when the requirement is applied to dynamics and to wave motion in later chapters. Here we settle for a rather modest example, re-deriving the radioactive decay law for uniformly moving particles.

Example 7.6. It is observed that particles of a certain kind at rest (relative to a frame S') decay according to Rutherford's law (see, for example, Section 4.4.1), so that the mean number surviving at time t' is

$$N = N(0) \exp\left(-t'/T_0\right). \tag{7.28}$$

What is the decay law for uniformly moving particles?

Solution. The strategy is (i) to re-express (7.28) in a manifestly covariant form, in this case as (four-scalar) = (four-scalar); and then (ii) to assert that since this covariant equation is true in S' it is true in all inertial frames. The tools are the coordinate and velocity four-vectors $\vec{X}$ and $\vec{U}$ of the particles.

(i) Start from the fact that N is a four-scalar. To see this, consider the event of counting the particles: the result is the same number for all observers. Hence the exponential must likewise be a four-scalar, and our first task is to express it, i.e. to express t', in a form making this obvious. With respect to S' one has $t' = X_0'/c$ and $(U_0', \mathbf{U}') = (c, \mathbf{0})$; whence, evaluating the four-scalar product $\vec{U} \cdot \vec{X}$ in S', one obtains

$$\vec{U} \cdot \vec{X} \equiv U_0' X_0' - \mathbf{U}' \cdot \mathbf{X}' = U_0' X_0' = ct'. \tag{7.29}$$

Thus the variable t', which refers explicitly to the rest frame S', may be replaced by the four-scalar $\vec{U} \cdot \vec{X}/c$, and the decay law may be re-expressed *in manifestly covariant form* as

$$N = N(0) \exp(-\vec{U} \cdot \vec{X}/cT_0). \tag{7.30}$$

(ii) Equation (7.30) makes no reference to any particular frame, and automatically embodies the decay law in any and every inertial frame. To answer questions asked in some particular frame S, we need merely evaluate the exponent in S. If the velocity of S with respect to S' is $-\mathbf{u}$, then relative to S the particle has velocity $\mathbf{u}$ and position $\mathbf{r} = \mathbf{u}t$, so that

$$\vec{U} \cdot \vec{X} \equiv U_0 X_0 - \mathbf{U} \cdot \mathbf{X} = \gamma(u)(c \cdot ct - \mathbf{u} \cdot \mathbf{r}) = \gamma(c^2 - u^2)t = t/\gamma.$$

Substituted into (7.30) this yields $N = N(0) \exp\left(-t/T_\beta\right)$ with $T_\beta \equiv \gamma(u)T_0$, the result already familiar from time dilation. ■

7.5 Notes

1. On the other hand, in principle one should never put a prime on $\vec{A}$ itself, because, as explained above, vectors are physical quantities in their own right, without any basic affinities with particular inertial frames. Nevertheless, in practice it can become very tempting to use $\vec{A}'$ merely as a shorthand for the set of primed components on the far right of (7.6), with $\vec{A}$ understood similarly as a shorthand for the set of unprimed components. Of course this shorthand would then have to be paid for by remembering that $\vec{A}$ and $\vec{A}'$ are actually the same vector.

2. Since the words scalar and invariant mean the same thing, they can be used interchangeably. Generally we shall prefer to speak of scalars rather than of invariants, so as to avoid, for instance, peculiar-sounding (though technically correct) references to invariant variables.
3. In other words, we assert that $A_0 B_0 - \mathbf{A} \cdot \mathbf{B}$ is invariant if and only if $(B_0, \mathbf{B})$ too is a four-vector. The "if" part is already established; its converse is the "only if" part to be proved. Logicians call an "if" condition sufficient, and an "only if" condition necessary. But they tend to name them in reverse order: "if and only if" is commonly called "necessary and sufficient".
4. Even though with S and S' in standard configuration $\mathbf{v}$ always points in the x direction, it is useful to put some expressions into forms valid for arbitrary directions of $\mathbf{v}$ and of $\boldsymbol{\beta} = \mathbf{v}/c$.
5. There are other possibilities, featuring tensors or spinors, but they are beyond our scope here. Tensors are central to covariant formulations of electrodynamics, and spinors are needed in quantum mechanics to describe spinning particles like electrons, protons, and neutrons.

8 Four-Acceleration

This chapter provides useful practice with four-vectors and proper times, but is not required elsewhere in the book except in Appendix C.

For handling Lorentz transformations of particle accelerations $\mathbf{a} \equiv \mathrm{d}\mathbf{u}/\mathrm{d}t$, it is convenient to define the associated four-vector

$$\vec{A} \equiv \frac{\mathrm{d}\vec{U}}{\mathrm{d}\tau} = \gamma(u)\frac{\mathrm{d}}{\mathrm{d}t}\frac{(c, \mathbf{u})}{\sqrt{1 - u^2/c^2}}. \tag{8.1}$$

That *the four-acceleration $\vec{A}$ is indeed a four-vector* is evident, since $\mathrm{d}\vec{U}$ is a four-vector while the increment $\mathrm{d}\tau$ in the proper time is a scalar: the definition of $\vec{A}$ in terms of $\vec{U}$ is patterned on the definition of $\vec{U}$ in terms of $\vec{X}$. The object is to make it simpler to derive and apply the rules for $\mathbf{a}$ already known from Section 4.2.5.

Since $\mathrm{d}(u^2)/\mathrm{d}t = 2(\mathbf{u} \cdot \mathrm{d}\mathbf{u}/\mathrm{d}t) = 2(\mathbf{u} \cdot \mathbf{a})$, the differentiations yield

$$A_0 = \frac{(\mathbf{u} \cdot \mathbf{a})/c}{(1 - u^2/c^2)^2}, \qquad \mathbf{A} = \frac{\mathbf{a}}{(1 - u^2/c^2)} + \frac{\mathbf{u}(\mathbf{u} \cdot \mathbf{a})/c^2}{(1 - u^2/c^2)^2}. \tag{8.2}$$

Conversely,

$$\mathbf{a} = (\mathbf{A} - \mathbf{u}A_0/c)(1 - u^2/c^2) \tag{8.3}$$

expresses $\mathbf{a}$ in terms of $\vec{A}$. Note that $\mathbf{A}$ is not parallel to $\mathbf{a}$ unless $\mathbf{u}$ is. For motion in a straight line (8.2) reduces to

$$\mathbf{A} = \mathbf{a}/(1 - u^2/c^2)^2, \qquad (\mathbf{a} \text{ parallel to } \mathbf{u}). \tag{8.4}$$

Differentiating $\vec{U} \cdot \vec{U} = c^2$ one finds

$$\vec{U} \cdot \mathrm{d}\vec{U}/\mathrm{d}\tau = \vec{U} \cdot \vec{A} = U_0 A_0 - \mathbf{U} \cdot \mathbf{A} = 0: \tag{8.5}$$

because their four-scalar product vanishes, one says that $\vec{U}$ and $\vec{A}$ are *orthogonal* to each other.

The reader should now re-derive the transformation rules for $\mathbf{a}$, using the strategy outlined in Example 7.2, and the standard Lorentz transformation of the four-vector

$\vec{A}$. We quote the result for convenience:

$$a'_x = \frac{(1 - v^2/c^2)^{3/2}}{(1 - vu_x/c^2)^3} a_x, \tag{8.6}$$

$$a'_{y,z} = \frac{(1 - v^2/c^2)}{(1 - vu_x/c^2)^3} \{(1 - vu_x/c^2)a_{y,z} + (vu_{y,z}/c^2)a_x\}. \tag{8.7}$$

Inversely,

$$a_x = \frac{(1 - v^2/c^2)^{3/2}}{(1 + vu'_x/c^2)^3} a'_x, \tag{8.8}$$

$$a_{y,z} = \frac{(1 - v^2/c^2)}{(1 + vu'_x/c^2)^3} \{(1 + vu'_x/c^2)a'_{y,z} - (vu'_{y,z}/c^2)a'_x\}. \tag{8.9}$$

Observe how the nonrelativistic limit ($v/c \to 0$ and $\mathbf{u}/c \to 0$) entails $\mathbf{a} \to \mathbf{a}'$, i.e. the Galilean invariance of acceleration.

The rest frame of an accelerating particle is obviously not inertial; but it is worth noting the components of $\vec{A}$ with respect to the *instantaneously co-moving inertial frame*, call it S', where, momentarily, $\mathbf{u}' = \mathbf{0}$. We call the acceleration in this frame *the proper acceleration*, and denote it by the special symbol $\boldsymbol{\alpha}$:

$$A'_0 = 0, \qquad \mathbf{A}' = \mathbf{a}' \equiv \boldsymbol{\alpha}, \qquad (\text{if } \mathbf{u}' = 0). \tag{8.10}$$

Using S' to evaluate the invariant associated with $\vec{A}$ one finds

$$\vec{A} \cdot \vec{A} = A_0'^2 - \mathbf{A}'^2 = -\alpha^2 < 0, \tag{8.11}$$

showing that $\vec{A}$ is spacelike. In S' the orthogonality (8.5) is obvious: $A'_0 U'_0 - \mathbf{A}' \cdot \mathbf{U}' = 0$ because A'_0 and $\mathbf{U}'$ vanish.

In applications it is generally best to choose axes that put S' into standard configuration with S: in other words we choose the x axis along $\mathbf{u}$, and note that $v = u$. But (8.8), (8.9) with $\mathbf{u}' = 0$ and $v = u$ yield

$$a_x = (1 - u^2/c^2)^{3/2}\alpha_x, \qquad a_{y,z} = (1 - u^2/c^2)\alpha_{y,z} \qquad (\text{if } \mathbf{u}' = 0, \text{ and with } \mathbf{a}' \equiv \alpha). \tag{8.12}$$

Especially interesting is motion in a straight line with constant α; Appendix C shows that this is the fate of a charged particle starting from rest under the action of an electric field that is constant and homogeneous with respect to the initial rest frame (the "laboratory" frame) S. The problem is to determine $u(t)$ and $x(t)$ relative to S. It is worth solving in two different ways. The obvious but rather tedious method works with the three-acceleration a, and starts by calculating the velocity $u(t)$ directly as a function of the laboratory time t. The more devious but far easier method starts by determining the four-vector $\vec{U}(\tau)$ and the time $t(\tau)$ as functions of the proper time τ, and $u(t)$ only afterwards.

Example 8.1. For rectilinear motion with constant proper acceleration α, determine (a) the laboratory velocity and position $u(t)$ and $x(t)$, given $u(0) = 0$, $x(0) = 0$; and (b) the proper time $\tau(t)$ of the particle, all as functions of laboratory time t.

Solution. (a) It turns out quickest to calculate with $\beta \equiv \beta(u) = u/c$. Equation (8.8) with $u' = 0$ and $v = u$ gives the laboratory acceleration as $a \equiv c\,\mathrm{d}\beta/\mathrm{d}t = (1 - \beta^2)^{3/2}\alpha$. Hence

$$\int \frac{\mathrm{d}\beta}{(1 - \beta^2)^{3/2}} = \frac{\beta}{(1 - \beta^2)^{1/2}} = \alpha t/c,$$

with no integration constant because $\beta(0) = 0$. (The integral is elementary though tedious; but the result is readily checked by differentiation.) Squaring, and solving for β, one finds

$$\beta(t) = u(t)/c = \frac{\alpha t/c}{\sqrt{1 + (\alpha t/c)^2}}. \tag{8.13}$$

Part (b) will require

$$1/\gamma = \sqrt{1 - \beta^2} = 1/\sqrt{1 + (\alpha t/c)^2}. \tag{8.14}$$

The position follows straightforwardly:

$$x(t) = \int_0^t \mathrm{d}t\, u(t) = \frac{c^2}{\alpha}\sqrt{1 + (\alpha t/c)^2}\,\Big|_{t=0}^{t} = \frac{c^2}{\alpha}\left\{\sqrt{1 + (\alpha t/c)^2} - 1\right\}, \tag{8.15}$$

where the integral is easy after changing the integration variable to $(\alpha t/c)^2$.

Initially, while $(\alpha t/c)^2 \ll 1$, we can expand the square root in powers of $(\alpha t/c)^2$; keeping only the first two terms one recovers the familiar Galilean formula for constant acceleration, $x(t) \approx \alpha t^2/2$. On the other hand, for large times one has

$$x(t) = \frac{c^2}{\alpha}\left\{(\alpha t/c)\sqrt{1 + (c/\alpha t)^2} - 1\right\}$$

$$= \left\{ct\left(1 + \frac{1}{2}(c/\alpha t)^2 + \cdots\right) - \frac{c^2}{\alpha}\right\} \approx ct - \frac{c^2}{\alpha} + \cdots, \tag{8.16}$$

where the dots stand for terms that vanish as $t \to \infty$. Since ct is the position of a light signal dispatched together with the particle, we see that in the limit the particle would keep pace with such a signal, running behind it by a constant distance c^2/α. Position and speed are shown in Figure 8.1.

(b) The proper time is given by

$$\tau(t) = \int \mathrm{d}t\sqrt{1 - \beta^2(t)} = \int \frac{\mathrm{d}t}{\sqrt{1 + (\alpha t/c)^2}} = \frac{c}{\alpha}\log\left\{\alpha t/c + \sqrt{1 + (\alpha t/c)^2}\right\}$$

$$= \frac{c}{\alpha}\sinh^{-1}(\alpha t/c). \tag{8.17}$$

For $(\alpha t/c)^2 \ll 1$, τ reduces to t as it should. For $(\alpha t/c)^2 \gg 1$ on the other hand

$$\tau(t) \approx \frac{c}{\alpha}\log\left\{\frac{2\alpha t}{c}\right\}. \tag{8.18}$$

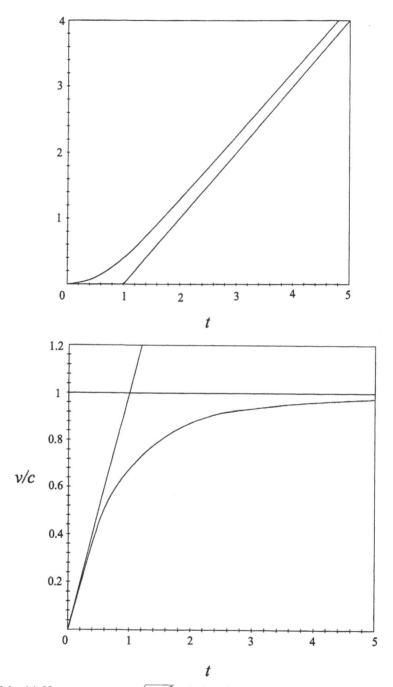

Figure 8.1 (a) Upper curve: $x = \sqrt{1 + t^2} - 1$, i.e. the position of a particle with constant acceleration relative to its instantaneous rest frame, starting at time $t = 0$ from rest at $x = 0$: equation (8.15) on a scale where $\alpha = 1 = c$. Lower curve: the large-time approximation on the same scale, i.e. $x = t - 1$, the rightmost expression in (8.16). (b) The velocity v of the particle moving as described in (a); the vertical axis shows $\beta = v/c$, equation (8.13). The straight line shows the Newtonian velocity $v = \alpha t$.

The proper time goes on increasing with t, though only slowly (logarithmically), because the velocity goes on increasing without ever actually reaching c: if it did, τ would remain fixed from then on. ∎

Example 8.2. For the same motion as in Example 8.1, (a) determine $\vec{U}(\tau)$; (b) from $\vec{U}(\tau)$ determine $t(\tau)$ and finally $u(t)$. (Problem 8.2 shows that in fact this approach works equally well when $\alpha(\tau)$ is not constant, but an arbitrary function of τ.)

Solution. (a) Start from the Lorentz transformation from S' to S written as in Section 7.3, but with $\vec{U}$ instead of $\vec{V}$, so that β and γ are now functions of u:

$$dU_0/d\tau = A_0 = \frac{1}{c}\left(U_0 A_0' + U_1 A_1'\right) = (\alpha/c)U_1, \tag{8.19}$$

$$dU_1/d\tau = A_1 = \frac{1}{c}\left(U_0 A_1' + U_1 A_0'\right) = (\alpha/c)U_0. \tag{8.20}$$

These equations apply to any rectilinear motion, whether α is constant or not. But, on differentiating (8.19) again, we can for constant α use (8.20) to eliminate $dU_1/d\tau$, and obtain

$$d^2 U_0/d\tau^2 = (\alpha/c)^2 U_0, \qquad \text{(constant } \alpha\text{)}. \tag{8.21}$$

Since the initial condition (in S) entails $U_0(\tau = 0) = c$, this can be solved by mere inspection:

$$U_0(\tau) = c\cosh(\alpha\tau/c). \tag{8.22}$$

Equation (8.19) then yields

$$U_1(\tau) = (c/\alpha)dU_0/d\tau = c\sinh(\alpha\tau/c). \tag{8.23}$$

(b) From (7.17) we have $dt = d\tau U_0/c = d\tau\cosh(\alpha\tau/c)$. Hence

$$t = \int d\tau\cosh(\alpha\tau/c) = (c/\alpha)\sinh(\alpha\tau/c), \tag{8.24}$$

tallying with (8.17). Thus

$$\beta(t) = U_1/U_0 = \tanh(\alpha\tau/c) = \frac{\sinh(\alpha\tau/c)}{\sqrt{1 + \sinh^2(\alpha\tau/c)}} = \frac{\alpha t/c}{\sqrt{1 + (\alpha t/c)^2}}, \tag{8.25}$$

reproducing (8.13) ∎

Finally a caution. For systems with several particles, one must bear in mind that each has its own proper time. For instance, given two particles, call them (1) and (2), one often needs to work with the scalar product $\vec{U}^{(1)} \cdot \vec{U}^{(2)}$ of their four-velocities. But if these vary, then one must be careful to write their four-accelerations as $\vec{A}^{(1)}(\tau^{(1)}) = d\vec{U}^{(1)}(\tau^{(1)})/d\tau^{(1)}$ and $\vec{A}^{(2)}(\tau^{(2)}) = d\vec{U}^{(2)}(\tau^{(2)})/d\tau^{(2)}$. Otherwise one might be tempted to write equations like $d(\vec{U}^{(1)} \cdot \vec{U}^{(2)})/d\tau = \vec{A}^{(1)} \cdot \vec{U}^{(2)} + \vec{U}^{(1)} \cdot \vec{A}^{(2)}$, which is nonsense because there exists no common variable "τ" for both $\vec{U}^{(1)}$ and $\vec{U}^{(2)}$ to be functions of.

The four-acceleration of speed-c particles would evidently require special treatment, just as their four-velocity does. We omit this, because it will not be needed.

Part III
Momentum and Energy

9 Particle Dynamics: Momentum and Energy

9.1 The Form-Invariant Conservation Laws: Redefinition of Momentum and Energy

We have seen how a finite invariant speed c combined with the relativity principle enforces a complete revision of Newtonian *kinematics*, by substituting the Lorentz for the Galilean transformations of event coordinates $(t, \mathbf{r})$ and of the velocities and accelerations of particles. It would be reasonable to expect that Newtonian *dynamics* needs a comparably drastic overhaul, and presently we shall show that it does. The best way to see this reverses the usual priorities of elementary mechanics, and starts by reconsidering not Newton's second law, i.e. not

$$d(\text{momentum})/dt = \text{force}, \tag{9.1}$$

but the conservation laws of momentum and energy as they apply to collisions and reactions between otherwise freely moving particles. As explained in Sections 2.4.2 and 2.4.4, these laws take the form

$$\Delta \mathbf{P} = 0, \qquad \Delta E = 0, \tag{9.2}$$

the differences Δ being defined as

$$\Delta \mathbf{P} = \sum_i \mathbf{p}_i \bigg|_{\text{after}} - \sum_j \mathbf{p}_j \bigg|_{\text{before}}, \tag{9.3}$$

and similarly for the energy. One can elucidate the conservation laws without being drawn into an unavoidably sophisticated discussion of the status of force laws in general and of relativistic force laws in particular, questions beyond the scope of this book, though Appendix C touches on them briefly.

In need of revision, as it turns out, are not the conservation laws (9.2), (9.3) as such, but the Newtonian expressions for the conserved quantities, namely $\mathbf{p} = m\mathbf{u}$ and $\varepsilon = mu^2/2 = p^2/2m$ for a particle's momentum and energy in terms of its velocity $\mathbf{u}$

and of a constant (i.e. velocity-independent) and invariant (i.e. frame-independent) mass m. With *these* expressions the conservation laws are not form-invariant under Lorentz transformations: in other words, even if they held in one inertial frame S they would fail in other such frames S', whence they fail to qualify as laws of nature under the relativity principle.

Since a single counter-example suffices to invalidate a would-be general rule, we choose the special case where, relative to S, two equal-mass particles with speeds $\pm u$ along the x axis collide, separating afterwards with speeds $\pm u$ along the y axis, as shown in Figure 9.1(a). That this scenario satisfies the Newtonian conservation laws with respect to S is evident by inspection. Figure 9.1(b) shows the initial and final velocities dictated by the Lorentzian velocity-combination rule for the same collision viewed from S', which moves relative to S with velocity v in the x direction. In S', P'_y is evidently zero both before and after, so that $\Delta P'_y = 0$ compatibly with (9.2); but Newtonian momentum conservation in the x' direction, namely $\Delta P'_x = 0$, demands

$$m\left\{ \frac{u-v}{1-uv/c^2} - \frac{u+v}{1+uv/c^2} \right\} \stackrel{?}{=} -2mv \quad \Rightarrow \quad \frac{-2mv(1-u^2/c^2)}{1-(u^2/c^2)(v^2/c^2)} \stackrel{?}{=} -2mv. \quad (9.4)$$

That (9.4) is evidently false makes our point: the Newtonian rule $\Delta \sum m_i \mathbf{u}_i = 0$ cannot be universally true, because it fails for the special case just considered. A similar but more laborious calculation shows that the rule $\Delta \sum m_i u_i^2/2 = 0$ also fails.

If momentum and energy are conserved, then the correct expressions for them must be *guessed*, and verified by experiment afterwards. Guesses are guided by two requirements. First, the conservation laws must be form-invariant under Lorentz transformations, i.e. the unprimed and primed versions must each imply the other, as discussed in Section 7.4 (and before that, albeit for Galilean physics, in Section 2.3). Second, if with respect to a given inertial frame S all the particle speeds u_i are

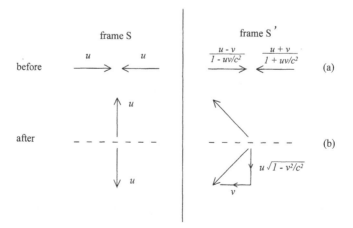

Figure 9.1   (a) Relative to the inertial frame S, two equal-mass particles collide with equal and opposite velocities, which are deviated through 90° by the collision. (b) The same collision as seen from another inertial frame S' moving with velocity v in the positive x direction with respect to S. The descriptions relative to S and to S' are connected by the standard Lorentz-transformation rule for velocities. *Caution: directions as shown by the arrows; the expressions indicate speeds, not velocities.*

low enough for terms proportional u_i^2/c^2 to be neglected (i.e. to justify the approximation $1 - u_i^2/c^2 \to 1$), then with respect to S the laws must reduce to their familiar Newtonian forms.

Experimental evidence backing the conjectures we shall adopt is presented in Section 9.3.

The search for admissible candidates for the momentum-conservation law is focused by recalling (i) the comments in Section 7.4 on the virtues of writing form-invariant equations as *four-scalar = four-scalar*, or *four-vector = four-vector*; and (ii) the association in Section 7.2 of four-velocities $\vec{U}$ with ordinary particle velocities **u**. Arguably the simplest guess is that the particle momentum as featured in (9.3) is the space part **p** of a four-vector, called the *momentum four-vector*, which we now define by[1]

$$\vec{p} = (p_0, \mathbf{p}) \equiv m\vec{U} = m(U_0, \mathbf{U}), \tag{9.5}$$

so that

$$p_0 = mU_0 = \gamma(u)mc, \quad \mathbf{p} = m\mathbf{U} = \gamma(u)m\mathbf{u}. \tag{9.6}$$

Here, as everywhere in this book, m is a four-scalar (i.e. an invariant under Lorentz transformations), characteristic of the particle in question, and (as in Newtonian physics) independent of its velocity. Following the usage universal in elementary-particle physics, we call m simply *the mass* of the particle; it is measured, at least in principle, as either the inertial or the gravitational mass of the particle in its instantaneous rest frame, these two quantities being equal as a matter of empirical fact. The reader is warned that some books (especially older ones) call our m the "rest"-mass, denote it by "m_0", and define also another velocity-dependent "mass" which we would write as $\gamma(u)m$, but which they write as "$m = \gamma(u)m_0$". We repeat that, by contrast, *in this book the word "mass" and the symbol m always refer to the constant four-scalar defined above, equal to the "rest mass" of writers who introduce masses of other kinds as well.*

The symbol **p** *is henceforth reserved for the three-vector defined in (9.6)*, rightfully called *the three-momentum*, though we may refer to it as "the momentum" where no confusion can result. We shall write

$$\mathbf{p} = (p_x, p_y, p_z), \quad \mathbf{p}^2 \equiv p_x^2 + p_y^2 + p_z^2 \equiv p^2. \tag{9.7}$$

The old combination $m\mathbf{u}$, if needed, will always be qualified as the "Newtonian" momentum.

At low speed one approximates $\mathbf{p} = \gamma(u)m\mathbf{u} = m\mathbf{u}/\sqrt{1 - u^2/c^2}$ by expanding $1/\sqrt{}$ in powers of u^2/c^2:

$$\mathbf{p} = m\mathbf{u}\left(1 + \frac{1}{2}\frac{u^2}{c^2} + \frac{3}{8}\frac{u^4}{c^4} + \cdots\right). \tag{9.8}$$

The first term is just the Newtonian momentum, as it should be. The other terms are corrections, small when $u^2/c^2 \ll 1$; they show that $|\mathbf{p}|$ rises with u faster than it would according to Newton.

We are now tooled up to state and to understand the implications of the proposed Lorentz-covariant four-momentum conservation law, which reads[2]

$$\Delta\vec{P} \equiv \Delta\sum_i \vec{p}_i = \Delta\sum_i (p_{i0}, \mathbf{p}_i) = 0. \tag{9.9}$$

Crucially, and as explicated in Section 7.4, the relativity principle dictates that all four components of this rule stand or fall together: any one of them entails the other three. Thus the correct version of three-momentum conservation, embodied in the space components of (9.9), inevitably entails a fourth conservation law given by the time component. By hindsight we define

$$\varepsilon \equiv cp_0 = \gamma(u)mc^2 = mc^2/\sqrt{1 - u^2/c^2}, \tag{9.10}$$

so that

$$\vec{p} = (\varepsilon/c, \mathbf{p}); \tag{9.11}$$

and we re-display the time component of (9.9) after multiplying it by c:

$$c\Delta P_0 \equiv \Delta \sum_i cp_{i0} = \Delta \sum_i \varepsilon_i = \Delta \sum_i \gamma(u_i)m_i c^2 = 0. \tag{9.12}$$

In other words, by virtue of the relativity principle the quantity $\sum \varepsilon$ is conserved in the same sense as the three-momentum $\sum \mathbf{p}$.

For guidance regarding the physics of ε, we approximate it at low speeds by expanding $1/\sqrt{}$ as in (9.8):

$$\varepsilon = mc^2 \left(1 + \frac{1}{2}\frac{u^2}{c^2} + \frac{3}{8}\frac{u^4}{c^4} + \cdots \right) = mc^2 + \frac{1}{2}mu^2 + \frac{3}{8}m\frac{u^4}{c^2} + \cdots. \tag{9.13}$$

In the nonrelativistic regime where $u^2/c^2 \ll 1$, the third and all higher terms are negligible. The second term $mu^2/2$ (whence c has cancelled) is just the Newtonian kinetic energy. Following this lead, we conjecture that ε is the *total energy of the particle, so that (9.12) is in fact the energy-conservation law*. The first term mc^2 is just an additive constant; these contributions cancel from (9.12) provided all particles retain their identity, i.e. provided the same numbers of particles of each type enter and leave the reaction. Accordingly we interpret mc^2 as the internal energy of the particle, namely its total energy when it is at rest, by analogy with the internal energy[3] U discussed in a Newtonian/Galilean context in Section 2.4.4. Commonly mc^2 is called *the rest energy* of the particle.

It is crucially important that on the normal scale of atomic physics and of chemistry these internal energies are enormous. Compare for instance the rest energy $m_e c^2$ of the electron (the lightest of all elementary particles) with the binding energy of the electron in the ground state of hydrogen, called the Rydberg energy Ry. From $m_e \approx 0.91 \times 10^{-30}$ kg one calculates

$$m_e c^2 = \left(0.91 \times 10^{-30} \right) \times (3 \times 10^8)^2 = 8.2 \times 10^{-14} \text{ J} = 0.51 \times 10^6 \text{ eV},$$

while

$$Ry = \frac{m_e}{2} \left(\frac{e^2}{4\pi\varepsilon_0\hbar} \right)^2 \approx 13.6 \text{ eV}.$$

The ratio[4] is

$$\frac{m_e c^2}{Ry} = \frac{2}{(e^2/4\pi\hbar c)^2} \approx \frac{2}{(1/137)^2} \sim 4 \times 10^4. \tag{9.14}$$

The *kinetic energy*, i.e. the energy needed to accelerate a particle from rest to speed u, is

$$K \equiv \varepsilon - mc^2. \tag{9.15}$$

(Low speed $u^2/c^2 \ll 1$ evidently entails $K \ll mc^2$, and validates the Newtonian approximation $K \approx mu^2/2 \approx mp^2/2$.) The energy conservation law $\Delta \sum \varepsilon = 0$ can be re-expressed as

$$\left\{ \left(\sum_i K_i \right)_{\text{after}} - \left(\sum_j K_j \right)_{\text{before}} \right\} = c^2 \left\{ \left(\sum_i m_i \right)_{\text{before}} - \left(\sum_j m_j \right)_{\text{after}} \right\} \equiv Q, \tag{9.16}$$

where the difference Q is often called the *energy release* in the reaction, though it would be more exact to call it the release of kinetic energy. When many reactions proceed at random, Q is experienced as a heat output, analogous to the heat of reaction in chemical processes; this makes (9.16) into one of the most fateful equations of applied physics, as we shall see in Section 9.3.

To recapitulate, under the relativity principle the conservation laws for three-momentum and for energy combine into a single inseparable four-vector equation

$$\Delta \sum_i \vec{p}_i = 0, \qquad \vec{p} = (p_0, \mathbf{p}) = (\varepsilon/c, \mathbf{p}) \equiv m\vec{U} = \gamma(u)m(c, \mathbf{u}). \tag{9.17}$$

Correspondingly, $\vec{p}$ is commonly called *the energy-momentum four-vector*.

That these four laws are closely linked is by no means peculiar to Einsteinian physics: Section 2.4.5 showed that the Galilean relativity principle also connects them. Of course there are differences too. The most spectacular difference is the loss of mass conservation: instead of the five conservation laws of Galilean physics, Einsteinian physics has only four. (Section 9.3.4 will consider why mass conservation had such a long innings.) By way of compensation, since the four quantities that continue to be conserved are now components of a quintessentially simple object (a four-vector), their connection has become completely standardized, and far easier to exploit than it was in Chapter 2.

9.2 Basics of the Energy-Momentum Four-Vector

9.2.1 Finite-Mass Particles

Equation (7.12) showed that $\vec{U} \cdot \vec{U} = c^2$ for any particle regardless of its speed u (provided only $u < c$). Hence (9.17) immediately gives the invariant squared length of $\vec{p}$ as

$$\vec{p} \cdot \vec{p} \equiv p_0^2 - \mathbf{p}^2 = \varepsilon^2/c^2 - \mathbf{p}^2 = m^2 \vec{U} \cdot \vec{U} = m^2 c^2, \tag{9.18}$$

showing that $\vec{p}$ is positive timelike (positive because, ε being the energy of a particle, $p_0 = \varepsilon/c > 0$). This yields the crucial expressions connecting energy and momentum:

$$\varepsilon = \sqrt{m^2 c^4 + c^2 p^2}, \qquad p = \sqrt{\varepsilon^2/c^2 - m^2 c^2}. \tag{9.19}$$

In the nonrelativistic regime where $p \ll mc$, one expands in ascending powers of p/mc:

$$\varepsilon = mc^2 \sqrt{1 + \left(\frac{p}{mc}\right)^2} = mc^2 \left\{ 1 + \frac{1}{2}\left(\frac{p}{mc}\right)^2 - \frac{1}{8}\left(\frac{p}{mc}\right)^4 + \cdots \right\}$$

$$= mc^2 + \frac{p^2}{2m} - \frac{1}{8}\frac{p^4}{m^3 c^2} + \cdots \qquad (p/mc \ll 1). \qquad (9.20)$$

Contrast this with (9.13): the kinetic energy $K = \varepsilon - mc^2$ rises faster than it does in Newtonian physics when expressed in terms of the speed u, but more slowly when expressed in terms of the momentum p.

In the opposite extreme-relativistic regime $p \gg mc$ the square root expands in ascending powers of mc/p:

$$\varepsilon = cp \sqrt{1 + \left(\frac{mc}{p}\right)^2} = cp \left\{ 1 + \frac{1}{2}\left(\frac{mc}{p}\right)^2 - \frac{1}{8}\left(\frac{mc}{p}\right)^4 + \cdots \right\}$$

$$= cp + \frac{1}{2}\frac{m^2 c^3}{p} - \cdots \qquad (p/mc \gg 1). \qquad (9.21)$$

The ordinary velocity **u** is given in terms of $\vec{p}$ by appeal to (7.13):

$$\beta(u) \equiv \mathbf{u}/c = \mathbf{U}/U_0 = \mathbf{p}/p_0 = c\mathbf{p}/\varepsilon = \mathbf{p}/\sqrt{m^2 c^2 + p^2}. \qquad (9.22)$$

This is the inverse of (9.6), which expresses $\vec{p}$ in terms of **u**; it is important in applications, where generally one calculates with ε and **p**, but where the results are sometimes required in terms of **u**.

Finally, (9.10) yields $\gamma(u)$ too in terms of p:

$$\gamma(u) = \varepsilon/mc^2 = \sqrt{1 + (p/mc)^2}. \qquad (9.23)$$

Example 9.1. A neutral K meson (mass M, such that $Mc^2 \approx 500$ MeV) at rest decays into two charged π mesons (masses m, such that $mc^2 \approx 140$ MeV), whose three-momenta are easily measured through the curvature of their tracks in a magnetic field (see Appendix C, Section C.2.2). Calculate the magnitudes of these momenta.

Solution. By momentum conservation, the three-momenta of the daughters are equal and opposite, each of magnitude p, say. Hence their energies are also equal, energy conservation yields $Mc^2 = 2\varepsilon$, and from (9.19) we have

$$p = \sqrt{\varepsilon^2/c^2 - m^2 c^2} = \sqrt{(Mc/2)^2 - m^2 c^2} \quad \Rightarrow \quad p/mc = \sqrt{(M/2m)^2 - 1} \approx 1.5.$$

This is probably the cleanest way of displaying the result. The energy release is $Q = (M - 2m)c^2$, and as a check on our algebra we note that $Q \rightarrow 0$ would entail $p \rightarrow 0$, i.e. zero kinetic energy for the reaction products.

Finally, given 1 MeV $\approx 1.6 \times 10^{-13}$ J, we have $m \approx 140 \times (1.6 \times 10^{-13})/(3 \times 10^8)^2 = 2.5 \times 10^{-28}$ kg, whence

$$p \approx 1.5mc \approx 1.5 \times (2.5 \times 10^{-28}) \times (3 \times 10^8) = 1.1 \times 10^{-19} \text{ kg} \cdot \text{m/s.} \quad \blacksquare$$

9.2.2 Zero-Mass Particles

Equations (9.18), (9.19), (9.22), (9.23) continue to apply straightforwardly in cases where $m = 0$. For such zero-mass particles, like photons and neutrinos, one has, evidently,

$$m = 0 \Rightarrow \vec{p} \cdot \vec{p} = 0, \quad p_0 = p, \quad \varepsilon = cp, \quad \mathbf{u}/c = \mathbf{p}/p_0 = \hat{\mathbf{p}}, \quad u = c. \quad (9.24)$$

Indeed, even for nonzero mass equations (9.24) can prove useful as approximations if $p \gg mc$: for instance, the exact zero-mass relation $\varepsilon = cp$ is also the leading term on the right of (9.21). The only qualitative difference is that $m \to 0$ changes $\vec{p}$ from positive timelike to positive lightlike.

In the light of the remarks at the end of Section 7.2, one ought to be very surprised indeed that equations (9.24) are so simple, and that we have found them so easily. Readers who were not surprised should recall that for $m = 0$ the expression we started from, namely $\vec{p} \equiv m\vec{U}$, appears to fail on two counts: first because one of the factors on the right vanishes, and then because the other factor $\vec{U} \equiv d\vec{X}/d\tau$ makes no sense for particles moving at speed c, for which no proper time τ can be defined. This makes it all the more remarkable that (9.24) enables us to continue Lorentz-transforming velocities $\mathbf{u}$ of magnitude c by reference to a four-vector (namely to $\vec{p}$), just when the original prescription based on $\mathbf{u} = U/U_0$ fails on account of the desertion of $\vec{U}$.

To appreciate these comments, observe that the startlingly simple relations (9.24) would have been much harder to derive if, instead of setting $m = 0$ outright in expressions interrelating the p_μ, one had tried to take the limit $m \to 0$ of the alternative expressions featuring $\mathbf{u}$, like (9.6). In a loose sense $m \to 0$ upstairs might appear to be compensated by the fact that $\sqrt{1 - u^2/c^2} \to 0$ downstairs, as we know by hindsight from (9.24); but it is not immediately obvious how these two limits should be combined. The key here is that $m \to 0$ must be implemented *at fixed* $\mathbf{p}$, which does validate (9.24) unexceptionably though in a somewhat roundabout manner. For example, judicious use of (9.23) leads to

$$\lim_{m \to 0} \varepsilon = \lim_{m \to 0} mc^2 \gamma(u) = \lim_{m \to 0} mc^2 \sqrt{1 + (p/mc)^2} = \lim_{m \to 0} mc^2 (p/mc) = cp,$$

the same conclusion as in (9.24).

Example 9.2. Neutral π mesons (mass m) in flight at speed $c\beta$ (with respect to the laboratory) decay into two photons. Calculate the energy of the photons emitted at a given angle to the flight path.

Solution. The decay is sketched in Figure 9.2. Let $\mathbf{p}, cp_0$ be the momentum and energy of the pion, and $\mathbf{k}^{(i)}, ck_0^{(i)}$ those of its daughter photons ($i = 1, 2$) emitted at angles $\theta^{(i)}$ to $\mathbf{p}$. Because photons have zero mass, $k_0^{(i)} = k^{(i)}$. Our strategy is to use energy and momentum conservation to determine say $k^{(1)}$ in the obvious way; a more efficient method will become available in Chapter 11.

Energy conservation yields

$$p_0 = k_0^{(1)} + k_0^{(2)} = k^{(1)} + k^{(2)}. \quad (i)$$

Three-momentum conservation along and at right angles to the flight path yields

$$p = k^{(1)} \cos \theta^{(1)} + k^{(2)} \cos \theta^{(2)}, \quad k^{(1)} \sin \theta^{(1)} = k^{(2)} \sin \theta^{(2)}. \quad (ii), (iii)$$

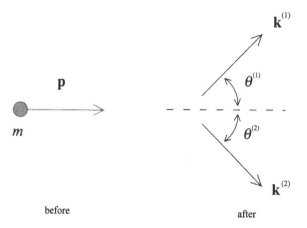

p

m

$\mathbf{k}^{(1)}$

$\theta^{(1)}$

$\theta^{(2)}$

$\mathbf{k}^{(2)}$

before   after

Figure 9.2 Neutral π meson (mass m) decaying into two photons (masses 0). Three-momenta and angles all relative to the laboratory frame.

Because we are not interested in $\theta^{(2)}$, we eliminate it from (ii), (iii):

$$k^{(2)2}(\sin^2\theta^{(2)} + \cos^2\theta^{(2)}) = k^{(2)2} = k^{(1)2}\sin^2\theta^{(1)} + (p - k^{(1)}\cos\theta^{(1)})^2$$
$$= p^2 + k^{(1)2} - 2pk^{(1)}\cos\theta^{(1)}. \qquad \text{(iv)}$$

But (i) leads to

$$k^{(2)2} = (p_0 - k^{(1)})^2 = p_0^2 + k^{(1)2} - 2p_0 k^{(1)}. \qquad \text{(v)}$$

Subtracting (iv) from (v) and then using $p_0^2 - p^2 = m^2 c^2$ we find

$$0 = p_0^2 - p^2 - 2p_0 k^{(1)} + 2pk^{(1)}\cos\theta^{(1)} = m^2 c^2 - 2k^{(1)}(p_0 - p\cos\theta^{(1)})$$
$$= m^2 c^2 - 2k^{(1)}(\gamma mc - \gamma mc\beta \cos\theta^{(1)}),$$

where $\gamma = 1/\sqrt{1 - \beta^2}$. Rearranging and multiplying by c we get the end-result:[5]

$$ck_0^{(1)} = ck^{(1)} = \frac{mc^2}{2\gamma(1 - \beta\cos\theta^{(1)})}. \qquad \text{(vi)}$$

Thus the photon energy falls with increasing angle. The maximum and minimum energies, realized for emission forward and back ($\cos\theta^{(1)} = \pm 1$) are

$$(ck_0^{(1)})_{\text{max}} = \frac{mc^2\sqrt{1 - \beta^2}}{2(1 - \beta)} = \frac{mc^2}{2}\sqrt{\frac{1 + \beta}{1 - \beta}}, \qquad (ck_0^{(1)})_{\text{min}} = \frac{mc^2}{2}\sqrt{\frac{1 - \beta}{1 + \beta}}. \qquad \text{(vii)}$$

■

9.2.3 Lorentz Transformations

Since $\vec{p}$ is a four-vector, its components Lorentz-transform according to the standard rules. For convenience we cite the transformation for standard configuration, and its inverse, with a reminder that they apply irrespective of the mass:

$$p_0' = \gamma(v)(p_0 - \beta(v)p_x), \qquad p_x' = \gamma(v)(p_x - \beta(v)p_0), \qquad p_{y,z}' = p_{y,z}; \qquad (9.25)$$

$$p_0 = \gamma(v)(p_0' + \beta(v)p_x'), \qquad p_x = \gamma(v)(p_x' + \beta(v)p_0'), \qquad p_{y,z} = p_{y,z}'. \qquad (9.26)$$

Recall $\beta(v) \equiv v/c$, $\gamma(v) \equiv 1/\sqrt{1 - \beta^2(v)}$, and $\varepsilon = cp_0$.

Directions transform as follows. Let **p** lie in the xy plane, at an angle ϕ to the x axis. Then **p**$'$ lies in the $x'y'$ plane, at an angle ϕ' to the x' axis, where

$$\tan \phi' = \frac{p_y'}{p_x'} = \frac{p_y}{\gamma(v)[p_x - \beta(v)p_0]}$$

$$= \frac{p \sin \phi}{\gamma(v)[p \cos \phi - \beta(v)p_0]} = \frac{(p/p_0)\sin \phi}{\gamma(v)[(p/p_0)\cos \phi - \beta(v)]}. \qquad (9.27)$$

Depending on the problem, it may or may not help to substitute $p/p_0 = u/c = \beta(u)$. Between the two values that (9.27) allows for ϕ' one must, as in all such cases, choose so as to fit the correct sign say of $\cos \phi'$, i.e. of p_x', which (9.25) shows to be the same as the sign of any of the combinations $[p_x - \beta(v)p]$, or $[p_x/p_0 - \beta(v)]$, or $[u_x - v]$.

Example 9.3. For *zero-mass particles* the Lorentz transforms of energy and momentum simplify considerably. Given two frames S, S' in standard configuration with relative velocity $c\beta$, and a photon with four-momentum components $(p_0, p_x, p_y, 0) = (p_0, p\cos \phi, p\sin \phi, 0)$ relative to S, determine the components p_μ' relative to S'.

Solution. The key fact is that for zero mass we have both $p_0 = p$ and $p_0' = p'$. All we need do is to substitute into the appropriate Lorent transforms (9.25):

$$p_0' = \gamma(p_0 - \beta p_x) = \gamma(p - \beta p \cos \phi) = \gamma p(1 - \beta \cos \phi), \qquad (9.28)$$

$$p_x' = \gamma(p_x - \beta p_0) = \gamma p(\cos \phi - \beta), \qquad p_y' = p_y = p \sin \phi, \qquad p_z' = p_z = 0. \qquad (9.29)$$

In the collinear case $\phi = 0$, these reduce to

$$p_0' = p' = p_x' = \frac{p(1 - \beta)}{\sqrt{1 - \beta^2}} = p\sqrt{\frac{1 - \beta}{1 + \beta}}; \qquad (9.30)$$

for $\phi = \pi$ one merely changes the sign of $\cos \phi$, or equivalently of β.

Also worth noting is the transformation rule for the angle ϕ thus defined:

$$\cos \phi' = \frac{p_x'}{p'} = \frac{(\cos \phi - \beta)}{(1 - \beta \cos \phi)}, \qquad \sin \phi' = \frac{p_y'}{p'} = \frac{\sin \phi}{\gamma(1 - \beta \cos \phi)}, \qquad (9.31)$$

$$\tan \phi' = \frac{\sin \phi}{\gamma(\cos \phi - \beta)}. \qquad (9.32)$$

The same trigonometry as in Section 4.2.4 then leads to

$$\tan \phi'/2 = \sqrt{\frac{1 + \beta(v)}{1 - \beta(v)}} \tan \phi/2. \tag{9.33}$$

These results for zero-mass (hence light-speed) particles can serve as shortcuts to the Doppler effect and the aberration of light waves, which will be considered in Chapter 13. ∎

9.3 Evidence

We require experimental confirmation that $\mathbf{p}$ and $\varepsilon \equiv cp_0$ defined by (9.6) and (9.10), (9.19) are indeed the momentum and the energy of a particle. The most compelling evidence comes from charged particles responding to the forces exerted on them by the electromagnetic fields in modern accelerators, which would simply not function otherwise. But this book does not deal with electromagnetic forces, except briefly in Appendix C, whence we settle for demonstrating only that the identifications are correct for freely moving particles, and that ε and $\mathbf{p}$ are the quantities conserved in collisions and reactions, as conjectured in (9.9), (9.12). Regarding ε such evidence is perfectly direct. But direct measurements of momentum need forces, whence the evidence we shall cite regarding $\mathbf{p}$ is indirect, relying on the fact that calculated *energies* after a collision depend on the conservation of momentum as well as of energy. The Compton effect (to be described in Chapter 11) will furnish a typical example, which is interesting also for other reasons.

9.3.1 *Energy as a Function of Speed*

We sketch a famous demonstration experiment[6] at the Massachusetts Institute of Technology, designed to make its interpretation clear and immediate, rather than to maximize accuracy. The idea is to verify (9.10) by direct measurements both of the kinetic energy K and of the speed u.

The layout is sketched in Figure 9.3. Successive well-localized bunches of electrons (mass m) are accelerated through a known potential difference of W volts, emerging with $K = eW$ eV. (There are 120 bunches per second, each lasting around 3×10^{-9} s.) They then coast down an evacuated drift tube, and are absorbed in a block of metal, called a Faraday cup, where K is converted into heat Q. The electric current j from the cup to earth is measured, and determines the average number $dn/dt = j/e$ of electrons per second (such averages being taken over many bunches). The temperature rise of the cup serves as a check on the average rate $dQ/dt = K dn/dt = Kj/e$ at which the electrons supply energy: this may reassure purists unwilling at this stage to accept $K = eW$, on the grounds that it is inferred from the behaviour of high-speed electrons in electric fields, an interaction about which we have not yet cited any evidence. Since the thermally measured energy agrees (within its 10% accuracy) with eW, we shall stretch a point and continue to use $K = eW$. Given $mc^2 = 0.51$ MeV, while eW is varied from 0.5 to 15 MeV, we see that K/mc^2 ranges from 1 to 30, so that the electrons range from the just-relativistic to the extreme-relativistic.

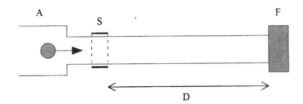

Figure 9.3 Sketch (not to scale!) of the MIT demonstration experiment described in Section 9.3.1. A: electrostatic accelerator, feeding bunches of electrons (small blob) into the drift tube. S: short sleeve around the drift tube, serving to indicate the passage of a bunch. F: Faraday cup, which absorbs the electrons. The distance $D = 8.4$ m is the flight path over which the bunches are timed.

To measure u, a short (about 0.1 m) metal sleeve is placed around the head of the drift tube, at a distance $D = 8.4$ m from the Faraday cup. The charge of every successive bunch of electrons induces a small voltage pulse in the sleeve as it passes through, and then another pulse in the cup; these pulses are displayed on an oscilloscope, allowing the time lag T between them to be determined. Since u does not in fact exceed c, the flight times D/c are at most of order 10^{-8} s, not particularly difficult to measure. Evidently $u = D/T$.

The results are displayed by plotting $\beta^2 \equiv (u/c)^2$ against K/mc^2. Theory gives

$$K = \varepsilon - mc^2 = \frac{mc^2}{\sqrt{1-\beta^2}} - mc^2 \Rightarrow \beta^2_{\text{theory}} = 1 - \left(\frac{1}{1 + K/mc^2}\right)^2,$$

while in Newtonian physics one would have $\beta^2_N = 2K/mc^2$, i.e. a linear rise.

The data from the paper are quoted in the following table:

K/mc^2	1	2	3	9	30
β^2	0.752	0.828	0.9220	0.964	1.0
β^2_{theory}	0.750	0.889	0.938	0.990	0.999

Figure 9.4 plots both β^2 and β^2_N against K/mc^2, and shows four of the data points (the fifth point is too far off to the right). The Newtonian expression is obviously a nonstarter, while agreement between theory and experiment is satisfactory within the accuracy claimed (which from the way the results are given appears to be a few percent).

9.3.2 The Cockroft-Walton Reaction

The first-ever transmutation of nuclei induced by artificially accelerated projectiles is the reaction

$$p + {}^7\text{Li} \rightarrow \alpha + \alpha, \tag{9.34}$$

observed in 1932 by Cockroft and Walton[7] in Cambridge. In terms, notionally, of the corresponding atoms (it is the atomic masses that are used in the analysis) this would read ${}^1\text{H} + {}^7\text{Li} \rightarrow {}^4\text{He} + {}^4\text{He}$. The target lithium atom is at rest; the protons from the

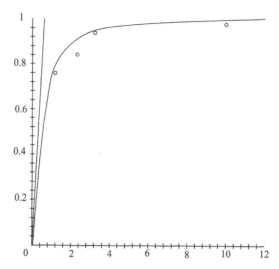

Figure 9.4   Horizontal scale: K/mc^2, where K is the kinetic energy. Vertical scale: $\beta^2 = (v/c)^2$. The straight line is the Newtonian relation $\beta^2 = 2K/mc^2$. Circles: measured values, from the table in Section 9.3.1; the last point is not shown.

accelerator had known kinetic energies K_p, up to about 0.5 MeV. (At least 0.1 MeV is needed to give protons a reasonable chance of reaching the Li nucleus.) The experiment detected α particles emitted at right angles to the protons, and measured their range in materials of known density. By calibration the range determines K_α, to within a percent or so.

Cockroft and Walton seem to have disregarded both K_p and the proton momentum, presumably because K_p is much smaller than the energy Q liberated in the reaction. Accordingly they assumed (and roughly verified) that their pairs of α particles emerged in approximately opposite directions with practically equal energies, in which case $Q \approx 2K_\alpha$. Equation (9.16) then predicts that $2K_\alpha$ should equal the mass defect times c^2. The atomic masses were known from measurements with mass spectrometers; the values available at the time were

$$m(^7\mathrm{Li}) = 7.0104 \pm 0.003, \quad m(^1\mathrm{H}) = 1.0072, \quad m(^4\mathrm{He}) = 4.0011, \quad (^{16}\mathrm{O\ units}),$$

expressed in the then atomic mass units.[8] At this accuracy the difference between old and new amu is irrelevant; and we convert from amu to MeV through (1 amu)$c^2 \approx$ 931 MeV. The dominant error was that in the lithium mass. The prediction reads

$$Q = 2K_\alpha = (1.0072 + 7.0104 - 2 \times 4.0011)\ \mathrm{amu} \times c^2$$

$$= (0.0154 \pm 0.003)\ \mathrm{amu} \times c^2 = (0.0154 \pm 0.003) \times 931\ \mathrm{MeV} = (14.3 \pm 2.7)\ \mathrm{MeV},$$

agreeing well enough with the measured $2K_\alpha \approx 17.2$ MeV.

Some comments are in order.

(i) The check depends critically on the phenomenal accuracy of the mass spectrometer, because Q is such a small difference between very much larger quantities.

(ii) It is only just good enough to disregard the momentum and energy of the proton, and only because the experiment detects α's at right angles. Problem 9.12 calls for a better calculation.

(iii) Newer measurements furnish more accurate confirmation. However, modern compilations of atomic masses take (16) for granted, and are generally adjusted to take account of various measured reaction energies, so that one must beware of using mass values derived by assuming the relation one means to check. Fortunately there exists at least one set of tables[9] enabling one to avoid doing this. They give

$$m(^7\text{Li}) = 7.016004, \quad m(^1\text{H}) = 1.007825, \quad m(^4\text{He}) = 4.002603, \quad (^{12}\text{C units}),$$

predicting

$$Q = (1.0078252 + 7.016004 - 2 \times 4.002603) \text{ amu} \times c^2 = 0.018623 \text{ amu} \times c^2$$

$$= 0.018623 \times 931.48 \text{ MeV} = 17.347 \text{ MeV}.$$

This is to be compared with an independently measured 17.373 ± 0.006 MeV, whose quoted error would appear to be simply over-optimistic, rather than to cast serious doubts on the theory.[10]

(iv) Many other reactions between nuclei (and neutrons) could of course supply evidence of essentially the same kind. Figure 9.5 is a qualitative reminder of the importance of the heat produced by some of them.

9.3.3 Positron annihilation

Positrons (e^+) are positive electrons, with masses equal to the electronic mass m. (We shall presently need $mc^2 = 0.51000$ MeV.) When a positron meets an electron, they can annihilate each other, creating photons, most commonly just two: $e^+ + e^- \to \gamma + \gamma$, with the γ's here denoting the photons, since these have energies in the gamma-ray region. A photon, a particle of zero mass, is a quantum of electromagnetic radiation; its energy ε is linked to the (circular) frequency $\omega = 2\pi\nu$ of the radiation by the Einstein relation $\varepsilon = \hbar\omega$, while frequency and wavelength λ are related by $\lambda = c/\nu = 2\pi c/\omega$, appropriately to waves with velocity c.

The experiment to be described[11] uses positrons from the beta-decay of short-lived ^{64}Cu, produced by radio-activating a sample of solid copper in a nuclear reactor. Collisions in the material slow down most of the positrons to thermal kinetic energies (of the order of $k_B T$) before they annihilate; when they do annihilate, both they and the electrons can be treated as effectively at rest, since both $k_B T$ and the kinetic energies of the electrons in copper are negligible compared with the pertinent energies mc^2. Thus the total momentum of the annihilating pair is zero; hence the momenta of the two photons are equal and opposite, and their energies are equal. Energy conservation and the Einstein relation then dictate that

$$\hbar\omega = mc^2 \implies \lambda_{\text{theory}} = \frac{2\pi c}{\omega} = \frac{2\pi\hbar}{mc} \equiv 2\pi\lambdabar_c = 2.42631 \times 10^{-12} \text{ m}, \tag{9.35}$$

where $\lambdabar_c \equiv \hbar/mc$ is called the Compton wavelength. Since $\hbar, m, c$ are all known beyond six significant figures, so is $\lambdabar_c$.

Figure 9.5   Some secondary effects of the heat output from a chain of certain nuclear reactions. "*When you sacrifice, pray the gods that it may never fall to your lot, either to suffer it, or to do it*". (Mary Renault, *The last of the wine*, Longmans, London, 1956). However, the scientist cannot count on never having to make such choices, nor on choosing right by conditioned reflex. (Reproduced by permission of the Negative Library, Ministry of Defence.)

The test consists in a direct measurement of the wavelength λ of the annihilation radiation. This is achieved by observing the angle of diffraction from a quartz crystal shaped into a lamina measuring $1 \times 50 \times 70$ mm, and bent into a segment of a cylinder of radius 2 m. (The curvature increases the accuracy by focusing the diffracted beam.) The experiment is a bravura piece: in ordinary X-ray diffraction the wavelengths are comparable to the lattice spacing (of the order of a few times 10^{-10} m), while here they are smaller by a factor of a hundred. Consequently the angles of incidence and of diffraction have to be very small, and the calibration of the instrument, with other X-rays of known wavelengths, is a *tour de force* in itself.

The reported result is $\lambda_{\text{exp}} = (2.4271 \pm 0.0010) \times 10^{-12}$ m, agreeing with λ_{theory} within the indicated statistical errors of about 1 part in 2000. The test is especially pleasing in that the theoretical expression features nothing but fundamental constants.

9.3.4 Why Was Mass Thought to be Conserved in Chemistry?

In Newtonian physics mass is conserved, a fact we discussed and exploited in Section 2.4. Perhaps the most searching evidence comes from the chemical balance; indeed it is unthinkable that modern chemistry could have developed until the fact was established.

Since mass conservation is contrary to (9.16) and to the evidence just discussed, the apparent contradiction must be resolved. It stems from the simple fact that the internal energy changes in chemical reaction are too small for the consequent mass differences to be detected by weighing. Such changes are typically somewhat smaller than the ~ 10 eV binding energy of the hydrogen atom, while molecular masses cannot be smaller than the mass M of the hydrogen atom, $Mc^2 \sim 1000$ MeV $= 10^9$ eV. Thus fractional mass changes in chemistry are at most of the order of $10/10^9 = 10^{-8}$, while the chemical balance can do no better than about 10^{-5}.

By contrast, nuclear reaction energies involving medium–heavy nuclei (say $A \sim 100$) are typically of the order of or higher than 0.1 MeV (in light nuclei sometimes considerably higher), while the masses, namely AM, are essentially the same as in chemistry (since almost all the atomic mass resides in the nucleus). Thus fractional mass changes are typically of order $0.1/1000A \sim 10^{-6}$, well within the accuracy of the mass spectrometer, which nowadays can detect fractional mass differences as small as 10^{-9}.

9.4 Notes

1. We use lower-case $\vec{p}$ and p_μ in spite of our habitual capitals for four-vectors, because P has been pre-empted for the total momentum of a *system* of particles.
2. The suffixes i label the particles, and must not be misread as vector indices: thus p_{i0} stands for the time component (suffix 0) of the four-vector $\vec{p}_i$. Similarly we might write p_{ix}, etc.
3. This Newtonian "U" is unrelated to the four-velocity $\vec{U}$ or to any of its components.
4. Remarkably, this ratio is independent of m_e. The combination $\alpha \equiv e^2/4\pi\hbar c \approx 1/137$ is a pure number, called the fine-structure constant for historical reasons from atomic spectroscopy (see Appendix C, Section C.2.3).
5. That the denominator in (vi) never vanishes, i.e. that (vi) makes sense for all θ from 0 to π, confirms that all θ can occur in such decays. This could have been foreseen from the

velocity-combination rule: since the speed $c\beta$ of the parent relative to the laboratory is inevitably less than the speed c of the daughters relative to the parent, daughters emitted forward or back in the parent rest frame move forward or back in the laboratory frame as well. By contrast, with finite-mass daughters, as for example in the decay $K \rightarrow \pi + \pi$, the range of laboratory angles is restricted if the rest-frame speed of the daughters is less than $c\beta$. See Problem 11.4.

6. W. Bertozzi, "Speed and kinetic energy of relativistic electrons", *American Journal of Physics* (1964), volume **32**, page 551.

7. J. D. Cockroft and E. T. S. Walton, "*Experiments with high velocity positive ions. II. – The disintegration of elements by high velocity protons*", *Proceedings of the Royal Society* (1932), volume A **137**, page 229. This is the work for which the authors later received the Nobel prize.

8. Originally, atomic mass units (amu) were defined so that the mass of one atom of ^{16}O was exactly 16 amu. More recently they have been re-defined so that the mass of one atom of ^{12}C is exactly 12 amu. (When using older tables in accurate calculations, one needs to look out for the difference.) The modern amu equals 1.660540×10^{-27} kg, so that (1 amu) $\times c^2 = 931.481$ MeV. The masses of atoms include the masses of the electrons, and also the mass equivalent of their binding energy. The latter is negligible to the accuracies of interest here; the former are not negligible, but cancel from the requisite differences, since in a reaction between atoms the total number of electrons is the same before and after.

9. J. H. E. Mattauch *et al.*, *Nuclear Physics* (1965), volume **67**, pages 1 to 120.

10. But for the discrepancy, the measured Q would presumably have been used in establishing the tabulated mass values, depriving us of our example.

11. J. W. M. DuMond *et al.* (California Institute of Technology), "Precision measurement of the wavelength and spectral profile of the annihilation radiation from Cu^{64} with the two-meter focusing curved crystal spectrometer", *Physical Review* (1949), volume **75**, page 1226.

10 Natural Units, and the Prevalence of MeV

In many calculations it is an extravagant waste of time to try to get all the c's right at every step. Such waste can be avoided in either of two ways, (a) or (b), which in practice are equivalent.

(a) Choose new units of length and/or time, called natural units (n.u.), so as to make the numerical value of c equal to unity: $c = 1$ n.u. (The "n.u." is universally omitted.) For definiteness we keep the second as the unit of time, and for the new unit of length adopt the light-second, namely the distance travelled in one second by "light", i.e. by any signal that moves with the invariant speed:

$$1 \text{ light-second} = c \times (1 \text{ second}) = 2.99792458 \times 10^8 \text{ metres.}$$

(b) Except perhaps in the end-result, simply drop all factors of c wherever they might occur.

In either scheme, one can if necessary restore the missing c's at the very end by mere inspection, making dimensions match in a way we shall illustrate presently.

In scheme (b) restoration amounts to correcting errors that algebraically though not physically speaking would have been trivial. We shall return to scheme (b) at the end of this chapter.

In scheme (a), restoring c's amounts to reverting from natural to conventional (say to SI) units.

For reasons unconnected with logic, the prospect of transcribing expressions under a change of units is widely experienced as frightening; that is why we shall spell out several examples at length. Actually however transcription is easy, depending as it does on nothing but the familiar fact that physical quantities (as distinct from pure numbers) can only be and always are given as (number of units) × (size of units): ten thousand dollars makes you a millionaire in cents. Given two different system of units, call them A and B, this fact reads

(physical quantity)

$$= \text{(numerical measure in A)} \times \text{(size of unit in A)}$$

$$= \text{(numerical measure in B)} \times \text{(size of unit in B)}$$

$$\Rightarrow \frac{\text{(numerical measure in B)}}{\text{(numerical measure in A)}} = \frac{\text{(size of unit in A)}}{\text{(size of unit in B)}}.$$

The change to natural *units* evidently changes, in practice, the number of independent physical *dimensions*, and thereby the scope of dimensional analysis. Length and time now assume the same dimension: $[L] = [T]$. Hence kinematics has only one physical dimension left, which we may choose to be time $[T]$. In particular, $[\text{velocity}] = [LT^{-1}]$ becomes dimensionless, so that $u = 1/2$ for instance signifies half the speed of light; by the same token $\beta(v) = v$. Acceleration has dimension $[LT^{-2}] = [T]^{-1}$.

Dynamics with natural units has two dimensions, say time $[T]$ and mass $[M]$. Then the natural dimensions of mass, $[\text{momentum}] = [M][\text{velocity}]$, and $[\text{energy}] = [M][\text{velocity}]^2$ are the same, namely $[M]$. For example, the dimensions of Planck's constant[1] are those of $[\text{angular momentum}] = [\text{momentum}][\text{length}] = [M][T]$. Most important, because overwhelmingly convenient, we are licensed to write $p_0 = \varepsilon = \sqrt{m^2 + p^2}$, and $u = p/\varepsilon$. Accordingly, in natural units the solution of Example 9.1 for instance takes, immediately, the simple form $p = \sqrt{(M/2)^2 - m^2}$.

Theory of course is indifferent to what units are chosen for measuring the surviving dimensions. We keep the second as the unit for $[T]$. For $[M]$ one could choose the conventional SI unit of mass, i.e. the kilogram; then one n.u. of mass would equal 1 kg, while one n.u. of energy would equal $(1 \text{ kg}) \times c^2 = 1 \times (3 \times 10^8)^2 = 9 \times 10^{16}$ J. Alternatively one could adopt for $[M]$ the SI unit of energy, namely the joule; then one n.u. of mass would equal $1/(3 \times 10^8)^2 = 1.111 \times 10^{-17}$ kg, while one n.u. of energy would equal 1 J.

However, in nuclear and high-energy physics the natural unit generally chosen for $[M]$ is the MeV, equal to 1.602177×10^{-13} J. For example, tables of particle properties often give the mass m_p of the proton as "938.272 MeV" (sometimes, and equivalently, as "938.272 MeV/c^2)"; expressed in conventional units this means that $m_p c^2 = 938.272$ MeV, whence

$$m_p \approx 938.272 \times (1.602177 \times 10^{-13})/(2.997925 \times 10^8)^2 = 1.67262 \times 10^{-27} \text{ kg},$$

which last would be a perversely inconvenient way of feeding the information into most calculations in high-energy physics. For instance, high-energy physicists would always write the final numerical result in Example 9.1 as $p \approx 1.5m = 1.5 \times 140$ MeV.

Example 10.1. According to the so-called Bohr theory, in the ground state of the hydrogen atom the orbital speed of the electron is $u = e^2/4\pi\varepsilon_0\hbar$. What is its value in natural units?

Solution. The question is trivial (though the answer is instructive). "The speed expressed in natural units" is just the pure number u/c, or in other words the speed in any units divided by c in the same units. Thus $u \rightarrow u/c = e^2/4\pi\varepsilon_0\hbar c \approx 1/137$: see Chapter 9, note 4. Since relativistic corrections (e.g. to the kinetic energy) are of order $(u/c)^2$, we see that in the hydrogen atom they will be of the order of 1 part in 10^4; that is why, to this accuracy, it may be treated nonrelativistically,[2] whether according to Bohr or quantum-mechanically according to the Schrödinger equation. ■

Example 10.2. What is the numerical value of Planck's constant $\hbar = 1.055 \times 10^{-34}$ kg·m^2/s, expressed in natural units with the unit of $[M]$ chosen to be 1 MeV?

Solution. Start by rearranging the units as follows:

$$1 \text{ kg} \cdot \text{m}^2/\text{s} = 1 \text{ kg} \cdot (1 \text{ m/s})^2 \cdot (1 \text{ s}) = (1 \text{ kg} \cdot c^2) \left(\frac{1 \text{ m/s}}{c}\right)^2 (1 \text{ s})$$

$$= \left(\frac{1 \text{ kg} \cdot c^2}{1 \text{ MeV}}\right) \left(\frac{1 \text{ m/s}}{c}\right)^2 (1 \text{ MeV})(1 \text{ s}).$$

The conversion factors are the pure numbers $(1 \text{ kg} \cdot c^2/1 \text{ MeV}) = (9 \times 10^{16})/(1.602 \times 10^{-13})$ and $(1 \text{ m/s})/c = 1/(3 \times 10^8)$. Hence

$$\hbar = (1.05 \times 10^{-34}) \left(\frac{9 \times 10^{16}}{1.602 \times 10^{-13}}\right) \left(\frac{1}{3 \times 10^8}\right)^2 \text{ MeV} \cdot \text{s} = 6.58 \times 10^{-22} \text{ MeV} \cdot \text{s}. \quad ■$$

Return now to scheme (b) for reverting from natural to conventional units. At first this sheme might induce discomfort, since it invites one to make any number of errors (to drop all c's), which are never corrected individually, though the cumulative error is corrected right at the end simply by matching dimensions. Nevertheless, experience suggests that after some practice scheme (b) is quicker and provokes fewer fruitless questions.

The undoctored outcome of a scheme-(b) calculation is, by prescription, an expression void of c's. What to do next (and finally) is best shown by example.

- Under the ubiquitous square roots $\sqrt{1 - v^2}$, the "1" is a pure number, i.e. dimensionless; therefore the addend v^2 must also be dimensionless; and since v is a speed, v^2 is made dimensionless by dividing it by c^2, or in other words by the replacement $v \rightarrow v/c \equiv \beta$. Thus we recover the familiar $\gamma(v) = 1/\sqrt{1 - v^2} \rightarrow 1/\sqrt{1 - v^2/c^2}$.
- Without c's, the Lorentz transformations of event coordinates read

$$t' = \gamma(t - vx), \qquad x' = \gamma(x - vt). \qquad \text{(i), (ii)}$$

We have already dealt with the dimensionless γ's. On the right of (i), the second factor must have dimension $[T]$. Thus the entry t is correct as it stands. The addend vx has dimensions $[LT^{-1}][L]$, and to turn this into the requisite $[T]$ we divide it by $[LT^{-1}]^2$, i.e. by c^2, which recovers the correct conventional rule $t' = \gamma(t - vx/c^2)$. On the right of (ii) the second factor must have dimension $[L]$; since both entries taken literally have this dimension already, no action is needed, and the conventional rule reads $x' = \gamma(x - vt)$.

• Without c's, equation (9.11) for $\vec{p}$ reads

$$\vec{p} = (\varepsilon, \mathbf{p}), \quad \varepsilon = \gamma(u)m \Rightarrow \vec{p} \cdot \vec{p} = \varepsilon^2 - p^2 = m^2. \tag{iii), (iv), (v)}$$

We take these relations in turn, recalling our convention that all components of any given four-vector have the same dimensions, and that we have chosen $\vec{p}$ to have the dimensions $[M][\text{velocity}]$ of momentum.

Since ε is an energy (dimension $[M][\text{velocity}]^2$), it must in (iii) be divided by a velocity, i.e. by c, in order to reduce the dimensions to those of momentum; hence the expression on the right becomes $(\varepsilon/c, \mathbf{p})$.

In (iv), $\gamma(u) = 1/\sqrt{1 - u^2}$ becomes $1/\sqrt{1 - u^2/c^2}$ by the same argument as we used above for $\gamma(v)$. Further, and again because ε is an energy, m on the right must be multiplied by a velocity squared, i.e. by c^2, in order to make the dimensions the same on both sides; thus we recover $\varepsilon = mc^2/\sqrt{1 - u^2/c^2}$.

Finally, in (v) all the entries on the right must have the dimensions $[M]^2[\text{velocity}]^2$ of the squared momentum on the left, and we achieve this by replacing $\varepsilon \to \varepsilon^2/c^2$ and $m^2 \to m^2 c^2$.

• Without c's, the Lorentz transformations for energy and momentum read

$$\varepsilon' = \gamma(v)(\varepsilon - vp_x), \qquad p'_x = \gamma(v)(p_x - v\varepsilon),$$

where γ as always is dimensionless. Since $[v][p_x] = [\text{velocity}][M \cdot \text{velocity}] = [M][\text{velocity}]^2 = [\text{energy}]$, the first relation is conventionally correct as it stands. In the second relation, we must doctor the term $v\varepsilon$ so as to give it the dimensions of momentum; this is achieved by the replacing it with $(v/c)(\varepsilon/c)$, which yields the correct rule in conventional units, namely $p'_x = \gamma(v)(p_x - v\varepsilon/c^2)$.

Of course, the results of this procedure are always arbitrary to within some power of c common to both sides of any equation. For example, faced with $\vec{p} = (\varepsilon, \mathbf{p})$, one might choose to assign to every term the dimensions not (as above) of momentum p but of energy ε; in that case the result in conventional units would read $c\vec{p} = (\varepsilon, c\mathbf{p})$, just as correct as, trivially convertible to, and in some problems marginally more useful than $\vec{p} = (\varepsilon/c, \mathbf{p})$.

Example 10.3. Relative to an inertial frame S a particle of mass m moves along the x axis with energy $\varepsilon = 2m$. Relative to another frame S' (in standard configuration with S) its energy is $\varepsilon' = 3m$. What is the velocity v of S' relative to S?

Solution. We answer the question, as it is put, in natural units. It is easier and clearer to do the calculation with $\varepsilon = am$ and $\varepsilon' = a'm$, substituting $a = 2$, $a' = 3$ only at the end. We shall need $p_x = p = \sqrt{\varepsilon^2 - m^2} = m\sqrt{a^2 - 1}$. Then the Lorentz transformation

$$\varepsilon' = a'm = (\varepsilon - vp)/\sqrt{1 - v^2} = m\left(a - v\sqrt{a^2 - 1}\right)/\sqrt{1 - v^2}$$

rearranges, after squaring, into a quadratic equation for v:

$$[a^2 + a'^2 - 1]v^2 - [2a\sqrt{a^2 - 1}]v + [a^2 - a']^2 = 0,$$

whence

$$v = \frac{a\sqrt{a^2 - 1} \pm a'\sqrt{a'^2 - 1}}{a^2 + a'^2 - 1} = \frac{2\sqrt{3} \pm 3\sqrt{8}}{12} = (0.99578, \quad -0.418).$$

Since v is a velocity, one multiplies these numbers by c to get the answers in conventional units:

$$v = (0.99578, \ -0.418)c = (2.97268, \ -1.25) \times 10^8 \text{ m/s}.$$

(What is likely to matter about the positive root is not just that it is close to $c = 2.9979$ m/s, but how close: hence we are likely to need many decimal places in the positive though maybe not in the negative root. Observe how one's attention is focused on this point much more readily by natural than by conventional units.) ∎

Though working with natural units under either scheme needs some practice, and is obviously optional in principle, it is almost essential for the avoidance of intolerable tedium in the course of any but the most trivial calculations: the reader is strongly urged to acquire the habit. In the next chapter we shall calculate with natural units whenever convenient, though end results will often be displayed in conventional units as well.

Notes

1. This way of remembering its dimensions is suggested by the so-called Bohr theory of the hydrogen atom, or by quantum mechanics generally, which make orbital angular momentum into an integral multiple of $\hbar$. Alternatively one can think of the Planck–Einstein relation $\hbar\omega = $ (photon energy), whence $[\hbar] = [\text{energy}]/[\text{frequency}] = [ML^2T^{-2}]/[T^{-1}] = [ML^2T^{-1}] = [MT]$.
2. Thus, in atomic (unlike high-energy) physics relativistic effects are small corrections, and our natural units are not natural at all. Accordingly, atomic physicists often use atomic units, defined so that the numerical values of $\hbar$, of the electron mass, and of the Bohr radius are all equal to 1. In these units the value of the speed of light is evidently $\simeq 137$, i.e. large though not quite infinite.

11 Systems of Particles: Four-Momentum Conservation using Invariants

In applying the conservation laws (9.9), (9.12), (9.16) for energy and momentum, it is far too easy to be led into formidably tedious algebra until, by precept or experience, one has learnt the methods specifically designed to avoid this. Such methods exploit three main ideas: (i) systems of freely-moving particles with total momentum $\vec{P}$ may often be treated as if they were a single particle having this momentum, and mass $M = \sqrt{\vec{P} \cdot \vec{P}}/c$; (ii) we may evaluate Lorentz-invariant (i.e. four-scalar) combinations of four-momenta, like M^2, in any inertial frame, and equate the expressions found in different frames; (iii) often it is convenient to start, counter-intuitively, by segregating, on one side of a conservation law, the four-momentum, say $\vec{q}$, of a particle that one does *not* want to know about, and then to construct the invariant $\vec{q} \cdot \vec{q}$, a prescription we shall abbreviate as *segregate and square*.

Before we explore these ideas systematically, Section 11.1 uses the simplest possible example, namely two-body decay, to illustrate some of them in action, and the penalties for ignoring them. The full armoury is set up in Section 11.2, and used in Section 11.3 to solve the important problem of the energy threshold for particle production. Section 11.4 deals with Compton scattering (photon by electron), famous in history for establishing the existence of photons, and incidentally supplying evidence for three-momentum conservation, as promised in the preamble to Section 9.3. Largely as an exercise (which may be omitted), Section 11.5 considers, somewhat more generally, the transformation between the so-called laboratory and centre-of momentum frames for elastic scattering. Finally, Section 11.6 deals briefly with macroscopic systems whose masses can in effect change continuously.

To think that such developments are mere technicalities would miss the point of relativity theory altogether. The basic notions of invariants and of form-invariant equations are central to the physics not because they are pretty, but because they furnish efficient methods of describing how things actually work; physical insight depends on the ability to identify such methods and then to adapt one's thinking until they become intuitive; and progress towards this depends wholly on practice in applying any would-be concepts to real physical processes. Thus, implementing the programme sketched in points (i) to (iii) above is likely to reveal more about nature than any amount of artificially trivialized discussion about light-clocks or science-fiction space-ships.

11.1 The Simplest Example: Two-Body Decay

Consider the decay of an unstable but long-lived particle[1] of mass M into two daughter particles of masses m_1, m_2; for instance the decay $\Lambda \to p + \pi^-$ of the neutral hyperon Λ into a negative π meson and a proton p, whose masses are $(M, m_1, m_2) = (1116, 140, 938)$ MeV/c^2. We want to calculate the energy ε_1^* and the magnitude p_1^* of the momentum of daughter 1 in the rest frame S^* of the parent.

11.1.1 The Obvious Method

In S^*, because the parent is at rest, the total three-momentum is zero: $\mathbf{P}^* = \mathbf{p}_1^* + \mathbf{p}_2^* = 0$. Hence the three-momenta of the daughters are equal and opposite, with equal magnitudes p^*, say. By the same token the total energy is Mc^2, so that energy conservation reads

$$Mc^2 = \varepsilon_1^* + \varepsilon_2^* = \sqrt{m_1^2 c^4 + p^{*2} c^2} + \sqrt{m_2^2 c^4 + p^{*2} c^2}. \tag{11.1}$$

The kinetic energy release defined by (9.16) is

$$Q = c^2(M - m_1 - m_2). \tag{11.2}$$

At this point we adopt natural units, i.e. we drop all c's until the end.

The most obvious procedure starts by solving (11.1) for p^{*2}, which means eliminating the square roots. This must be done in two stages. (i) Square:

$$M^2 = m_1^2 + m_2^2 + 2p^{*2} + 2\sqrt{\cdot} \sqrt{\cdot}.$$

(ii) Rearrange so that the product of the roots stands alone, and square again:

$$(M^2 - m_1^2 - m_2^2 - 2p^{*2})^2 = 4(m_1^2 + p^{*2})(m_2^2 + p^{*2}).$$

After some tedious algebra (which it is essential that for this once the reader do in full) one finds, as one of many equivalent variants,

$$p^{*2} = \frac{1}{4M^2}[M^2 - (m_1 - m_2)^2][M^2 - (m_1 + m_2)^2]. \tag{11.3}$$

The last factor of (11.3) checks that $Q \equiv (M - m_1 - m_2) = 0$ entails $p^* = 0$, as it should.

Finally one substitutes from (11.3) into ε_1^{*2}, and finds

$$\varepsilon_1^{*2} = m_1^2 + p^{*2} = \frac{1}{4M^2}\left\{4M^2 m_1^2 + [M^2 - (m_1 - m_2)^2][M^2 - (m_1 + m_2)^2]\right\}$$

$$= \frac{1}{4M^2}\left\{4M^2 m_1^2 + [M^4 - 2M^2(m_1^2 + m_2^2) + (m_1^2 - m_2^2)^2]\right\}$$

$$= \frac{1}{4M^2}\left\{M^4 + 2M^2(m_1^2 - m_2^2) + (m_1^2 - m_2^2)^2\right\} = \frac{\{M^2 + m_1^2 - m_2^2\}^2}{4M^2},$$

$$\varepsilon_1^* = \left(M^2 + m_1^2 - m_2^2\right)/2M. \tag{11.4}$$

The other energy follows either through $\varepsilon_2^* = M - \varepsilon_1^*$, or, somewhat more easily, by interchanging the suffixes 1 and 2. Inserting the values of the masses in MeV, one obtains the ε directly in MeV: thus, the π meson energy in Λ decay is $\varepsilon_1^* \approx 173$ MeV. Alternatively, with the masses in kg and c in m/s, $\varepsilon_1^* = c^2\left(M^2 + m_1^2 - m_2^2\right)/2M$ gives the answer in joules.

11.1.2 The Obvious Method Improved

The proximate cause of the algebraic excesses of the preceding section is the decision to start with p^{*2}, since most of the labour was expended on deriving (11.3). They can be avoided if one tackles (11.1) by using $p^{*2} = \varepsilon_1^{*2} - m_1^2$ to manipulate the (initially unwanted) energy ε_2^* as follows:

$$\varepsilon_2^* = \sqrt{p^{*2} + m_2^2} = \sqrt{\varepsilon_1^{*2} - m_1^2 + m_2^2}.$$

This turns (11.1) into

$$M = \varepsilon_1^* + \sqrt{\varepsilon_1^{*2} - m_1^2 + m_2^2} \;\Rightarrow\; (M - \varepsilon_1^*)^2 = M^2 - 2M\varepsilon_1^* + \varepsilon_1^{*2} = \varepsilon_1^{*2} - m_1^2 + m_2^2,$$

whence (11.4) follows at once.

If necessary, one can now evaluate $p^{*2} = \varepsilon_1^{*2} - m_1^2$, which is far easier than finding it directly, as in the preceding section.

11.1.3 Four-Vectors: Segregate and Square

Although for two-body decay the improvement just described allows even the obvious approach to deliver the answer rather easily, there are few comparably effective improvements in more complicated problems. The basic cause of the awkwardness is that we have not concerted energy and momentum conservation in the best possible way.

In order to do so, we start from the four-vector form (9.9) of the conservation law, which now reads

$$\vec{P} = \vec{p}_1 + \vec{p}_2, \tag{11.5}$$

with the four-momentum of the parent on the left and the sum of the four-momenta of the daughters on the right. When questioned about particle 1, as we are here, we segregate the four-momentum of the *other* particle by writing

$$\vec{p}_2 = \vec{p}_1 - \vec{P}, \qquad (11.6)$$

and then construct the four-scalar product of each side with itself:

$$\vec{p}_2 \cdot \vec{p}_2 = m_2^2 = (\vec{p}_1 - \vec{P}) \cdot (\vec{p}_1 - \vec{P})$$

$$= \vec{p}_1 \cdot \vec{p}_1 + \vec{P} \cdot \vec{P} - 2\vec{P} \cdot \vec{p}_1 = m_1^2 + M^2 - 2(P_0 p_{10} - \mathbf{P} \cdot \mathbf{p}_1). \qquad (11.7)$$

The method turns on exploiting the fact that, in this relation between invariants, the four-scalar product $\vec{P} \cdot \vec{p}_1 = (P_0 p_{10} - \mathbf{P} \cdot \mathbf{p}_1)$ may be evaluated in any inertial frame. We choose the rest frame S^* of the parent, defined by $\mathbf{P}^* = \mathbf{0}$, which entails $P_0^* = M$ and therefore $\vec{P} \cdot \vec{p}_1 = M p_{10}^* = M \varepsilon_1^*$; substituting into (11.7) one gets

$$m_2^2 = m_1^2 + M^2 - 2M \varepsilon_1^* \Rightarrow \varepsilon_1^* = (M^2 + m_1^2 - m_2^2)/2M, \qquad (11.8)$$

as in (11.4). The superiority of this to the obvious method hardly needs comment.

11.2 Several Free Particles Treated as One

We proceed to construct some formulae that prove extremely useful in problem-solving.

Consider a system of N *freely moving* particles, indexed by a label n, so that the total energy and momentum read

$$\vec{P} = (P_0, \mathbf{P}) = (E/c, \mathbf{P}) = \sum_{n=1}^{N} \vec{p}_n = \sum_{n=1}^{N} (\varepsilon_n/c, \mathbf{p}_n). \qquad (11.9)$$

The limits on $\sum_n$ will be omitted for brevity. Our point is that it often pays to envisage $\vec{P}$ as the four-momentum of a single particle, whose mass M would then be given by

$$c^2 M^2 = \vec{P} \cdot \vec{P} = P_0^2 - \mathbf{P}^2 = E^2/c^2 - \mathbf{P}^2. \qquad (11.10)$$

Irrespective of how we choose to think of M, it is, by construction, an *invariant*, having the same value relative to all inertial frames. Moreover, it is *conserved* (because $\vec{P}$ is), so that it has the same value before and after any collisions or reactions between the particles. It is the combination of these two properties that makes M so important in practice.

The special inertial frame S^* where the total three-momentum vanishes ($\mathbf{P}^* = \mathbf{0}$) is called *the centre-of-momentum frame*[2] of the system (consistently with the notation in Section 11.1, where S^* is just the rest frame of the parent). Thus

$$\mathbf{P}^* = \mathbf{0}, \qquad E^* = \sum_n \varepsilon_n^* = \sum_n \sqrt{m_n^2 c^4 + p_n^{*2} c^2} = Mc^2, \qquad (11.11)$$

where m_n is the mass of particle number n, and p_n^* the magnitude of its three-momentum $\mathbf{p}_n^*$ relative to S^*.

What is the least possible value $M_{\min}$ of M? The answer is visible from the second of equations (11.11): in the sum, every entry ε_n^* is positive, and no smaller than $m_n c^2$.

Hence M is least when every $\mathbf{p}_n^*$ vanishes, so that relative to S^* every particle is at rest:

$$M \geq M_{\min} = \sum_n m_n. \tag{11.12}$$

In other words, the mass of our fictitious single particle cannot be less than the sum of the masses of the real particles in the system. Systems that realize $M_{\min}$ have a readily recognizable signature: since relative to the special frame S^* all the particles are at rest, they all have one and the same velocity relative to *any* inertial frame S, with $\mathbf{u}_n = c^2 \mathbf{p}_n / \varepsilon_n = \mathbf{v}$ for all n, where $\mathbf{v}$ is simply the velocity of S with respect to S^*. Conversely, any system where all the particles have the same velocity has $M = \sum_n m_n$.

Example 11.1. Two photons with energies 200 and 300 MeV travel in the positive x and y directions, respectively. If a single particle had their combined momentum and energy, what would be (a) its direction; (b) its speed; (c) its mass?

Solution. (a) Evidently the direction is that of $\mathbf{P}$, at an angle $\phi = \tan^{-1}(P_y/P_x) = \tan^{-1}(200/300) = 33.7°$ to the x axis. The other value of the inverse tangent is ruled out by the known signs of P_y and P_x.
(b) The speed is given by the standard formula

$$u = c\frac{|\mathbf{P}|}{P_0} = c\frac{\sqrt{200^2 + 300^2}}{200 + 300} = c\frac{13}{5} = 2.6c.$$

(c) Photons have zero mass ($\vec{p}_n \cdot \vec{p}_n = 0$), whence M is determined by

$$c^2 M^2 = (\vec{p}_1 + \vec{p}_2)^2 = \vec{p}_1 \cdot \vec{p}_1 + \vec{p}_2 \cdot \vec{p}_2 + 2\vec{p}_1 \cdot \vec{p}_2 = 0 + 0 + 2\vec{p}_1 \cdot \vec{p}_2 = 2(p_{10}p_{20} - \mathbf{p}_1 \cdot \mathbf{p}_2)$$

$$= 2p_{10}p_{20} = 2 \times (200 \text{ MeV}/c) \times (300 \text{ MeV}/c) = 120\,000 \, (\text{MeV}/c)^2.$$

Thus $M \approx 346 \text{ MeV}/c^2$. ∎

The relation used in part (c) above is a special case of a general formula resulting from (11.9) and (11.10), namely

$$c^2 M^2 = \left(\sum_i \vec{p}_i\right) \cdot \left(\sum_j \vec{p}_j\right)$$

$$= \sum_i \vec{p}_i \cdot \vec{p}_i + \sum\sum_{i \neq j} \vec{p}_i \cdot \vec{p}_j = c^2 \sum_i m_i^2 + 2\sum\sum_{i>j} \vec{p}_i \cdot \vec{p}_j. \tag{11.13}$$

As one immediate consequence of (11.12), combined with the fact that M is conserved, it is impossible for a single isolated photon (zero mass, $M = 0$), however energetic, to convert spontaneously into any other system containing finite-mass particles: the expression on the right of (11.12) shows that $M > 0$ if any of the final-particle masses m_n are nonzero. For instance, no free photons can ever decay into an electron–positron pair.

11.3 Production Thresholds

High-energy physics progresses mainly through producing increasingly massive particles, in collisions initiated by projectiles from increasingly energetic accelerators.

Hence one must be able to calculate production thresholds, i.e. the minimum projectile energies needed in order to produce new particles (one or more) of given mass. We start by doing this calculation for collisions with targets that are stationary in the laboratory; reasons for using, instead, collisions between oppositely directed beams are considered later in the chapter. In both cases, the initial state consists of just two particles.

What makes the calculation straightforward is the recognition (which we repeat from Section 11.1 for emphasis) that (a) the total mass M defined in the preceding Section is conserved in the collision, whence it may be evaluated equally before the collision or after; and that (b) M is a four-scalar, whence it may be evaluated equally in the laboratory frame S^L, or in the centre-of-momentum frame S^*. Components of four-vectors relative to these frames will be labelled by superfixes L and $*$ respectively. We use natural units.

11.3.1 Stationary Target

The initial state (before the collision) consists of a projectile (usually a proton or an electron) of mass m_1 and four-momentum $\vec{p}_1 = (\varepsilon_1^L, \mathbf{p}_1^L)$, plus a stationary target particle[3] of mass m_2 and four-momentum $\vec{p}_2 = (\varepsilon_2^L, \mathbf{p}_2^L) = (m_2, \mathbf{0})$. The final state is to contain a set of particles having masses μ_f; they may, but need not, include the two initial particles. Figure 11.1 serves to visualize this scenario.

The problem is to calculate the minimum possible value of ε_1^L. The answer is not obvious because not all of the projectile's energy can be converted into mass: the projectile inevitably supplies three-momentum as well, which must be accommodated in the final state, some of whose energy must therefore be kinetic.

One proceeds by considering the conserved total four-momentum $\vec{P}$ and the consequently conserved invariant $\vec{P} \cdot \vec{P} = M^2$. Having studied the final state relative

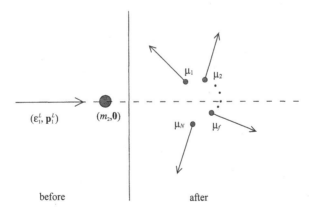

Figure 11.1 An inelastic collision, represented symbolically in the laboratory frame. Initial state on the left, indicating four-momenta. The target (three-momentum $\mathbf{p}_2^L = \mathbf{0}$) is at rest, so it certainly has mass, symbolized by the solid blob. The projectile (three-momentum $\mathbf{p}_1^L$), whose motion is symbolized by the arrow, may or may not have mass. Final state on the right. It may contain any number N of particles, with masses $\mu_f (f = 1, \ldots, N)$ that may or may not be zero. The arrows indicate that in general they will be moving (though some of them may happen to be at rest).

to S^*, we already know from (11.12) that $M \geq M_{\min} = \sum_f \mu_f$. On the other hand, from the initial state we infer

$$M^2 = \vec{P} \cdot \vec{P} = (\vec{p}_1 + \vec{p}_2) \cdot (\vec{p}_1 + \vec{p}_2) = \vec{p}_1 \cdot \vec{p}_1 + \vec{p}_2 \cdot \vec{p}_2 + 2\vec{p}_1 \cdot \vec{p}_2 = m_1^2 + m_2^2 + 2\vec{p}_1 \cdot \vec{p}_2,$$
(11.14)

a special case of (11.13). The crucial step is to evaluate the remaining four-scalar product in the frame where this is easy, namely in S^L (defined by $\mathbf{p}_2^L = 0$):

$$\vec{p}_1 \cdot \vec{p}_2 = \varepsilon_1^L \varepsilon_2^L - \mathbf{p}_1^L \cdot \mathbf{p}_2^L = \varepsilon_1^L m_2 \Rightarrow M^2 = m_1^2 + m_2^2 + 2\varepsilon_1^L m_2.$$
(11.15)

Substituting into (11.14) and using the minimum value of M we find the desired threshold condition[4]:

$$M^2 = m_1^2 + m_2^2 + 2\varepsilon_1^L m_2 \geq \left(\sum_f \mu_f \right)^2 \Rightarrow \varepsilon_1^L \geq \left\{ \frac{\left(\sum_f \mu_f \right)^2 - m_1^2 - m_2^2}{2m_2} \right\}.$$
(11.16)

One should remember that ε_1^L is the total energy: what one pays for is only the kinetic energy $K_1^L = \varepsilon_1^L - m_1$, supplied by the accelerator. However, with modern accelerators producing protons of order 10^6 MeV compared with their mass of 938 MeV (and electrons of order 10^5 compared to 0.5 MeV), the difference in cost is negligible.

Example 11.2. Antiprotons ($\bar{\text{p}}$) have the same mass $m \approx 938$ MeV as protons (p), but opposite charge. They can be made only as members of p$\bar{\text{p}}$ pairs, and were first made in Berkeley (California) in the reaction

$$\text{p} + \text{p} \rightarrow \text{p} + \text{p} + \text{p} + \bar{\text{p}}$$

between projectile protons from the Bevatron and almost stationary protons in the nuclei of a solid copper target. Neglecting motion in the target, determine the minimum *kinetic* energy K_1^L of the projectiles,[5] and compare it with the mass $2m$ of the newly made proton–antiproton pair.

Solution. One need merely substitute appropriately into (11.16), and then convert from ε_1^L to K_1^L. In this reaction $m_1 = m_2 = m$, and $\sum_f \mu_f = 4m$. Hence $\left(\varepsilon_1^L \right)_{\min} = (16 - 1 - 1)m/2 = 7m$, and $\left(K_1^L \right)_{\min} = (7 - 1)m = 6m \approx 5628$ MeV. Since this is three times the total mass of the newly made pair, you must pay for at least three times what you get. ∎

The example illustrates numerically an unfortunate fact evident from (11.16): once $\sum_f \mu_f$ is well above m_1 and m_2, further increases require accelerator energies rising in proportion not to $\sum_f \mu_f$ but to $\left(\sum_f \mu_f \right)^2$. Since the cost of accelerators rises roughly in proportion to their energy (it is a triumph of engineering that it rises no faster), this means that say a tenfold increase in $\sum_f \mu_f$ multiplies the cost by a hundred, which would halt progress dependent on stationary targets in very short order.

11.3.2 Colliding Beams

The way out of the impasse just described is to accumulate accelerated charged particles in a storage ring, around which they are forced to circulate by a magnetic field, and then to make them collide head-on with other particles of equal mass moving with the same energy ε but in the opposite direction and therefore with opposite three-momentum. Then the total three-momentum of the colliding pair is zero (so that the centre-of-momentum and the laboratory frames are now one and the same, allowing one to dispense with the superfixes $*$ and L); thus there is nothing to prevent them from converting all their joint energy into rest energy of the reaction products, because nothing now constrains the products to possess kinetic energy as well.

To express this in the language we used above, start by noting that now one has

$$(\varepsilon_1, \mathbf{p}_1) = (\varepsilon, \mathbf{p}), \qquad (\varepsilon_2, \mathbf{p}_2) = (\varepsilon, -\mathbf{p}),$$

so that

$$M^2 = (\varepsilon_1 + \varepsilon_2)^2 - (\mathbf{p}_1 + \mathbf{p}_2)^2 = (\varepsilon + \varepsilon)^2 - (\mathbf{p} - \mathbf{p})^2 = (2\varepsilon)^2,$$

which confirms that all the incident energy is indeed convertible into rest mass: $2\varepsilon = M \geq \sum_f \mu_f$.

Thus the energies and therefore the cost of accelerators coupled with storage rings rise only in proportion to the masses of the particles to be created. This is the solution now universally adopted by high-energy laboratories; its disadvantage is a great diminution of intensity, i.e. of the number of collisions per unit time, because the density of particles circulating in a storage ring is far below the densities available in fixed targets.

11.4 The Compton Effect

11.4.1 Theory

The scattering of electromagnetic radiation (originally of hard X-rays or of γ-rays) by electrons is called Compton scattering. The Compton effect (we shall derive it presently) is the increase $\Delta\lambda$ in the wavelength of the radiation scattered through an angle θ by initially stationary electrons:

$$\Delta\lambda \equiv \lambda_f - \lambda_i = \lambda_c(1 - \cos\theta), \qquad \lambda_c \equiv h/mc \approx 2.427 \times 10^{-12} \text{ m}, \quad (11.17)$$

where the suffixes i, f stand for *initial* (incident) and *final* (scattered), and m is the electron mass. The combination λ_c is called the Compton wavelength; we have already met it in a different context in Section 9.3.3. (In particle physics one commonly uses the reduced Compton wavelength $\lambda_c \equiv \lambda_c/2\pi = \hbar/mc$.) Most remarkably, $\Delta\lambda$ is independent of λ_i.

Historically[6] the effect was crucial as evidence for the existence of photons, i.e. for the granularity of the energy carried by light: the point is that according to classical theory electromagnetic waves should be scattered without change of frequency or wavelength (as shown retrospectively by the fact that $\Delta\lambda$ would vanish if Planck's constant were zero). Here however we shall take photons for granted, treating them simply as zero-mass particles with four-momentum $\vec{q} = (q_0, \mathbf{q})$, i.e. as having energy cq_0 and three-momentum $\mathbf{q}$, with $q_0 = q$. The connection with the corresponding radiation will be made only at the end, through the Planck–Einstein relation

$$cq_0 = cq = \hbar\omega, \qquad \lambda = 2\pi c/\omega = 2\pi\hbar/q = h/q, \quad (11.18)$$

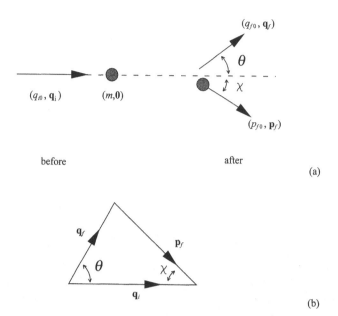

$(q_{f0}, \mathbf{q}_f)$

θ

$(q_{i0}, \mathbf{q}_i)$ $(m, 0)$ χ

$(p_{f0}, \mathbf{p}_f)$

before   after

(a)

$\mathbf{q}_f$ $\mathbf{p}_f$

θ χ

$\mathbf{q}_i$

(b)

Figure 11.2 Compton scattering. All quantities are referred to the laboratory frame. (a) Symbolic representation of the four-momenta. Initial state on the left: the electron (nonzero mass symbolized by the solid blob) is at rest. Final state on the right: the final photon and the recoiling electron travel at angles θ and χ to the incident direction. (The angle χ does not feature in the Compton formula, but is asked for in Problem 11.5.) For the photons (zero mass), $q_{i0} = q_i$ and $q_{f0} = q_f$. (b) Three-momentum conservation: the vector sum of the three-momenta of the final photon ($\mathbf{q}_f$) and the recoiling electron ($\mathbf{p}_f$) equals the three-momentum of the initial photon ($\mathbf{q}_i$). This diagram would look exactly the same in Newtonian physics.

where ω and $k = 2\pi/\lambda$ are the circular frequency and the wavenumber. We shall also take it for granted that the photons travel in the same direction as the waves, so that θ is equally the angle between incident and scattered photons.

Figure 11.2(a) sketches the process in terms of the momenta, under the pretence that initially the target electron is at rest, i.e. disregarding its motion inside the atom or solid where in fact it is bound.[7] In the original experiments the scattered electron was not observed: thus it proves convenient that (11.17) refers only to the photons.

To derive (11.17) one must first express q_f in terms of q_i and of θ. Once again we do this in two ways, the obvious and the efficient. For brevity we use the standard notation by dispensing with the superfix L for vector components with respect to the laboratory frame.

(i) *The obvious method* starts from energy and three-momentum conservation written down separately. The zero-mass relations $q_0 = q$ are exploited without comment. We use natural units. Energy conservation yields

$$q_i + m = q_f + \sqrt{m^2 + p_f^2};$$

the unobserved and therefore unwanted p_f^2 is displayed by segregating the root, squaring, and rearranging:

$$m^2 + p_f^2 = (q_i + m - q_f)^2 \Rightarrow p_f^2 = q_i^2 + q_f^2 - 2q_iq_f + 2m(q_i - q_f).$$

Three-momentum conservation is embodied in the triangle shown in Figure 11.2(b); through the cosine rule it yields

$$p_f^2 = q_i^2 + q_f^2 - 2q_i q_f \cos \theta.$$

Eliminating p_f^2 one finds

$$mc(q_i - q_f) = q_i q_f (1 - \cos \theta); \tag{11.19}$$

this is the simplest form of the requisite connection between the momenta. In the last line we have restored the factor c needed in conventional units so as to give the left-hand expression the same dimensions of momentum squared as appear on the right.

(ii) *The efficient method* ("segregate and square", as explained in Section 11.1.3) starts directly from four-momentum conservation, segregates the unwanted $\vec{p}_f$, and squares, using $\vec{p}_i \cdot \vec{p}_i = m^2 = \vec{p}_f \cdot \vec{p}_f$ and $\vec{q}_i \vec{q}_i = 0 = \vec{q}_f \vec{q}_f$:

$$\vec{p}_f \cdot \vec{p}_f = (\vec{p}_i + \vec{q}_i - \vec{q}_f) \cdot (\vec{p}_i + \vec{q}_i - \vec{q}_f),$$

$$m^2 = m^2 + 0 + 0 - 2\vec{q}_i \cdot \vec{q}_f + 2\vec{p}_i \cdot (\vec{q}_i - \vec{q}_f) \;\Rightarrow\; \vec{p}_i \cdot (\vec{q}_i - \vec{q}_f) = \vec{q}_i \cdot \vec{q}_f. \tag{11.20}$$

But $\vec{p}_i = (m, \mathbf{0})$, whence the left-hand side reduces to $m(q_i - q_f)$, while on the right

$$\vec{q}_i \cdot \vec{q}_f = q_i q_f - \mathbf{q}_i \cdot \mathbf{q}_f = q_i q_f (1 - \cos \theta).$$

Substitution into (11.20) yields (11.19).

Finally we re-express (11.19) in terms of wavelengths, setting $q = h/\lambda$ according to (11.18):

$$mch \left(\frac{1}{\lambda_i} - \frac{1}{\lambda_f} \right) = h^2 \left(\frac{1 - \cos \theta}{\lambda_i \lambda_f} \right) \;\Rightarrow\; \lambda_f - \lambda_i = \frac{h}{mc}(1 - \cos \theta) = \lambda_c (1 - \cos \theta),$$

as anticipated in (11.17). This rearranges into the relation directly accessible to experiment:

$$\lambda_f = \lambda_i + 2\lambda_c \sin^2(\theta/2). \tag{11.21}$$

11.4.2 Evidence

Compton's original data[8] were obtained with γ-rays from Radium C, having $\lambda_i = 2.2 \times 10^{-12}$ m. They are given in the following table:

$\theta°$	$\lambda(\exp)$	$\lambda(\text{theory})$
45	3.0	2.9
90	4.3	4.7
135	6.8	6.3

where $\lambda(\text{theory})$ is just λ_f from (11.21), and wavelengths are measured in 10^{-12} m. The chief difficulties seem to have been (a) the low intensity of the scattered radiation, lowest at high angles (as $\theta \to \pi$) where $\Delta\lambda$ is largest; and (b) the fact that the wavelengths were too short to be measured conveniently, then, by the standard method of

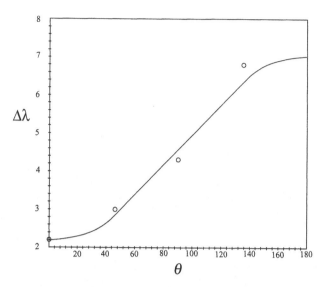

Figure 11.3 The Compton effect. Horizontal scale: the scattering angle θ in degrees. Vertical scale: the scattered wavelength λ_f in units of 10^{-12} m. The circles are Compton's measured values, listed in the table. The curve is the theoretical result (11.21) with $\lambda_i = 2.2 \times 10^{-12}$ m.

crystal diffraction. In fact the energies cq were determined by absorption, i.e. from the known relation between energy and the amount of material per unit area needed to stop the rays. Figure 11.3 compares the data points with (11.21).

A later experiment[9] in Compton's laboratory did use crystal diffraction, with the K_α X-rays from molybdenum ($\lambda_i = 70.8 \times 10^{-12}$ m), at angles close to $161°$ such that their cosines averaged to -0.948 ± 0.003. The result was

$$\Delta\lambda(\theta = 161°) = (4.721 \pm 0.003) \times 10^{-12} \text{ m}, \qquad (11.22)$$

to be compared with 4.728×10^{-12} m from (11.17). For very accurate modern tests one looks in vain: at low energies λ_f is appreciably affected by the momentum distribution of the target electrons (which in fact motivates current measurements), while really high-energy accelerators have more urgent uses.

11.5 Elastic Scattering

Any two particles colliding with enough energy can create others, but the simplest possibility is that they do not. This process is called elastic scattering, Compton scattering being just one example. One studies the angular distribution (called the differential scattering cross-section), namely the proportion of the incident particle flux deviated (scattered) per unit solid angle, as a function of the scattering angle and of the incident energy. This distribution contains information about the interaction between the particles, and often about their structure. (The analysis generally requires quantum mechanics: for instance, to explore very small distances a one needs particles with momenta p high enough to yield de Broglie wavelengths $\lambda = 2\pi\hbar/p < a$.) Important examples include the scattering of low-energy alpha particles by nuclei, which led to Rutherford's planetary model of the atom, and also gave the first estimates of nuclear radii through deviations from the Coulomb repulsion between point charges; the scattering of high-energy electrons from nuclei, which determines the detailed charge distributions of the latter; and electron–electron scatter, which confirms that electrons behave as point charges down to the smallest distances accessible as of now. Discussions can be found in books on nuclear and particle physics.

To describe scattering, one generally uses at least two reference frames: the laboratory frame S (for brevity we dispense with superfixes L in this section), where one of the colliding particles, called the target (say particle 2), is at rest; and the centre-of-momentum (CM) frame S^*, where the total momentum is zero. Observations are made in S, but theories are simplest in S^*. The Lorentz transformation of cross-sections in general is rather intricate, and beyond our scope; here we consider only the transformation of energies and angles between S and S^*, largely as an exercise in the practicalities of four-momenta.

Figure 11.4(a) symbolizes elastic scattering in terms of the four-momenta, with inward- and outward-pointing arrows indicating the initial and final particles. Thus, in natural units,

$$\vec{p}_1 + \vec{p}_2 = \vec{p}_3 + \vec{p}_4, \qquad \vec{p}_1 \cdot \vec{p}_1 = \vec{p}_3 \cdot \vec{p}_3 = m_1^2, \qquad \vec{p}_2 \cdot \vec{p}_2 = \vec{p}_4 \cdot \vec{p}_4 = m_2^2. \qquad (11.23)$$

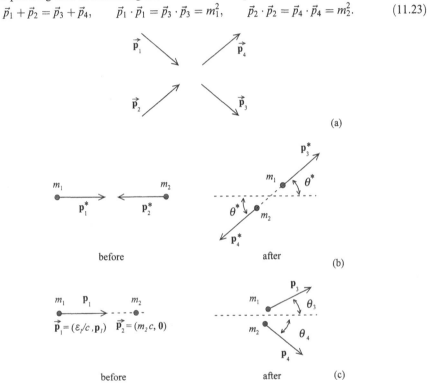

(a)

(b)

(c)

Figure 11.4 Various ways of symbolizing elastic scattering. (a) The *four-momenta* are indicated, without reference to any particular inertial frame. Ingoing and outgoing arrows serve merely to identify the initial and final particles. The $\vec{p}$'s are subject only to constraints (11.23). When the masses are unequal (and even if they are equal but the particles are nevertheless distinguishable), particle 1 emerges as 3 and particle 2 emerges as 4. (b) The process referred to the centre-of-momentum frame S^*. The arrows indicate *three-momenta*, and the blobs indicate the masses (one or both of which may be zero). All the **p***'s have the same magnitude p^*. The two initial momenta are collinear; the two final momenta are likewise collinear, but have been deviated from the incident direction (the broken line) through the scattering angle θ^*. Energies are not indicated on the figure; in fact $\varepsilon_1^* = \varepsilon_3^*$ and $\varepsilon_2^* = \varepsilon_4^*$, and in the equal-mass case $m_1 = m_2$ all the ε^*'s are equal. (c) The same process referred to the laboratory frame S, where the target particle 2 is at rest; superfixes L omitted, as in the text. The arrows indicate three-momenta, and the blobs indicate masses (m_1 may be zero, but m_2 may not). Relative to the incident direction (the broken line) the projectile (mass m_1) is deviated through the scattering angle θ_3, and the target (mass m_2) recoils at an angle θ_4. As a reminder, the initial (though not the final) four-momentum components are also shown.

Relative to S^* the process is sketched in Figure 11.4(b). We write p^* for the common magnitude of all the three-momenta, while the arrows indicate their directions. Since $\mathbf{P}^* = \mathbf{p}_1^* + \mathbf{p}_2^* = \mathbf{0}$, the only change allowed by four-momentum conservation is the change of direction through the scattering angle[10] θ^*. The same process relative to the laboratory frame is sketched in Figure 11.4(c); our task is to connect the entries in (b) and (c). The velocity of S^* with respect to S we call $\mathbf{w}$. Obviously the velocity of the target with respect to S^* is $-\mathbf{w}$.

The total four-momentum $\vec{P} = \vec{p}_1 + \vec{p}_2$ defines the invariant $\vec{P} \cdot \vec{P} = M^2$. The CM energies and momenta are found in terms of M exactly as in Section 11.1: ε_1^* is given by (11.4), ε_2^* by interchanging the suffixes 1 and 2, and p^* by (11.3).

The laboratory energy ε_1 of the projectile is found by evaluating

$$M^2 = (\vec{p}_1 + \vec{p}_2) \cdot (\vec{p}_1 + \vec{p}_2) = \vec{p}_1 \cdot \vec{p}_1 + \vec{p}_2 \cdot \vec{p}_2 + 2\vec{p}_1 \cdot \vec{p}_2 = m_1^2 + m_2^2 + 2\vec{p}_1 \cdot \vec{p}_2$$

in S, where $\vec{p}_1 \cdot \vec{p}_2 = \varepsilon_1 m_2$, whence

$$\varepsilon_1 = (M^2 - m_1^2 - m_2^2)/2m_2,$$

$$p_1^2 = \varepsilon_1^2 - m_1^2 = \frac{[M^2 - (m_1 + m_2)^2][M^2 - (m_1 - m_2)^2]}{4m_2^2}. \tag{11.24}$$

(The expression for ε_1 should be compared with the threshold condition (11.16): the points of view are different but the calculation is the same.)

The speed w is accessible in two ways. First, consider the fictitious single particle consisting of the two actual particles regarded as one; it is at rest with respect to S^* (because $\mathbf{P}^* = \mathbf{0}$), while its velocity with respect to S is (as for any particle) just the ratio $\mathbf{P}/P_0$ of its momentum to its energy; hence

$$w = P/P_0 = p_1/(\varepsilon_1 + m_2). \tag{11.25}$$

Alternatively, because the target is at rest with respect to S, while relative to S^* its velocity is $-p^*/\varepsilon_2^*$, we have

$$w = p^*/\varepsilon_2^*. \tag{11.26}$$

(One should check explicitly that these two expressions agree!)

We consider angles only in the special case of *equal masses*. Then $\varepsilon_1^* = \varepsilon_2^* \equiv \varepsilon^*$; the speed of both particles with respect to S^* is w; and we define $\gamma^* \equiv 1/\sqrt{1 - w^2}$, so that

$$M = \sqrt{2m(\varepsilon_1 + m)} = 2\varepsilon^* = 2m\gamma^*, \quad w = \sqrt{(\varepsilon_1 - m)/(\varepsilon_1 + m)}, \quad (m_1 = m_2 \equiv m). \tag{11.27}$$

The first expression for M quotes (11.15); the resulting γ^* then implies the expression for w.

The laboratory angles θ_3 and θ_4 are related to the CM angle θ^* by Lorentz-transforming explicitly from S^* to S in the now-familiar way. One finds (see equation (9.27))

$$\tan \theta_3 = \frac{p_{1y}}{p_{1x}} = \frac{p^* \sin \theta^*}{\gamma^*[p^* \cos \theta^* + w\varepsilon^*]} = \frac{\sin \theta^*}{\gamma^*[1 + \cos \theta^*]}, \tag{11.28}$$

where the last step divides top and bottom by ε^* and uses $p^*/\varepsilon^* = w$. Similarly one obtains

$$\tan \theta_4 = \frac{\sin \theta^*}{\gamma^*[1 - \cos \theta^*]}, \tag{11.29}$$

so that

$$\tan \theta_3 \tan \theta_4 = \frac{1}{\gamma^{*2}} = \frac{2m}{\varepsilon_1 + m} = \frac{2}{\gamma_1 + 1}, \tag{11.30}$$

where for γ^* we have used the first of equations (11.27), and have defined $\gamma_1 \equiv \varepsilon_1/m = 1/\sqrt{1 - u_1^2}$.

The detailed application of such rules to elastic scattering cross-sections is just the start of a vast industry[11] devoted to scattering and production processes generally. Its products are

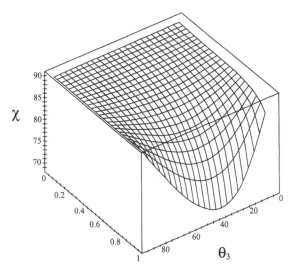

Figure 11.5 Elastic scattering of equal-mass particles in the laboratory frame S. Three-dimensional plot of the opening angle χ (the angle between the three-momenta of the outgoing particles) as a function (11.31) of the incident speed u and of the scattering angle θ_3.

better pursued when needed, rather than as illustrations of special relativity. However, (11.30) suggests that it is of some general interest to study the laboratory *opening angle* $\chi \equiv \theta_3 + \theta_4$, bearing in mind that if the product of two tangents is unity, i.e. if one is the reciprocal of the other, then the sum of the angles is a right angle. But the nonrelativistic limit $c \to \infty$ does entail $1/\gamma^{*2} = 1 - w^2/c^2 \to 1$, reminding us that in Newtonian physics $\chi = \pi/2$, regardless of θ^* and regardless of the incident velocity u_1. To see what actually happens we choose to express χ in terms of the laboratory variables θ_3 and γ_1:

$$\tan \chi = \frac{\tan \theta_3 + \tan \theta_4}{1 - \tan \theta_3 \tan \theta_4} = \frac{\tan \theta_3 + 1/\gamma^{*2} \tan \theta_3}{1 - 1/\gamma^{*2}} = \frac{(\gamma_1 + 1) \tan \theta_3 + 2/\tan \theta_3}{\gamma_1 - 1}. \qquad (11.31)$$

Because projectile and target have the same mass, one cannot distinguish dynamically between the scattered and the recoiling target particles; hence χ must remain unaffected on replacing θ_3 with θ_4, and one can easily check from (11.30) that it does.

Figure 11.5 shows χ as a function of u_1 and of θ_3. An early experiment[12] tested four-momentum conservation through (11.31), by taking stereoscopic cloud-chamber photographs of the charged-particle tracks in events where beta rays (electrons) from Radium E collided with an effectively stationary atomic electron. The angles were determined directly from the pictures, while u_1 was found from the curvature of the incident track in a known magnetic field. (Beta decay produces electrons with all energies up to a maximum, and events near the maximum were selected because they give χ's further below $\pi/2$ and thereby a better test of relativistic effects.) With θ_3 and χ to about $\pm 1°$, and u_1 to about 1%, the results are given in the following table:

Event	1	2	3	4	5	6	7	8	9	10	11	12	13	14
u_1	0.85	0.83	0.83	0.82	0.85	0.83	0.84	0.90	0.88	0.85	0.91	0.93	0.85	0.82
θ_3°	20	26.6	31.4	22	22.2	22.4	23.4	24.5	35.4	21.1	36.9	29.6	21.8	36.9
χ°(theory)	82	82	81	83	82	82	82	78	78	82	75	73	82	81
χ°(exp)	84	81	81	84	82	82	83	80	77	83	75	73	82	81

11.6 Discrete Masses and Continuous Masses

11.6.1 Discrete Masses: Bound States and Elementary Particles

In Section 11.2 it was only by choice, and for convenience, that we treated several particles as if they were one, since the particles in question were taken to move freely; which is also the reason why the effective mass of the fictitious single particle can assume any value above the threshold $\sum_f m_f$. This is in contrast to bound multiparticle complexes, like molecules, atoms, or nuclei, whose component particles interact with each other. Interactions are always mediated by fields, though in near-nonrelativistic systems the fields can often be ignored, and their effects approximated well enough by forces acting directly between the constituents. In any case, (a) the constituent particles do not move freely, and (b) the resultant mass M of the complex cannot change continuously, because quantum mechanics admits only certain discretely spaced values of the binding energy B, and thereby of the total rest energy $Mc^2 = (\sum_f m_f c^2) - B$.

However, in principle it is perhaps more satisfactory to start by regarding *all* systems with reasonably well-defined masses as particles on an equal footing. From this point of view, different bound states of a given atom or nucleus are different particles. It so happens that for the hydrogen atom we have a theory that views all such states as proton–electron complexes, and can predict their properties on that basis to any presently verifiable accuracy; for nuclei we have theories that are somewhat similar though far less accurate and less well warranted; and latterly we try to treat all the observed strongly-interacting so-called elementary particles as bound states of other unobserved particles called quarks, on circumstantial evidence that is plausible but not yet compelling.

Such a common approach to bound states and to provisionally elementary particles has the merit of reminding us that we may indeed apply four-momentum conservation exactly as we have been doing, without any need to ask questions about the internal structure of the particles. For instance, when calculating energies in the decay $K \rightarrow \pi + \pi$ or the photoproduction $p + \gamma \rightarrow N^*$, it simply does not matter in what proportions the masses of these particles might be made up of the rest energy and of the kinetic energies of their constituents (if any), and of the field energies or potential energies responsible for their cohesion. Exactly the same is true say for the decay $H_{3p}^* \rightarrow H_{2s}^* + \gamma$ of the 3p into the 2s states of hydrogen, and for the photoexcitation process $H_{1s} + \gamma \rightarrow H_{2p}^*$ of the ground to the 2p (the first excited) states.

11.6.2 Continuously Variable Mass

So far we have considered only discrete-mass particles. For atomic or subatomic bound states, the discreteness is dictated by quantum mechanics, and one hopes that in some deeper theory the same will be true for all the so-called elementary particles (except perhaps for quarks if they exist, though history inspires caution regarding them too). But the masses that quantum mechanics allows to *macroscopic bodies* are so closely spaced that on the relevant scales they can in effect be changed continuously, for instance by heating, or by the shedding or accretion of very much smaller particles, like dust (see Problem 11.10). Unfortunately, it is far beyond our scope to study the Lorentz transformation of macroscopic thermodynamic quantities[13] like pressure, temperature, and entropy. Hence we settle for just three pieces of physics. (a) Appendix E calculates how the Planck distribution of black-body radiation appears to an observer moving uniformly with respect to the frame where the distribution is isotropic. The result is important because this is how observations on the universal thermal background radiation measure our speed relative to the cosmos. (b) The example just below considers variable temperatures and masses in a simple case of approach to equilibrium. (c) Finally, the next subsection considers the hitherto fictive relativistic rocket.

Example 11.3. Thermodynamic equilibrium. Consider two solid bodies ($i = 1, 2$), made of the same material. The mass of each, i.e. its total energy in its rest frame, is given by $m_i = n_i\mu[1 + (s/c^2)\theta_i]$, where n_i is the number of molecules, θ_i is the Celsius temperature measured in the rest frame of the body, and μ, s are constants characteristic of the material. This is a reasonable approximation over a limited temperature range. Evidently s is the specific heat. The bodies start at rest-frame temperatures θ_i, and with velocities $\mathbf{u}_i$ relative to some inertial frame S. They can exchange heat and exert frictional forces on each other, by mechanisms we need not specify. When they reach mutual equilibrium, they have a common velocity $\mathbf{u}$ relative to S, and a common temperature θ^* relative to their common rest frame. Determine $\mathbf{u}$ and θ^*. (See also Problem 11.16.)

Solution. The total four-momentum $\vec{P} = \vec{p}_1 + \vec{p}_2$ is conserved, and so therefore is $M^2 c^2 \equiv \vec{P} \cdot \vec{P}$. The number of molecules is also conserved, so that the final number is $n = n_1 + n_2$.

The final velocity follows immediately from

$$\mathbf{u} = c\frac{\mathbf{P}}{P_0} = \frac{m_1\gamma_1\mathbf{u}_1 + m_2\gamma_2\mathbf{u}_2}{m_1\gamma_1 + m_2\gamma_2},$$

where $\gamma_i \equiv \gamma(u_i)$. Define also the four-velocities $\vec{U}_i = \gamma_i(c, \mathbf{u}_i)$.

To find θ^*, we equate M^2 before and after:

$$M^2 = m_1^2 + m_2^2 + 2\vec{p}_1 \cdot \vec{p}_2/c^2 = m_1^2 + m_2^2 + 2m_1m_2\vec{U}_1 \cdot \vec{U}_2/c^2$$

$$= m_1^2 + m_2^2 + 2m_1m_2\gamma_1\gamma_2\left(1 - \mathbf{u}_1 \cdot \mathbf{u}_2/c^2\right) = n^2\mu^2\left[1 + (s/c^2)\theta^*\right]^2.$$

The last line determines θ^*, because all other entries are known.

This calculation is straightforward only because it has managed to avoid asking what temperatures should be assigned to the bodies by observers relative to whom they are moving. There could be no such let-out if we asked about the entropy increase in the process. ∎

11.6.3 The Relativistic Rocket

This nonexistent but instructive rocket moves in a straight line through space free of gravitational forces. We require its velocity u relative to the inertial frame S where it starts from rest. Its instantaneously co-moving inertial frame is called S'. In a small proper-time interval $\delta\tau$ the engine consumes propellant of mass $|\delta m|$; when this is burnt, (or when it has undergone some other reaction that liberates kinetic energy), all the products are ejected with velocity w relative to the rocket; w is a design parameter remaining constant throughout. Since mass is not conserved, the expelled mass, call it $\delta\mu$, must be calculated: it need not and does not equal $|\delta m|$. Meanwhile, relative to S'

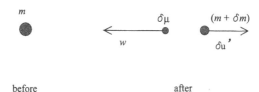

before   after

Figure 11.6   The relativistic rocket referred to its instantaneously co-moving frame S' at some value τ of its proper time ("before"), and to the same frame at proper time $\tau + \delta\tau$ ("after"). In the interval $\delta\tau$, the mass of the rocket has changed by the (negative) amount δm; the used-up propellant has been expelled leftwards with speed w, and with mass $\delta\mu$; and the remainder of the rocket has gained speed $\delta u'$ to the right. The arrows represent *velocities*.

(co-moving at the start of the interval), the velocity u' of the remaining part of the rocket increases from zero to $\delta u'$. This elementary process is indicated in Figure 11.6.

To determine u relative to S as a function of time t relative to S can be quite awkward (Problems 11.11 and 11.12 consider some relatively simple cases), and here we determine it only as a function of the remaining mass m. The initial mass is m_i. We use natural units.

The strategy is (i) to find the velocity gain $\delta u'$ relative to S'; (ii) Lorentz transform to S so as to find the corresponding increment δu in u; and (iii) integrate in order to find u itself.

(i) Over the small interval in question, momentum and energy conservation referred to S' read

$$0 = \frac{(m + \delta m)\delta u'}{\sqrt{1 - (\delta u')^2}} - \frac{\delta\mu w}{\sqrt{1 - w^2}}, \tag{11.32}$$

$$m = \frac{m + \delta m}{\sqrt{1 - (\delta u')^2}} + \frac{\delta\mu}{\sqrt{1 - w^2}}. \tag{11.33}$$

(Remember that the mass change δm of the rocket is negative.) We work only to first order in the small quantities. Then (11.33) yields

$$\delta\mu = -\delta m\sqrt{1 - w^2}; \tag{11.34}$$

substituting for $\delta\mu$ in (11.33) we obtain

$$m\delta u' = -w\delta m. \tag{11.35}$$

This contains the essence of the dynamics, giving the velocity change $\delta u'$ in terms of δm, which is controlled by the operator. Equations (11.34) and (11.35) are re-derived somewhat more elegantly at the end of this section.

(ii) Since S' has velocity u with respect to S, the velocity-transformation rule (combining $\delta u'$ with u) approximated to first order gives

$$u + \delta u = \frac{u + \delta u'}{1 + u\delta u'} \Rightarrow (1 - u^2)\delta u' = \delta u = -(1 - u^2)w\delta m/m, \tag{11.36}$$

where the last step has used (11.35).

(iii) Integration yields

$$-w \int_{m_i}^{m} \frac{dm}{m} = w \log\left(\frac{m_i}{m}\right) = \int_0^u \frac{du}{1-u^2} = \frac{1}{2}\log\left(\frac{1+u}{1-u}\right).$$

Hence we get, restoring c's,

$$\left(\frac{m}{m_i}\right)^{2w/c} = \frac{1-u/c}{1+u/c} \;\Rightarrow\; \frac{u}{c} = \frac{1-(m/m_i)^{2w/c}}{1+(m/m_i)^{2w/c}}. \tag{11.37}$$

It is instructive to check this in the limit $c \to \infty$. We use the first version. On the left

$$\lim_{c\to\infty}\left(\frac{m}{m_i}\right)^{2w/c} = \lim_{c\to\infty}\exp\left\{\frac{2w}{c}\log\frac{m}{m_i}\right\} = 1 + \frac{2w}{c}\log\frac{m}{m_i}+\cdots,$$

while on the right

$$\lim_{c\to\infty}\frac{1-u/c}{1+u/c} = 1 - 2u/c +\cdots.$$

Equating these expressions one does indeed recover the familiar nonrelativistic relation $u = w\log(m_i/m)$.

Observe that (11.37) is independent of the burn rate, and that u is greatest when $w = c$, i.e. when all the used-up propellant is expelled at light speed. Of course this cannot be done by literally burning fuel; but it might be done, in principle, if the propellant is some kind of stored electromagnetic energy convertible into light.

Example 11.4. A **photon rocket** ($w = c = 1$) has used up most of its initial mass as propellant, so that $m/m_i \ll 1$. Show that the surviving portion has energy $\varepsilon \approx m_i/2$.

Solution. We need the square root in $\varepsilon = m/\sqrt{1-u^2}$ when u is just below unity. From (11.37) with $m/m_i \ll 1$ we approximate

$$u \approx 1 - 2(m/m_i)^{2w} \;\Rightarrow\; 1 - u^2 = (1+u)(1-u) \approx 2(1-u) \approx 4(m/m_i)^{2w},$$

$$\varepsilon \approx m(m_i/m)^w/2.$$

For $w < 1$ the energy vanishes as $m \to 0$. But for $w = 1$ the energy tends to $m_i/2$, becoming independent of m (and thus also of u): eventually a finite amount of energy piles into vanishing mass. Note that the first conclusion is <u>not</u> obvious and the second is not paradoxical. At the end of the burn both m and $\sqrt{1-u^2}$ vanish, and until one has actually calculated their ratio ε one cannot reasonably foresee whether in the limit it tends to zero or to some finite value: from Section 9.2.2 we are perfectly familiar with zero-mass but light-speed objects carrying nonzero energy. ∎

The crucial equation (11.35) can be derived more succinctly by envisaging the infinitesimal burn of a mass $|\delta m|$ of fuel as the decay, into two daughters, of an initial particle of mass m at rest with respect to S'. One daughter is the ejected fuel, having mass $\delta\mu$, momentum $\delta p'$, energy $\delta\varepsilon'$, and speed $w = \delta p'/\delta\varepsilon'$. The other daughter is the remainder of the rocket, having mass $(m + \delta m)$, momentum $-\delta p'$, energy $(m - \delta\varepsilon')$, and therefore speed $\delta u' = \delta p'/(m - \delta \varepsilon') \simeq \delta p'/m$, where the approximation retains only terms of first order of smallness. The same approximation yields

$$m - \delta \varepsilon' = \frac{(m + \delta m)}{\sqrt{1 - (\delta u')^2}} \simeq m + \delta m \Rightarrow -\delta m = \delta \varepsilon' = \frac{\delta \mu}{\sqrt{1 - w^2}},$$

confirming (11.34). Finally

$$\delta u' = \delta p'/m = w\delta \varepsilon'/m = -w\delta m/m,$$

confirming (11.35).

At first sight the basic equation $m\delta u' = -w\delta m$ might well deceive one into thinking that it has been derived nonrelativistically. Newtonian dynamics would reason thus: in S', the rocket is initially at rest, and then ejects mass $|\delta m|$ backwards with speed w, and therefore with momentum $-w|\delta m|$; hence its own momentum rises by just the opposite amount, and its velocity rises by $\delta u' = w|\delta m|/m$. It is instructive to convince oneself that this argument contains two errors, which just happen to cancel each other out: the ejected mass is not $|\delta m|$ but $\delta \mu = |\delta m|\sqrt{1 - w^2}$, and the momentum of the true ejected mass $\delta \mu$ at speed w is not $-w\delta \mu$ but $-w\delta \mu/\sqrt{1 - w^2}$.

11.7 Notes

1. Quantum mechanics shows that if a particle has mean life T_0 in its rest frame, then its mass is indeterminate to within $\Delta M \sim \hbar/c^2 T_0$. (see also Problem 9.12.) We want T_0 to be long enough to make $\Delta M/M$ negligible.
2. Although $S*$ is a relativistic generalization of the Newtonian centre-of-mass frame, this connection sheds little useful light on $S*$, because the relativistic analogue of the Newtonian centre-of-mass *coordinate* is hardly ever relevant to the physics.
3. Of course the target particles are never strictly stationary: they undergo thermal agitation, and also execute their quantum-mechanically determined motions as constituents of the atoms, molecules, solids, or nuclei that the material of the real target consists of. However, the kinetic energies arising from such motions are usually negligible or almost negligible on the energy scales typical of the collision. Moreover, we consider only cases where the projectile interacts with just one of the many actual target particles: this is what normally happens if the $p*$ are large enough for the associated de Broglie wavelengths $2\pi\hbar/p*$ to be much shorter than the mean spacing between the latter.
4. In order to revert the conventional units the rightmost expression should be multiplied by c^2. This would however be extremely silly, because such masses are universally quoted in MeV, and inserting these values yields ε_1^L directly in MeV, which is what one normally wants.
5. In other words: how energetic did the accelerator have to be in order to induce this reaction and thus elicit a Nobel prize? The experiment is reported by O. Chamberlain *et al.*, *Physical Review* (1955), volume **100**, page 947. It is also described in many textbooks of particle physics.
6. A riveting account is given by R. H. Stuewer, *The Compton effect*, Science History Publications, New York, 1975.
7. By contrast, nowadays the departures from this approximation are measured precisely in order to study such motions.
8. A. H. Compton, "A quantum theory of the scattering of X-rays by light elements", *Physical Review* (1923), volume **21**, page 483. This is the work that earned Compton the Nobel prize.
9. N. S. Gingrich, "An analysis of scattered X-rays with the double crystal spectrometer", *Physical Review* (1930), volume **36**, page 1050.

10. In scattering problems one always chooses the polar axis, conventionally the z axis, along the incident direction, and calls scattering angles θ, as we also do (and already did for Compton scattering in Section 11.4). Unfortunately, with this choice of axes the velocity of S^* relative to S points in the z direction, and not in the x direction as it should for S^* and S in standard configuration. The difference should cause no confusion, since all it means is that we use symbols θ even for angles that might be measured in the xy plane, where ordinary spherical polar coordinates would have led one to use symbols ϕ instead.

11. See for instance H. Goldstein, *Classical mechanics*, Addison-Wesley, Reading, 1980, second edition, Section 7.7; and R. Hagedorn, *Relativistic kinematics*, Benjamin, New York, 1964.

12. F. C. Champion, "On some close collisions of fast β-particles with electrons, photographed by the expansion method", *Proceedings of the Royal Society* (1932), volume A **136**, page 630. Champion calculates χ not from (11.31), but from another equivalent formula, whose equivalence is not however obvious from mere inspection.

13. Although some advanced texts do discuss these problems, the writer knows of only one intelligible treatment unlikely to mislead, namely N. G. van Kampen, "Relativistic thermodynamics of moving systems", *Physical Review* (1968), volume **173**, page 295; and "Lorentz-invariance of the distribution in phase space", *Physica* (1969), volume **43**, page 244.

Part IV
Waves

12 Plane Waves

12.1 Introduction

This book considers only periodic plane waves. The simplest such are simple-harmonic waves described by

$$\psi(t, \mathbf{r}) = A \cos[\phi(t, \mathbf{r})], \qquad \phi \equiv \omega t - \mathbf{k} \cdot \mathbf{r}, \tag{12.1}$$

with the symbols as explained in Section 2.5.1. For the moment the physical nature of the waves, i.e. of the amplitude A, is irrelevant: it affects only the dispersion relation $\omega(\mathbf{k})$ for the frequency. In particular, we need pay no attention to any partial differential equation that ψ might be subject to.[1] Moreover, everything in this chapter applies equally even if (12.1) is generalized to $\psi(t, \mathbf{r}) = F(\phi)$, with F *any* periodic function of the phase ϕ, i.e. any function such that $F(\phi + 2\pi) = F(\phi)$. The main reason why simple-harmonic waves dominate most discussions is that the dispersion relation is rarely simple for any other kind.

Like the description of a freely moving particle by means of its energy and momentum, the description of a plane wave by means of its frequency ω and of its wave-vector $\mathbf{k}$ depends on the inertial frame of the observer. In other words ω and $\mathbf{k}$ are subject to Lorentz transformations, and our first task, in Section 12.2, is to determine these. One finds that $(\omega/c, \mathbf{k}) \equiv \vec{K}$ constitutes a four-vector; accordingly, waves could be classified through the value of the invariant $\vec{K} \cdot \vec{K} \equiv \kappa^2$. In empty space, where there are no parameters other that $\vec{K}$, electromagnetic and other light-speed waves have $\kappa^2 = 0$, while the quantum wavefunctions of free particles have $\kappa^2 = (mc/\hbar)^2$, where $\hbar$ is Planck's constant and m is the mass. By contrast, for waves through matter κ^2 generally depends both on the four-velocity of the undisturbed medium and on the wave-vector in its rest frame, so that the expressions for κ^2 can become quite intricate. In fact we shall not organize the discussion around κ^2, but shall proceed from the simpler questions to the more complicated, choosing examples as simple as possible at each stage.

Accordingly, Sections 12.3 and 12.4 find the transformation rules for group and phase velocities, which stem directly from the basic rules for ω and $\mathbf{k}$. Some of the experimental evidence is outlined in Section 12.5. Finally, Section 12.6 describes how relativistic invariance determines the quantum-mechanical assignment of wavefunctions to freely moving particles, through the Einstein and the de Broglie relations

$\varepsilon = \hbar\omega$ and $\mathbf{k} = \mathbf{p}/\hbar$. The underlying wave equations are considered briefly in Appendix D; to understand them one must first find the Lorentz transforms of the partial derivatives $\partial/\partial t$ and ∇, which are not required elsewhere in this book.

For the next and final chapter, on light in vacuo, Sections 12.3 to 12.6 may be skipped; only Section 12.2 is essential.

We defer to the next chapter details of waves with speed c in empty space, characterized by the dispersion relation $\omega = ck$. For brevity we call them light waves, though everything we say about them applies equally to electromagnetic waves of any frequency, and to any other type of wave propagating with the same speed. This special case is by far the most important in practice, because light is so widely used to gather information about its sources. It so happens that the transformation rules specific to light could have been anticipated by treating a monochromatic light wave *in vacuo* simply as if it were a stream of photons, i.e. of zero-mass particles ($\varepsilon = cp$ corresponding to $\omega = ck$). While such arguments are unobjectionable as short cuts, they are not substitutes for a proper understanding of the relativistic physics of waves in their own right. But we do appeal to this correspondence in Appendix E, in order to Lorentz-transform the Planck distribution for black-body radiation.

From Section 2.5 on Galilean-relativistic[2] waves, our only absolute need is for Section 2.5.1, and for the introductory comments in Section 2.5.2. But comparisons with the Galilean theory can prove illuminating throughout, especially regarding the differences between the views that the old theory and the new take of waves in what we now consider empty space, calling it simply the vacuum. Inevitably, such waves travel with different velocities $\mathbf{u}$ and $\mathbf{u}'$ (and in general with different speeds) relative to different inertial frames S and S'; but the Lorentz transformations between $\mathbf{u}$ and $\mathbf{u}'$ feature only one other velocity, namely the velocity $\mathbf{v}$ of S' relative to S. By contrast, Newtonian physics regarded even light waves propagating through the vacuum as vibrations of some medium (the "aether") with its own uniquely privileged rest frame $S^{(0)}$; then the speed of light would be c only relative to $S^{(0)}$; relative to any other uniformly moving frame S, the dispersion relation and the speed would necessarily depend on the velocity $-\mathbf{w}$ of S with respect to $S^{(0)}$; and the transformation between two such frames S and S' would depend not only on $\mathbf{v}$, but also on their separate velocities $-\mathbf{w}$ and $-\mathbf{w}'$ with respect to $S^{(0)}$. This earlier (and as it turns out incorrect) theory for light in empty space was rather like the theory given in Section 2.5 for sound in air, with $S^{(0)}$ analogous to the rest frame of the air.[3] Thus, quite apart from being right, by dispensing with the additional parameters $\mathbf{w}$ the new theory also turns out to be far simpler to apply than the old.

12.2   Lorentz Transformations: The Four-Vector $\vec{K} = (\omega/c, \mathbf{k})$

The reasoning in this section applies to all types of physically measurable[4] waves, irrespective of their dispersion relation, whether they propagate through empty space or through a medium, and if through a medium, then irrespective of its motion relative to the observer.

We start from the fact that the phase ϕ of any such disturbance is an invariant (ie a four-scalar), assigned the same numerical value by all inertial observers. The reason why was explained at the start of Section 2.5.2. To exploit the fact, we write ϕ, with hindsight, as

$$\phi = (\omega/c)(ct) - \mathbf{k} \cdot \mathbf{r} = K_0 X_0 - \mathbf{K} \cdot \mathbf{X} = \vec{K} \cdot \vec{X}; \qquad (12.2)$$

here $\vec{X} \equiv (ct, \mathbf{r})$ is the familiar coordinate four-vector specifying the event when and where the phase is to be determined, and we have defined a new four-vector

$$\vec{K} = (K_0, \mathbf{K}) \equiv (\omega/c, \mathbf{k}). \qquad (12.3)$$

That $(K_0, \mathbf{K})$ so defined does indeed transform as a four-vector follows from the theorem proved in Example 7.1, because $\vec{X}$ transforms as one (by definition!), while $K_0 X_0 - \mathbf{K} \cdot \mathbf{X}$ is invariant.

- Accordingly, the Lorentz transformation for ω and $\mathbf{k}$ between two inertial frames in standard configuration (relative velocity $\mathbf{v} = c\boldsymbol{\beta}$) reads

$$\omega'/c = \gamma(\omega/c - \beta k_x) \quad \Rightarrow \quad \omega' = \gamma(\omega - v k_x), \qquad (12.4)$$

$$k'_x = \gamma(k_x - \beta(\omega/c)) \quad \Rightarrow \quad k'_x = \gamma(k_x - v\omega/c^2), \quad k'_{y,z} = k_{y,z}, \qquad (12.5)$$

where $\gamma = 1/\sqrt{1 - v^2/c^2}$. The inverses follow from the familiar rule: interchange primed and unprimed, and reverse the sign of $\mathbf{v}$.
- In the nonrelativistic limit $c \to \infty$ one has $\gamma \to 1$ and $v\omega/c^2 \to 0$, so that (12.4), (12.5) reduce to $\omega' = \omega - \mathbf{k} \cdot \mathbf{v}$ and $\mathbf{k}' = \mathbf{k}$, which tally with the Galilean (2.38).
- The invariant associated with the new four-vector is $\vec{K} \cdot \vec{K} = \omega^2/c^2 - \mathbf{k}^2$.
- For the special case of light waves *in vacuo*, as for all waves whose speed is the invariant speed c of relativity theory, we have $\omega/k = c$ and therefore $\vec{K} \cdot \vec{K} = 0$.
- There is an important analogy between the ratio $\mathbf{K}/K_0 = c\mathbf{k}/\omega$ of the space to the time components of $\vec{K}$, and the ratio $\mathbf{U}/U_0 = \mathbf{u}/c$ constructed from the velocity four-vector $\vec{U}$ for particles. Since numerators and denominators, respectively, transform alike in the two cases, so do the ratios; *consequently $c^2 \mathbf{k}/\omega$ for any wave transforms in exactly the same way as do particle velocities $\mathbf{u}$*.

Example 12.1. Relative to the rest frame $S^{(0)}$ of a transparent dielectric, light inside the medium obeys the dispersion relation $\omega^{(0)} = (c/n^{(0)})k^{(0)}$, where $n^{(0)} > 1$ is called *the refractive index*. (a) Taking $n^{(0)}$ as a constant independent of $\mathbf{k}^{(0)}$ (a reasonable approximation in many problems), derive the dispersion relation in a form valid relative to an inertial frame S with respect to which the medium has velocity $\mathbf{v}$. (b) Thence determine the refractive index $n(\mathbf{k}, \mathbf{v})$ relative to S, *defined by* $\omega = (c/n)k$. (c) Approximate n to first order in v/c (i.e. appropriately to $v \ll c$).

Solution. (a) The strategy is the same as for the analogous Galilean problem in Section 2.5.3: (i) write the dispersion relation known to be correct in $S^{(0)}$ as a relation between Lorentz invariants (i.e. between four-scalars), which must then be true in all inertial frames because it is both form-invariant and true in one such frame, namely in $S^{(0)}$; (ii) re-express this form-invariant equation in terms of quantities referred to S.
   (i) The problem features only two four-vectors, namely $\vec{K}$ and the four-velocity $\vec{V}$ of the medium. The components of $\vec{V}$ with respect to S are $(V_0, \mathbf{V}) = \gamma(c, \mathbf{v})$, where $\gamma \equiv 1/\sqrt{1 - v^2/c^2}$; with respect to $S^{(0)}$ (where by definition $\mathbf{v}^{(0)} = \mathbf{0}$) they are $(V_0^{(0)}, \mathbf{V}^{(0)}) = (c, \mathbf{0})$. The only invariants one can construct from $\vec{K}$ and $\vec{V}$ are $\vec{K} \cdot \vec{K}$,

$\vec{K} \cdot \vec{V}$, and $\vec{V} \cdot \vec{V} = c^2$; the third can play no role because its value does not depend on **v**; hence the invariant form of the dispersion relation can feature only $\vec{K} \cdot \vec{K}$ and $\vec{K} \cdot \vec{V}$. In $S^{(0)}$, evaluating $\vec{K} \cdot \vec{K}$ and using the dispersion relation we obtain

$$\vec{K} \cdot \vec{K} = \omega^{(0)2}/c^2 - k^{(0)2} = -(n^{(0)2} - 1)\omega^{(0)2}/c^2. \tag{12.6}$$

Likewise $\vec{V} \cdot \vec{K} = V_0^{(0)} K_0^{(0)} - \mathbf{V}^{(0)} \cdot \mathbf{K}^{(0)} = (\omega^{(0)}/c)c = \omega^{(0)}$. Substitution on the far right of (12.6) then yields the invariant dispersion relation we require, namely

$$\vec{K} \cdot \vec{K} = -\frac{(n^{(0)2} - 1)}{c^2}(\vec{V} \cdot \vec{K})^2. \tag{12.7}$$

Observe incidentally that in this case $\vec{K}$ is spacelike.

(ii) Expressing (12.7) with respect to S, where $\vec{V} \cdot \vec{K} = V_0 K_0 - \mathbf{V} \cdot \mathbf{K} = \gamma(\omega - \mathbf{v} \cdot \mathbf{k})$, one finds

$$\omega^2/c^2 - k^2 = -\frac{(n^{(0)2} - 1)}{c^2}\gamma^2(\omega - \mathbf{v} \cdot \mathbf{k})^2. \tag{12.8}$$

This quadratic equation could be solved for ω, giving precisely the dispersion relation $\omega(\mathbf{k}, \mathbf{v})$ that in principle we require; but the result is quite complicated, and often it is better to work directly with (12.8).

(b) A similar quadratic equation for the refractive index emerges on substituting $\omega = ck/n$ and $\mathbf{v} \cdot \mathbf{k} = vk \cos\phi = v_\| k$, with ϕ the angle between **k** and **v**, and $v_\| \equiv \mathbf{v} \cdot \mathbf{k} = v \cos\phi$ the component of **v** along **k**:

$$n^2 - 1 = (n^{(0)2} - 1)\gamma^2(1 - nv_\|/c)^2 = (n^{(0)2} - 1)\frac{(1 - n(v/c)\cos\phi)^2}{1 - v^2/c^2}. \tag{12.9}$$

Note that ω now depends not only on the magnitude k of **k** but also, through $\cos\phi$, on the direction of **k** relative to **v**: in other words, with respect to S the material is no longer isotropic. But it is still nondispersive in the restricted sense that n does not depend on k.

(c) To order v/c in (12.9) we can replace $\gamma \to 1$. Also we can replace $nv_\|/c \to n^{(0)}v_\|/c$, because the (as yet unknown) difference between n and $n^{(0)}$ is itself of first order in $v_\|/c$, and can contribute at most proportionately to $(v_\|/c)^2$. Finally we can replace $(1 - n^{(0)}v_\|/c)^2 \simeq (1 - 2n^{(0)}v_\|/c)$, and find

$$n^2 \simeq 1 + (n^{(0)2} - 1)(1 - 2n^{(0)}v_\|/c) = n^{(0)2} - 2(n^{(0)2} - 1)n^{(0)}v_\|/c,$$

$$n \simeq \sqrt{n^{(0)2} - 2(n^{(0)2} - 1)n^{(0)}v_\|/c} = n^{(0)} + \delta n, \quad \delta n = -\frac{v_\|}{c}(n^{(0)2} - 1) + \cdots. \tag{12.10}$$

This is the prediction tested by the measurement described in Section 12.5. ■

12.3 Group Velocity

Recall from Section 2.5.1 the physics of the group velocity, and its definition

$$\mathbf{u}_g = \nabla_\mathbf{k}\omega. \tag{12.11}$$

We show that, irrespective of the dispersion relation, $\mathbf{u}_g$ Lorentz-transforms in exactly the same way as particle velocities do.

Since the definition involves differentiation with respect to the components of $\mathbf{k}$, one must consider small changes in these components (which are the independent variables of the problem), the consequent change

$$\delta\omega = \delta\mathbf{k}\cdot\nabla_\mathbf{k}\omega = \delta\mathbf{k}\cdot\mathbf{u}_g \tag{12.12}$$

of the dependent variable ω, and the Lorentz transforms of all these. We treat in turn the component $u_{gx} = \partial\omega/\partial k_x \equiv \partial_{\|}\omega = u_{g\|}$ parallel and the components $u_{gy,z} = \partial\omega/\partial k_{y,z} = \partial_\perp\omega = \mathbf{u}_{g\perp}$ normal to $\mathbf{v}$. The derivative with respect to each independent variable must be calculated with all the other independent variables held fixed.

To determine $u'_{g\|}$ one need vary only k_x; doing this in (12.4), (12.5) and using (12.12) we find

$$\delta\omega'/c = \gamma(\delta\omega/c - \beta\delta k_{\|}) = \delta k_{\|}\gamma(\partial_{\|}\omega/c - \beta) = \delta k_{\|}\gamma(u_{g\|} - v)/c,$$

$$\delta k'_{\|} = \gamma(\delta k_{\|} - \beta\delta\omega/c) = \delta k_{\|}\gamma(1 - \beta\partial_{\|}\omega/c) = \delta k_{\|}\gamma(1 - vu_{g\|}/c^2).$$

Multiplying the first equation by c and dividing it by the second then yields

$$u'_{g\|} = \lim\left(\frac{\delta\omega'}{\delta k'_{\|}}\right)_{\mathbf{k}'_\perp} = \frac{u_{g\|} - v}{1 - vu_{g\|}/c^2}; \tag{12.13}$$

here it is essential that, as the notation prescribes and as we have already stressed, the partial derivative on the left has indeed been calculated at constant $\mathbf{k}'_\perp$, i.e. appropriately to $\delta\mathbf{k}'_\perp = \delta\mathbf{k}_\perp = \mathbf{0}$.

By contrast, to determine $\mathbf{u}'_{g\perp}$ one needs to vary all three components of $\mathbf{k}$, in such a way that $\delta k'_{\|} = 0$ (rather than $\delta k_{\|} = 0$). Thus

$$0 = \delta k'_{\|} = \gamma(\delta k_{\|} - \beta\delta\omega/c) = \gamma(\delta k_{\|} - \beta\delta\mathbf{k}\cdot\nabla_\mathbf{k}\omega/c) = \gamma(\delta k_{\|} - v\delta\mathbf{k}\cdot\mathbf{u}_g/c^2)$$

$$\Rightarrow \delta k_{\|}[1 - vu_{g\|}/c^2] = v\delta\mathbf{k}_\perp\cdot\mathbf{u}_{g\perp}/c^2,$$

which entails

$$\delta\omega = \delta\mathbf{k}\cdot\mathbf{u}_g = \delta\mathbf{k}_\perp\cdot\mathbf{u}_{g\perp} + \delta k_{\|}u_{g\|} = \delta\mathbf{k}_\perp\cdot\mathbf{u}_{g\perp}\left[1 + \frac{vu_{g\|}/c^2}{1 - vu_{g\|}/c^2}\right] = \frac{\delta\mathbf{k}_\perp\cdot\mathbf{u}_{g\perp}}{1 - vu_{g\|}/c^2}.$$

Consequently, with (say) $\delta k_y = \delta k'_y \neq 0 = \delta k_z = \delta k'_z$ (so that $\delta\mathbf{k}_\perp\cdot\mathbf{u}_{g\perp} = \delta k_y u_{gy}$) one has

$$\frac{\delta\omega'}{\delta k'_y} = \frac{\gamma(\delta\omega - vk_{\|})}{\delta k_y} = \frac{\gamma}{\delta k_y}\left(\frac{\delta k_y u_{gy}}{1 - vu_{g\|}/c^2} - \frac{v^2\delta k_y u_{gy}/c^2}{1 - vu_{g\|}/c^2}\right) = \gamma u_{gy}\left(\frac{1 - v^2/c^2}{1 - vu_{g\|}/c^2}\right),$$

and similarly for the z component. Hence

$$\mathbf{u}'_{g\perp} = \lim\left(\frac{\delta\omega'}{\delta\mathbf{k}'_\perp}\right)_{k'_{\|}} = \frac{\mathbf{u}_{g\perp}}{\gamma(1 - vu_{g\|}/c^2)}. \tag{12.14}$$

But comparison shows that the transformation rules (12.13), (12.14) are indeed identical to the rules (4.15), (4.16) for particle velocities. We stress again that this is true for any wave, and irrespective of the dispersion relation.

Example 12.2. The simplest conceivable invariant dispersion relation is that for light *in vacuo*, namely $\vec{K} \cdot \vec{K} = 0$. The next simplest is $\vec{K} \cdot \vec{K} = \mu^2$, with μ^2 a constant; and it does in fact apply to several important kinds of wave (see, for example, Section 12.6, and the problems for this chapter). (a) Determine the group velocity of such waves; (b) check *explicitly* that it Lorentz-transforms in the right way; and (c) consider whether μ^2 need be positive.

Solution. (a) The dispersion relation yields $\omega^2 = c^2(k^2 + \mu^2)$; differentiating with respect to **k** we obtain

$$2\omega \mathbf{V_k}\omega = 2c^2\mathbf{k} \;\Rightarrow\; \mathbf{V_k}\omega = \mathbf{u}_g = c^2\mathbf{k}/\omega.$$

(b) As observed just above Example 12.1, the ratio $c^2\mathbf{k}/\omega$ does transform like a particle velocity, which is how a group velocity should. (c) Finally, since wave groups can carry signals, in unbounded field-free space their speed $u_g = ck/\sqrt{k^2 + \mu^2}$ cannot exceed c, whence under such conditions[5] one must indeed have $\mu^2 \geq 0$. ∎

12.4 Phase Velocity

Recall from Section 2.5.1 the definition of the phase velocity:

$$\mathbf{u}_p \equiv \hat{\mathbf{k}}\omega/k = \mathbf{k}\omega/k^2. \tag{12.15}$$

The Lorentz-transformation rules for $\mathbf{u}_p$ are deplorably complicated, and quite unlike those common to group and to particle velocities. But fortunately they do reduce to the rules for particle velocities in two important special cases, namely (i) for collinear motion; and (ii) for light speed ($u_p = c$), regardless of directions. We deal with these special cases first, and then with the general case mainly for completeness.

Everything in this as in the preceding section applies to waves of all types, regardless of the dispersion relation.

12.4.1 Collinear Motion

In this effectively one-dimensional scenario, u_p, k_x and the velocity v of S' relative to S, can each be positive or negative; and we have $u_p = \omega/k_x$. Then the general rules (12.4), (12.5) yield

$$u_p' = \frac{\omega'}{k_x'} = \frac{\omega - vk_x}{k_x - v\omega/c^2} = \frac{\omega/k_x - v}{1 - v\omega/k_xc^2} = \frac{u_p - v}{1 - vu_p/c^2}, \tag{12.16}$$

coinciding as promised with the rule for collinear particle and group velocities.

12.4.2 Light Speed

Phase speed c with respect to an inertial frame S means $\omega = ck$, whence

$$\mathbf{u}_p \equiv \hat{\mathbf{k}}\omega/k = \hat{\mathbf{k}}c = c^2\mathbf{k}/\omega = \mathbf{V_k}\omega = \mathbf{u}_g. \tag{12.17}$$

But $c^2\mathbf{k}/\omega$ does transform like a particle velocity, as explained just above Example 12.1, and as already used in Example 12.2.

Example 12.3. Find the Lorentz-transformation rule for wavelengths (a) in the collinear case, and (b) for light *in vacuo*.

Solution. (a) Collinearly, (12.5) now entails $k_x' = \gamma(k_x - v\omega/c^2) = \gamma k_x(1 - vu_p/c^2)$. Since $\lambda = 2\pi/k$ and $k = |k_x|$, we find

$$\lambda' = \frac{\lambda}{\gamma(1 - vu_p/c^2)} \qquad \text{(collinear case)}, \qquad (12.18)$$

with vu_p positive (negative) when $\mathbf{k}$ and $\mathbf{v}$ are parallel (antiparallel).

(b) For light one has $\omega' = \gamma(\omega - vk_x) = \gamma(\omega - v(\omega/c)\cos\phi) = \omega\gamma(1 - (v/c)\cos\phi)$. Since $\lambda = 2\pi/k = 2\pi c/\omega$, we find

$$\lambda' = \frac{\lambda}{\gamma(1 - (v/c)\cos\phi)} \qquad \text{(light)}, \qquad (12.19)$$

where ϕ is the angle between $\mathbf{k}$ and $\mathbf{v}$. For light in the collinear case, where $\cos\phi = \pm 1$ corresponds to $v \lessgtr 0$, this reduces to

$$\lambda' = \lambda\frac{\sqrt{1 - v^2/c^2}}{1 - v/c} = \lambda\sqrt{\frac{1 + v/c}{1 - v/c}} \qquad \text{(light, collinear case)}. \qquad (12.20)$$

Note that in both the collinear cases, (12.18) and (12.20), the ratio λ'/λ is totally different from the ratio $l'/l = 1/\gamma$ appropriate to the Lorentz contraction of the length l of a material object. Similarly, (12.19) yields $\lambda'/\lambda = 1/\gamma$ precisely in the transverse case $\phi = \pi/2$, $\cos\phi = 0$, where the Lorentz contraction vanishes. Recall also from Section 2.5.2 that under Galilean transformations $\mathbf{k}$ is invariant, implying that wavelengths $\lambda = 2\pi/k$ do not change at all. ■

12.4.3 Phase Velocity and Refractive Index in the General Case

This section may be skipped with impunity.

We consider separately the transformation first of the speed and then of the direction.

In order to transform phase speeds it proves convenient to start by evaluating the invariant $\vec{K} \cdot \vec{K}$ in both frames:

$$\vec{K} \cdot \vec{K} = \omega^2/c^2 - k^2 = \omega^2(1/c^2 - 1/u_p^2) = (\omega'^2/c^2)(1 - c^2/u_p'^2),$$

where ϕ_p' is the angle made by $\mathbf{v}$ with $\mathbf{k}'$ and $\mathbf{u}_p'$. On the other hand, the Lorentz transformation of frequencies yields

$$\omega = \gamma(\omega' + vk_x') = \gamma(\omega' + vk'\cos\phi_p') = \omega'\gamma(1 + (v/u_p')\cos\phi_p').$$

Equating the two expressions that result for ω^2/ω'^2 we obtain

$$\gamma^2(1 + (v/u_p')\cos\phi_p')^2 = \frac{(1 - c^2/u_p'^2)}{(1 - c^2/u_p^2)}. \qquad (12.21)$$

Of course this is not yet the end-result: rather it is an equation to be solved for u_p', given u_p and ϕ_p.

As an alternative to dealing with u_p, one can for any kind of wave and with respect to any inertial frame S *define* a refractive index by

$$n \equiv c/u_p = ck/\omega \iff \omega = ck/n;\qquad(12.22)$$

no generality is lost, but n may turn out to be quite a complicated function of $\mathbf{k}$, and n' a different but comparably complicated function of $\mathbf{k}'$. Then (12.21) yields

$$\gamma^2(1 + (n'v/c)\cos\phi'_p)^2 = \frac{(1 - n'^2)}{(1 - n^2)},\qquad \gamma^2(1 - (nv/c)\cos\phi_p)^2 = \frac{(1 - n^2)}{(1 - n'^2)},\qquad(12.23)$$

where the second equation is the inverse given by the familiar rule of thumb.

Example 12.4.   (a) For the moving medium in Example 12.1, derive (12.9) directly from (12.23).

Solution. There is next to nothing left to do. Identify S' with the rest frame $S^{(0)}$ of the medium, so that $n' \to n^{(0)}$. Since we want n for a given value of ϕ_p (rather than of $\phi_p^{(0)}$), we must start from the second of equations (12.23); but this is identically the same as (12.9).   ∎

Solving (12.21) for u_p we find

$$u_p^2/c^2 = \left\{1 + \frac{c^2 - u_p'^2}{\gamma^2(u_p' + v\cos\phi'_p)^2}\right\}^{-1},\qquad(12.24)$$

and construct, with hindsight, and after some tedious but straightforward algebra, the combination

$$1 - u_p^2/c^2 = \frac{(1 - u_p'^2/c^2)}{\gamma^2\{(1 - u_p'^2/c^2)/\gamma^2 + (u_p' + v\cos\phi'_p)^2/c^2\}}.\qquad(12.25)$$

The point is that this is quite unlike its analogue (4.16) for particle speeds u:

$$1 - u^2/c^2 = \frac{(1 - u'^2/c^2)}{\gamma^2(1 + (vu'/c^2)\cos\phi')^2}.\qquad(12.26)$$

It is reassuring to check that, as we already know, there is agreement in the special cases where one has either $u_p' = u' = c$, or $\cos\phi'_p = \cos\phi' = \pm 1$, or both.

As regards directions, using $\omega' = k'u_p'$ one finds

$$\tan\phi_p = \frac{k_y}{k_x} = \frac{k_y'}{\gamma(k_x' + v\omega'/c^2)} = \frac{k'\sin\phi'_p}{\gamma(k'\cos\phi'_p + vk'u_p'/c^2)} = \frac{\sin\phi'_p}{\gamma(\cos\phi'_p + vu_p'/c^2)}.\qquad(12.27)$$

Note that this too is quite unlike the relation (4.17)

$$\tan\phi = \frac{\sin\phi'}{\gamma(\cos\phi' + v/u')}\qquad\text{(particles)}\qquad(12.28)$$

that applies to particle velocities; though again one can check that there is agreement if $u_p' = u' = c$.

12.5 Evidence from the Fizeau Effect: The Refractive Index of a Moving Medium

12.5.1 Introduction

We describe a modern version of the classic Fizeau experiment[6] to determine, for light of given frequency $\omega = 2\pi f$, its inverse wavelength $1/\lambda = k/2\pi$ in a transparent medium moving with speed v. This amounts to a fairly direct test of the basic Lorentz-transformation rules (12.4), (12.5) for ω and $\mathbf{k}$. Alternatively, since the phase speed u_p in a medium is linked to the refractive index n through

$$u_p = f\lambda = \omega/k = c/n, \qquad 1/\lambda = nf/c, \tag{12.29}$$

the experiment is equivalent to measuring[7] n as a function of v.

For simplicity, we start by taking the dispersion relation in the rest frame $S^{(0)}$ of the medium to be $\omega^{(0)} = ck^{(0)}/n^{(0)}$, with strictly constant $n^{(0)}$. At the end of Section 12.5.2 we will estimate the corrections due to the variation of $n^{(0)}$ with $\omega^{(0)}$.

Presently practicable measurements are accurate only to first order in $v/c \ll 1$. Example 12.1 showed that to this order the theory predicts

$$n = n^{(0)} + \delta n, \qquad \delta n = -(v_\parallel/c)(n^{(0)2} - 1), \tag{12.30}$$

implying

$$u_p = \frac{c}{n^{(0)} + \delta n} = \frac{c}{n^{(0)}}\left(1 - \frac{\delta n}{n^{(0)}} + \cdots\right) = \frac{c}{n^{(0)}} + \alpha v_\parallel + \cdots,$$

$$\alpha \equiv \left(1 - \frac{1}{n^{(0)2}}\right).$$

The combination α is called the *Fresnel drag coefficient*. Since $c/n^{(0)}$ is the phase speed relative to the medium, α indicates what fraction of the medium velocity increments the phase velocity of the light: pre-Einstein theories related α to the question (now meaningless) of the drag exerted by the moving medium on the light-carrying aether. Section 12.5.3 will show that Galilean instead of Lorentz transformation of ω and k gives the so-called nonrelativistic expression

$$\delta n_{\mathrm{NR}} = -(v_\parallel/c)n^{(0)2}, \tag{12.31}$$

implying that nonrelativistically $\alpha \to \alpha_{\mathrm{NR}} = 1$.

Since v/c is very small, δn used to be difficult to measure. (In fact it is rare anywhere in physics to find truly Lorentzian corrections of first rather than of only second order in v/c, and Section 12.5.3 comments further on the provenance of this one.) The beauty of the Fizeau idea is that, provided δn can be measured at all, it can be distinguished from δn_{NR} without needing to be measured very accurately, because their ratio $\delta n/\delta n_{\mathrm{NR}} = (1 - 1/n^{(0)2})$ is not at all close to unity.

12.5.2 The Experiment

The essentials of the layout are sketched in Figure 12.1. Light from a laser is split into two beams which are led in opposite directions around the same closed rectangular path (roughly a square with 1 m sides), being reflected by mirrors fixed at the corners.

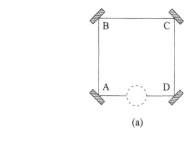

(a)

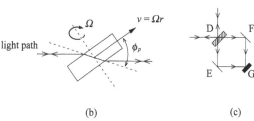

(b)   (c)

Figure 12.1 Sketch of the layout of the Fizeau experiment. For other details see the text. (a) Two mutually coherent light beams circulate in opposite directions around the square, reflected by mirrors A, B, C, D at the corners. The arrangements for producig the beams and getting them to circulate are not shown. The region inside the broken circle is shown in (b). (b) The quartz disk spins with angular velocity Ω around an axis at a distance r behind the plane of the diagram. Thus, at the light path (solid line), the velocity $\mathbf{v}$ of the material points along the arrow (in the plane of the figure and at right angles to the rotation axis), with $v = \Omega r$. Its components $v_\parallel$ along the beams are $\pm v \cos \phi_p$, where one allows for refraction on entry and exit. (c) One possible arrangement for extracting light from the two beams and to detect the beats. The mirror D is only partially silvered. A small fraction of the clockwise-circulating beam is transmitted and then reflected by the mirror E. The same fraction of the counterclockwise-circulating beam is also transmitted and is then reflected by the mirror F. These beams meet and interfere at G, where their resultant is detected on absorption by a photocathode (shown black).

Along one side they traverse, obliquely, a spinning disk of fused quartz, so that the velocity $v_\parallel$ of the material projected onto the beam direction is parallel to one light beam and antiparallel to the other, i.e. positive and negative respectively. As we shall explain presently, this difference forces the two beams to acquire slightly different frequencies $f(\pm v_\parallel) \equiv f_\pm = f(0) \pm \delta f$. Parts of both beams leak out through one of the mirrors which is only partially silvered, and are then brought together so as to inter-fere on absorption by a photocathode. The output current of the cathode therefore has a component that oscillates with the beat frequency $f_+ - f_- = 2\delta f$, which can be measured without difficulty. To allow interference, the frequencies must be very stable, and the light must be coherent over distances greater than the length of the circuit: that is why one needs a laser source.

The frequencies are governed, indirectly, by the fact that each beam settles to steady circulation around a closed path, which is possible only if one circuit accom-modates a whole number N of wavelengths. Since N is an integer it cannot change continuously, and therefore remains fixed under small changes of wavelength and frequency; in particular it is the same for both beams when the disk is at rest, and remains fixed as $v_\parallel$ increases from zero. We proceed to show how this condition links δf to δn.

The length of the circuit is L; a distance l lies inside the quartz (refractive index n), and $L - l$ in air (effectively *in vacuo*). Start by considering just one of the beams. The frequency f is the same everywhere along it, inside the material and out, by virtue of the continuity condition (say on the electric field of the wave) at the interfaces.[8] In air the wavelength is $\lambda_{out} = c/f$, and in the quartz it is $\lambda_{in} = c/nf$. Thus the standing-wave condition reads

$$N = \frac{L-l}{\lambda_{out}} + \frac{l}{\lambda_{in}} = \frac{f}{c}[(L-l) + nl] = \frac{f}{c}[L + (n-1)l], \qquad (12.32)$$

requiring from theory expressions for $1/\lambda$ or equivalently for n inside a moving medium.

Now consider both the beams. Inside the disk they experience different refractive indices, namely $n(\pm v_\parallel) \equiv n_\pm = n(0) \pm \delta n$ respectively, where $n(0) = n^{(0)}$ is the index for the material at rest, i.e. in its rest frame $S^{(0)}$. Accordingly we write[9] $n_\pm = n^{(0)} \pm \delta n$. On the other hand each beam satisfies (12.32) with the appropriate n but with the same N for both, whence they must have different frequencies $f_\pm = f(0) \pm \delta f$:

$$0 = [f(0) + \delta f][L + (n^{(0)} + \delta n - 1)l] - [f(0) - \delta f][L + (n^{(0)} - \delta n - 1)l].$$

The terms independent of δn and δf cancel. Further, since δn and δf are first-order small, we can drop terms proportional to $\delta n \delta f$ because they are second-order small. This leads to

$$0 = \frac{2\delta f}{c}[L + (n^{(0)} - 1)l] + \frac{f(0)}{c}[2l\delta n] \Rightarrow \frac{\delta f}{f(0)} = -\delta n \frac{l}{[L + (n^{(0)} - 1)l]}. \qquad (12.33)$$

Using (12.30) we find for the beat frequency that

$$\frac{2\delta f}{f(0)} = -\frac{2v_\parallel}{c} n^{(0)2} \frac{l(1 - 1/n^{(0)2})}{[L + (n^{(0)} - 1)l]}. \qquad (12.34)$$

The report on the experiment cites the following values: $\lambda(0) = 1.153\,\mu$, whence $f(0) = c/\lambda(0) = 2.6 \times 10^{14}$ Hz; $L = 4.08$ m; $l = 0.017$ m; and $n^{(0)} = 1.45$. Thus $(n^{(0)} - 1)l$ is negligible compared with L. The velocity $v_\parallel$ could be raised to around 3 m/s, i.e. to $v_\parallel/c \sim 10^{-8}$. Hence δf is expected to rise to the order of $(v/c) \times f(0) \times (l/L) \sim 10^{14} \times 10^{-8} \times 10^{-2} \sim 10^4$ Hz.

Figure 12.2 shows the results. The lower straight line is the relativistic prediction (12.34). The upper straight line is the "nonrelativistic" prediction using δn_{NR}, i.e. (12.34) divided by $(1 - 1/n^{(0)2})$. Clearly the data rule out the latter, and are quite close to (12.34). Judged by eye, there is agreement to within an experimental uncertainty of around 2%.

The variation of $n^{(0)}$ with the rest-frame frequency $\omega^{(0)}$ entails another contribution to δn, call it $\delta_2 n$, also of first order in v/c, but much smaller than the expression just derived. It allows for the fact that in the rest frame of the material the two beams have different frequencies, namely

$$\omega_\pm^{(0)} = \gamma(\omega_\pm \mp v_\parallel k_\pm) = \gamma\omega_\pm(1 \mp v_\parallel/c) \simeq \omega(1 \mp v_\parallel/c),$$

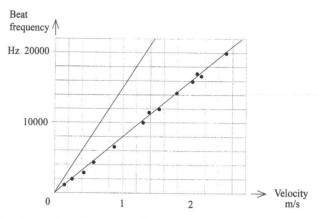

Figure 12.2 Experimental results on the Fizeau effect. Horizontal axis: the speed $v_\parallel$ of the material parallel to the light. Vertical axis: the beat frequency $2\delta f$. Lower sloping line: the theoretical prediction, equation (12.34). Upper sloping line: the "nonrelativistic" prediction, greater than the relativistic by the Fresnel drag coefficient $1/(1 - 1/n^{(0)2}) = 1.91$ (see Sections 12.5.1 and 12.5.3). This line has been superposed on the figure published in the paper. The dots are measured values. Reproduced from W. M. Macek *et al.*, "Measurement of Fresnel drag with the ring laser", *Journal of Applied Physics* (1964), volume **35**, page 2556, by permission of the American Institute of Physics.

where the ω's and k's are the frequencies and the vacuum wave-numbers in the laboratory frame. Hence the appropriate refractive indices are

$$n^{(0)}(\omega \mp \omega v_\parallel/c) \simeq n^{(0)}(\omega) \pm \delta_2 n, \qquad \delta_2 n = -(v_\parallel/c)\omega dn^{(0)}/d\omega,$$

and $\delta_2 n$ should be added to δn on the right of (12.33). But in quartz or glass at optical frequencies $|\omega dn^{(0)}/d\omega|$ is typically between 0.01 and 0.05, so that $|\delta_2 n/\delta n| = |\omega dn^{(0)}/d\omega|/(n^{(0)2} - 1) \sim |\omega dn^{(0)}/d\omega|$ is at most a 5% correction. For the measurement we have described this barely matters; but in more recent experiments claiming much higher accuracy $\delta_2 n$ must be taken into account.

12.5.3 Why is the Fizeau Effect not Second-Order Small?

Evidently one can determine $n = n^{(0)} + \delta n = c/u_p$ by finding the phase speed u_p, an approach that helps to illuminate how δn comes to be linear rather than merely quadratic in v/c. For simplicity we consider only the collinear case, even though it does not fully cover the experimental scenario of Section 12.5.2. From this point of view the experiment tests the Lorentz-transformation rule (12.16) for u_p.

Such tests for velocities of any kind have in common the basic difficulty that if both the speeds being combined are far below c, then specifically relativistic effects, the term vu_p/c^2 in this case, are second-order small, which normally is too small to observe. But in the Fizeau experiment, although one still has $v/c \ll 1$, the phase speed $u_p = c/n$ is (effortlessly so to speak) of order c, because the refractive index n of the medium is of order unity. Thus the crucial ratio $vu_p/c^2 = v/nc$ is only first-order small, and it is the presence of this term that the experiment confirms.

To trace this feature through the calculation, we put boxes around the terms present under Lorentz and absent under Galilean transformations:

$$u_p = \frac{u_p^{(0)} + v}{1 + \boxed{u_p^{(0)}v/c^2}} = \frac{c/n^{(0)} + v}{1 + \boxed{v/n^{(0)}c}} = \frac{(c/n^{(0)})(1 + vn^{(0)}/c)}{1 + \boxed{v/n^{(0)}c}}.$$

On the other hand, $u_p = c/n$. Upending, equating the two resultant expressions for $1/u_p$, and linearizing as indicated by the arrow, we find

$$\frac{n}{n^{(0)}} = \frac{1 + \boxed{v/n^{(0)}c}}{1 + vn^{(0)}/c} \rightarrow (1 + \boxed{v/n^{(0)}c})(1 - vn^{(0)}/c + \cdots)$$

$$= 1 - vn^{(0)}/c + \boxed{v/n^{(0)}c} + \cdots = 1 - \frac{v}{c}n^{(0)}(1 - \boxed{1/n^{(0)2}} + \cdots),$$

as before. One should ponder the fact that in the last line both the corrections are of order v/c but that only one is relativistic, and that without the boxes mere inspection would fail to reveal this.

12.6 Quantum Mechanics: The Wave Functions of Free Particles

Modern quantum mechanics started by associating plane waves with free particles, in a way that appeared to be uniquely determined by Lorentz invariance. The only other physical input was the much earlier Planck–Einstein relation $\varepsilon = \hbar\omega$ for the energy ε of a light quantum (photon) of frequency ω. Effectively if loosely this associates particles (photons) with waves, i.e. with the classically well-understood oscillations of the electromagnetic field.

Suppose now that one looks for the converse, i.e. for a way to associate plane waves with the free particles of pre-quantal physics. Eventually of course quantum mechanics must explore the precise physical significance of any such association, and its bearing on particles that are not free. But in any case one would like the equation relating wave and particle to be form-invariant under Lorentz transformations, in order to satisfy the relativity principle in the simplest conceivable way. If a plane wave is characterized by a unique four-vector $\vec{K} = (K_0, \mathbf{K}) = (\omega/c, \mathbf{k})$, then this requirement determines the issue, given that a free particle is characterized by its four-momentum $\vec{P} = (P_0, \mathbf{P}) = (\varepsilon/c, \mathbf{p})$. The only plausible form-invariant way to link these four-vectors is to make them parallel, with a constant of proportionality dictated (see Section 7.4) by the prior Planck–Einstein proportionality between their time components:

$$K_0 = P_0/\hbar \Rightarrow \vec{K} = \vec{P}/\hbar \Rightarrow \mathbf{K} = \mathbf{P}/\hbar \Rightarrow \mathbf{k} = \mathbf{p}/\hbar. \tag{12.35}$$

In view of $k = 2\pi/\lambda$, the last equation yields de Broglie's expression for the wavelength:

$$\lambda = 2\pi p/\hbar. \tag{12.36}$$

Thus the wave finally associated with the particle reads

$$\psi = A \exp(-i\vec{K} \cdot \vec{X}) = A \exp\{-i(\omega t - \mathbf{k} \cdot \mathbf{r})\} = A \exp\{-i(\varepsilon t - \mathbf{p} \cdot \mathbf{r})/\hbar\}, \tag{12.37}$$

where the minus sign in the exponent is conventional, and for our present purposes the amplitude A is irrelevant.

Then $\vec{P} \cdot \vec{P} = \varepsilon^2/c^2 - p^2 = m^2c^2$ leads to

$$\kappa^2 \equiv \vec{K} \cdot \vec{K} = \omega^2/c^2 - k^2 = \vec{P} \cdot \vec{P}/\hbar^2 = 1/\lambda_c^2, \qquad \lambda_c = \hbar/mc, \tag{12.38}$$

where λ_c is the (reduced) Compton wavelength for the particle (not necessarily an electron), familiar in another context from Section 11.4. This in turn entails the dispersion relation

$$\omega = c\sqrt{k^2 + \kappa^2}, \qquad (12.39)$$

which identifies the group and the phase velocities:

$$\mathbf{u}_g = \boldsymbol{\nabla}_k \omega = c^2 \mathbf{k}/\omega = c^2 \mathbf{p}/\varepsilon = \mathbf{u}, \qquad (12.40)$$

$$\mathbf{u}_p = \mathbf{k}\omega/k^2 = \mathbf{p}\varepsilon/p^2. \qquad (12.41)$$

The group velocity equals the particle velocity $\mathbf{u}$. This is essential: there could be no physics in any dispersion relation that allowed a particle to travel at velocities different from those of the localizable wavegroup associated with it. By contrast, the phase velocity points in the same direction but is faster than c; indeed $u_g u_p = c^2$ (see also Problem 12.4). However, the phase speed is by no means irrelevant: for given frequency it determines the wavelength through $\lambda = 2\pi u_p/\omega$, and λ in turn governs the diffraction and interference patterns that tell one *where* (rather than *how fast*) the particle is going.

Example 12.5. The proton radius R is of the order of 10^{-15} m. To study its charge distribution one measures the differences between the elastic scattering of electrons from protons and from (hypothetical) point charges. For such differences to show up, the de Broglie wavelength of the electrons must be comparable to or smaller than R. What is the least energy that the electrons must have?

Solution. Few words are needed:

$$\lambda = 2\pi/k = 2\pi\hbar/p \leq R \ \Rightarrow\ p \geq 2\pi\hbar/R$$

$$\Rightarrow \varepsilon = \sqrt{c^2 p^2 + m^2 c^4} \geq \sqrt{(2\pi c\hbar/R)^2 + m^2 c^4}.$$

Express ε in MeV: then

$$2\pi c\hbar/R = 2\pi(3 \times 10^8)(1.05 \times 10^{-34})(10^{-15})(1.6 \times 10^{-13}) = 1.2 \times 10^3 \text{MeV}.$$

Evidently the electron rest energy $mc^2 \approx 0.5$ MeV is negligible by comparison, whence we estimate $\varepsilon \geq 1000$ MeV. Obviously such extreme-relativistic electrons are expensive. ∎

In fact the initial argument for $\vec{K} = \vec{P}/\hbar$ is leaky. Quantum-mechanical wavefunctions are not directly observable, whence their phase need not be invariant, and cannot compel $(\omega/c, \mathbf{k})$ for plane waves of this kind to transform like a four-vector. It is not unusual in physics for correct conclusions to be drawn from arguments that might have failed more thorough scrutiny, or that become untenable as the subject develops. Our present example is of the second type. Appendix D shows that the underlying relativistic wave equation does make $(\omega/c, \mathbf{k})$ into a four-vector, without prior reference to the phase.

That the point is not a mere quibble is easily demonstrated by lapsing from Lorentzian to Galilean physics. "Nonrelativistic" just like relativistic quantum mechanics starts from the de Broglie relation $\mathbf{k} = \mathbf{p}/\hbar$. But then the relativity principle makes it *impossible* for the phase to be invariant: the principle requires $\mathbf{k}' = \mathbf{p}'/\hbar$, while an invariant phase requires $\mathbf{k}' = \mathbf{k}$, incompatibly with the transformation rule $\mathbf{p}' = \mathbf{p} - m\mathbf{v}$.

12.7 Notes

1. To the physics of any given type of waves the wave equation that they obey is of course vital: for instance, without it there would be no convenient way to link the plane waves considered here with others of different geometry, say with spherical waves.
2. Section 4.1 explained that we shall sometimes refer to Galilean (i.e. pre-Einstein) physics as "nonrelativistic", and to Einsteinian physics as "relativistic", without repeated apologies for this absurd but firmly entrenched way of describing their difference.
3. Many details of course differ. For instance, sound waves are unpolarized, while plane waves of light vibrate at right angles to the propagation direction. However, for our purposes this particular difference is irrelevant.
4. All waves in classical (nonquantum) physics are measurable in this sense, including for instance acoustic, elastic, electromagnetic, and (if they exist) gravitational waves. But the wave functions of quantum mechanics are not; they are discussed separately in Section 12.6. However, it turns out in the end that the phases of *relativistic* quantum wavefunctions are invariant too, though this is not self-evident, and needs proof.
5. The qualifications are essential: for instance, in some gravitational fields u_g for *light* can exceed c.
6. A scholarly discussion is given by I. Lerche, "The Fizeau effect: theory, experiment, and Zeeman's measurements", *American Journal of Physics* (1977), volume **45**, page 1154; but see also Section 4 in R. J. Cook, H. Fearn, and P. W. Milonni, "Fizeau's experiment and the Aharonov–Bohm effect", *American Journal of Physics* (1995), volume **63**, page 705. The experiment we describe is reported by W. M. Macek, J. R. Schneider, and R. M. Salamon, "Measurement of Fresnel drag with the ring laser", *Journal of Applied Physics* (1964), volume **35**, page 2556.
7. Unfortunately, it is often analysed instead as a measurement of u_p. Though legitimate in principle this is misleading, because (i) the experiment involves no direct measurement of velocity; and (ii) if it did, it would yield information not about $\mathbf{u}_p$ but about the group velocity, which transforms quite differently from $\mathbf{u}_g$ in general, and also under the (noncollinear) conditions of this particular experiment. Worse, and more often still, such measurements have been presented as verification of the velocity-combination rule for particles, which they are not, because the physics of the two cases is completely different: even for collinear motion it is only by pure coincidence that phase and particle velocities happen to transform alike.
8. This is true because the material moves parallel to its surface. For material advancing or receding, the light frequency would change on entry and exit: see Problem 12.10.
9. Remember that the superfix (0) as on $n^{(0)}$ is reserved for quantities defined with respect to the rest frame of the material, while the argument (0) of $f(0)$ and of $n(0)$ indicates zero value of a velocity $v_\parallel$ defined with respect to the laboratory frame.

13 Light Waves in Empty Space: Aberration and Doppler Effect

13.1 Notation and Agenda

This chapter applies the Lorentz transformations from Section 12.2 for $\vec{K} = (\omega/c, \mathbf{k})$ to light propagating through empty space, where its speed is c, and its group and phase velocities equal (with both written simply as $\mathbf{u}$):

$$\vec{K} \cdot \vec{K} = 0, \qquad \omega = ck, \qquad \mathbf{u}_p = \mathbf{u}_g \equiv \mathbf{u} = c\hat{\mathbf{k}} = c^2 \mathbf{k}/\omega. \qquad (13.1)$$

More specifically we are concerned with a monochromatic signal from an emitter E, detected by a receiver R. The signal is modelled as a single plane wave, which is adequate if it propagates like a ray governed by geometrical optics. In a first discussion it saves time and confusion to adopt a completely explicit (if somewhat laboured) notation,[1] as follows. The rest frames of E, R are $S^{(E)}$, $S^{(R)}$; superfixes (E, R) always specify the values of quantities (generally of four-vector components) referred to these frames. By contrast, suffixes E, R specify characteristics of emitter and receiver. For instance, $\vec{U}_E$, $\vec{U}_R$ are their four-velocities. The only pertinent transformations are those from $S^{(E)}$ to $S^{(R)}$ and back,[2] featuring their relative velocity which we now define as

$$\mathbf{v} \equiv \mathbf{u}_R^{(E)} = -\mathbf{u}_E^{(R)}. \qquad (13.2)$$

The proper frequency of E, i.e. its frequency in its own rest frame $S^{(E)}$, is $\omega_E^{(E)}$, while its frequency in $S^{(R)}$ is $\omega_E^{(R)}$.

Quantities without suffixes refer to the wave: thus $\omega^{(E)}$, $\omega^{(R)}$ are its frequencies with respect to $S^{(E)}$, $S^{(R)}$. With respect to $S^{(E)}$ the frequency of the wave is naturally the same as the frequency of the emitter, so that $\omega^{(E)} = \omega_E^{(E)}$. The frequency registered by the receiver (in its own rest frame $S^{(R)}$) is $\omega_R^{(R)}$; naturally it equals the frequency of the wave in this frame, so that $\omega_R^{(R)} = \omega^{(R)}$.

We have two tasks: to understand *aberration*, namely the difference between the wave directions $\hat{\mathbf{k}}^{(R)}$ and $\hat{\mathbf{k}}^{(E)}$; and the *Doppler effect*, namely the deviation of the ratio $\omega_R^{(R)}/\omega_E^{(E)}$ from unity. To this end, the mere Lorentz transformations of a given plane wave are necessary but not sufficient. Incomparably the most efficient method is to solve both problems jointly by using invariants, as in Section 13.4. However, we shall preface this with alternative arguments from first principles, for aberration in Section 13.2 and for the collinear Doppler effect in Section 13.3. In exchange for somewhat obscuring the structure of the general solution, these more primitive approaches reveal certain aspects of the physical input rather more readily; and they make it easier to compare the general expressions with the so-called nonrelativistic ones, derived without allowing for time dilation.

Closely related to aberration and the Doppler effect is the important problem of Lorentz-transforming the Planckian spectrum of black-body radiation. Unfortunately, to do this directly in terms of waves one needs also the transformation rules for intensity, which we have not considered. Instead, Appendix E will find the answer by reasoning in terms of photons, i.e. through the short-cut mentioned in Section 12.1.

13.2 Aberration, from First Principles

13.2.1 Theory

Strictly speaking the problem of Lorentz-transforming the direction $\hat{\mathbf{k}} = \hat{\mathbf{u}}$ is already solved by the proof, in Section 12.4.2, that the velocities $\mathbf{u}$ of light waves *in vacuo* transform exactly like particle velocities of magnitude $u = c$. Thus the rules for standard configuration are (4.15), (4.16); for directions they entail (4.21), namely

$$\tan\phi' = \frac{\sqrt{1 - v^2/c^2}\,\sin\phi}{\cos\phi - v/c}. \tag{13.3}$$

They simplify only in that light speed naturally entails $u' = u = c$, and allows (13.3) to be replaced by (4.24):

$$\tan(\phi'/2) = \sqrt{\frac{1 + v/c}{1 - v/c}}\,\tan(\phi/2). \tag{13.4}$$

However, the implications are best understood through an example, and we choose the discovery of stellar aberration by Bradley.[3]

13.2.2 Evidence: Stellar Aberration

Consider a star fixed in the rest frame S of the sun, far enough to allow us to disregard the difference between the directions from star to sun and from star to earth: in other words parallax across the earth's orbit is negligible. We call this direction $\hat{\mathbf{k}}$. Referred to the instantaneous rest frame S' of the earth,[4] it is $\hat{\mathbf{k}}'$: in order to see the star, a terrestrial telescope must point along $-\hat{\mathbf{k}}'$. What Bradley discovered was that $\hat{\mathbf{k}}'$ oscillates sinusoidally with a period of one year: Figure 13.2 plots the two sets of

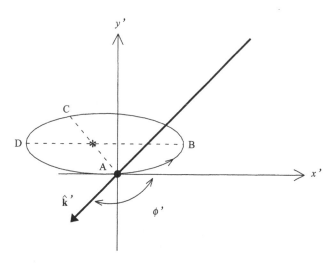

Figure 13.1 Stellar aberration. The earth's orbit around the sun *. Light from the very distant star arrives in the direction $\hat{\mathbf{k}}'$ relative to the inertial frame S' instantaneously co-moving with the earth at A, where its velocity **v** relative to the sun points directly towards the projection of the star onto the ecliptic. Note that the sine of the angle ϕ' is negative: the aberration angle $\delta\phi$ given by (13.5) is minimal at A and maximal at C.

observations that he reports in full. His measurements indicated that the amplitude of the oscillations is proportional to $\sin\phi$, where ϕ is the elevation of the star above the ecliptic (the plane of the earth's orbit); and that twice the amplitude approaches $(40.5 \pm 0.5)''$ as ϕ approaches $\pi/2$. (The currently accepted value is $40.94''$. The symbol $''$ stands for second of arc.) Our task is to explain this, and to connect $\hat{\mathbf{k}}$ with $\hat{\mathbf{k}}'$, i.e. the "true" direction with the "apparent".

Figure 13.1 shows the earth at A, at a time t_1 when its orbital velocity $\mathbf{v}(t_1)$ points directly towards the projection of the star onto the ecliptic. We take the x and x' axes along $\mathbf{v}(t_1)$, and the y and y' axes perpendicular to the ecliptic, so that at time t_1 the two frames are in standard configuration, and $v = v_x$. In fact both the polar angles of $\mathbf{k}'$ oscillate, but like Bradley we consider only the angle ϕ'.

Since $v/c \sim 10^{-4}$ is so small, one can approximate on the right of (13.3) to first order:

$$\tan\phi' \approx \frac{\sin\phi}{\cos\phi - v_x/c} = \frac{\sin\phi}{\cos\phi}\left\{1 + \frac{v_x/c}{\cos\phi} + \cdots\right\} = \tan\phi + \left(\frac{v_x}{c}\right)\frac{\sin\phi}{\cos^2\phi} + \cdots.$$

Furthermore, one is interested not directly in ϕ', but in the aberration $\delta\phi \equiv \phi' - \phi$ which likewise is very small. Hence we can approximate

$$\tan\phi' = \tan(\phi + \delta\phi) = \tan\phi - \frac{\delta\phi}{\cos^2\phi} + \cdots.$$

Equating the two expressions for $\tan\phi'$ we obtain $\delta\phi \approx (v_x/c)\sin\phi$. Into this formula v_x has entered as the component, towards the projection of the star onto the ecliptic, of the earth's velocity relative to the common rest frame S of sun and star. Evidently $\delta\phi$ oscillates with a one-year period, so that at time t one has

$$\delta\phi(t) \approx (v/c)\cos[2\pi(t - t_1)/(1\text{year})]\sin\phi. \tag{13.5}$$

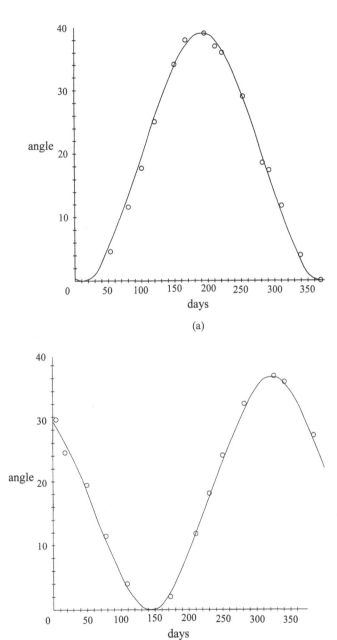

(a)

(b)

Figure 13.2 Stellar aberrations measured by Bradley. Horizontal axis: time *t* in days: day 1 is 1 September 1727. Vertical axis: the angle of aberration, in seconds of arc, relative to its minimum taken as zero. The circles are the measured values, governed by equation (13.5). Both stars are close to the zenith (ϕ' close to $\pi/2$). The curves have been drawn so as to pass through the measured point with the greatest deviation: they are not best fits. (a) The star γ Draconis (i.e. in the Dragon); the curve is $(39/2)\{1 + \cos[2\pi(t - 190)/365]\}$. (b) The star η Ursae Majoris (i.e. in the Great Bear); the curve is $(36/2)\{1 + \cos[2\pi(t - 322)/365]\}$.

That the observed $\delta\phi$ is not due to parallax[5] is seen from the absolute phase of its oscillations, i.e. from t_1: aberration is extremal at the points A, C of the orbit, while parallax would be extremal at B, D.

Bradley's value for the amplitude when $\sin\phi = 1$ implies that an angle of $2v/c$ radians is equal to $40.5''$. He presented this as a value for the time T needed by light to travel over a distance equal to the (mean) radius $\mathcal{R}$ of the earth's orbit. From $v = 2\pi\mathcal{R}/(1 \text{ year})$ we have

$$T = \frac{\mathcal{R}}{c} = \frac{(v/2\pi)(1 \text{ year})}{c} = \frac{1}{4\pi}\left(\frac{2v}{c}\right)(1 \text{ year})$$

$$= \frac{1}{4\pi}\left(\frac{40.5 \times 2\pi}{360 \times 60 \times 60}\right)(365 \times 24 \times 60 \times 60) = 493 \text{ s} = 8 \text{ minutes } 13 \text{ seconds.}$$

Currently accepted values yield $\mathcal{R}/c = (1.496 \times 10^{11} \text{ m})/(2.998 \times 10^8 \text{ m/s}) = 499 \text{ s}$.

13.2.3 The Galilean View: Euler's Paradox

The expression for $\delta\phi$ just derived and checked depends only on the relative velocity **v** of earth and star. In Newtonian days the problem was more subtle. A mere ten years after Bradley's paper Euler[6] already noticed that, if light propagates with speed c relative to an aether, then aberration would in principle be different if the star moved and the earth were at rest. However, as for the Doppler effect (Example 2.11) it turns out that the differences between the various expressions are only of second order in the pertinent speeds divided by c; to first order they all agree, which makes it difficult and often impossible to discriminate between them when, as in Bradley's case, all these ratios are or are assumed to be very small. We defer calculating second-order terms until we come to the Doppler effect, where they have been checked experimentally.

13.3 The Collinear Doppler Effect from First Principles

We proceed to derive the Doppler shift from first principles, for the special case where the relative motion of emitter E and receiver R is along the line of sight. In this method it is particularly easy to keep track of the consequences peculiar to time dilation. Experimental evidence is deferred to Section 13.5.1.

Recall the comments on notation in Section 13.1. It is instructive to do the calculation twice, first in the rest frame $S^{(R)}$ of R, and then in the rest frame $S^{(E)}$ of E. The frequency and period of E in its own rest frame are $\omega_E^{(E)} = 2\pi/T_E^{(E)}$.

In the rest frame of R, Figure 13.3a shows E moving with speed $|u_E^{(R)}| \equiv v$ directly *towards R.* By time dilation, the period and the frequency of the source are $T_E^{(R)} = T_E^{(E)}/\sqrt{1 - v^2/c^2}$ and $\omega_E^{(R)} = \omega_E^{(E)}\sqrt{1 - v^2/c^2}$. From this point the argument runs as in the small print just above Example 2.10. The distance between successive crests is $D^{(R)} = T_E^{(R)}(c - v)$, because in the time $T_E^{(R)}$ that it takes for a second crest to be emitted, the first has already progressed through $T_E^{(R)}c$, while the source has made up a distance $T_E^{(R)}v$. The crests travel at speed c, so they reach R at intervals of $T^{(R)} = D^{(R)}/c$. But this $T^{(R)}$ is just the period of the wave as perceived by R, so that $T^{(R)} = T_R^{(R)}$:

$$T_R^{(R)} = \frac{D^{(R)}}{c} = T_E^{(R)}(1 - v/c) = T_E^{(E)}\frac{1 - v/c}{\sqrt{1 - v^2/c^2}} = T_E^{(E)}\sqrt{\frac{1 - v/c}{1 + v/c}}. \tag{13.6}$$

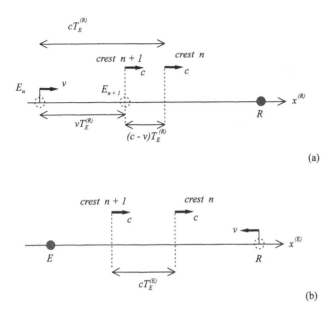

Figure 13.3   The collinear Doppler effect. For the argument see the text. The small arrows show speeds and directions. The relative magnitudes of the labelled separations (and only of these) are drawn to scale for $v/c = 3/4$. (a) In the receiver rest frame. E_n, E_{n+1}: positions of the emitter at the emission of crests numbers n and $n + 1$. Broken lines show the positions of these crests *when crest $n + 1$ is emitted*. (b) In the emitter rest frame. Broken lines show *simultaneous* positions of crests numbers n and $n + 1$. (The display must not be taken literally: small sources (as drawn) would produce spherical crests, while collimated waves (i.e. plane crests, also as drawn) could be produced only by sources extending sideways over many wavelengths).

Reciprocating, we get the end-result

$$\frac{\omega_R^{(R)}}{\omega_E^{(E)}} = \frac{T_E^{(E)}}{T_R^{(R)}} = \frac{\sqrt{1 - v^2/c^2}}{1 - v/c} = \sqrt{\frac{1 + v/c}{1 - v/c}}, \quad \text{collinear case:} \ \binom{\text{approach: } v > 0}{\text{recession: } v < 0}.$$

$$(13.7)$$

(These ratios could have been anticipated from Problems 5.6 and 5.7.)

For small v/c one expands the frequency ratio in a power series. Doing this in the third term of (13.7) yields

$$\frac{\omega_R^{(R)}}{\omega_E^{(E)}} = \left(1 - \frac{1}{2}\left(\frac{v}{c}\right)^2 + \cdots\right)\left(1 + \left(\frac{v}{c}\right) + \left(\frac{v}{c}\right)^2 + \cdots\right) = 1 + \left(\frac{v}{c}\right) + \frac{1}{2}\left(\frac{v}{c}\right)^2 + \cdots.$$

$$(13.8)$$

In the rest-frame of E, Figure 13.3(b) shows the same system with E fixed and the source approaching it at the same speed $|u_R^{(E)}| = v$. In $S^{(E)}$ the crests are emitted at time intervals of $T_E^{(E)}$; the distance between them is $D^{(E)} = T_E^{(E)}c$; the closing speed[7] between crests and receiver is $c + v$; hence the crests reach R at time intervals of $D^{(E)}/(c + v) = T_E^{(E)}/(1 + v/c)$. This is the interval measured in $S^{(E)}$; by time dilation the same interval measured in $S^{(R)}$ is[8]

$$T_R^{(R)} = \sqrt{1 - v^2/c^2}\, T_E^{(E)} \frac{1}{1 + v/c} = T_E^{(E)} \sqrt{\frac{1 - v/c}{1 + v/c}},$$

whence

$$\frac{\omega_R^{(R)}}{\omega_E^{(E)}} = \frac{1 + v/c}{\sqrt{1 - v^2/c^2}} = \sqrt{\frac{1 + v/c}{1 - v/c}}, \tag{13.9}$$

agreeing with (13.7) as it must.

Example 13.1. How fast would you need to drive towards a traffic light so that the red light appears green? Take the wavelengths of red and green as 6.5×10^{-7} and 5.6×10^{-7} m.

Solution. The wavelength ratio is given by $\omega^{(R)}/\omega^{(E)} = \lambda^{(E)}/\lambda^{(R)}$, because light speed *in vacuo* is the same in both frames: $c = \lambda^{(E)}\omega^{(E)}/2\pi = \lambda^{(R)}\omega^{(R)}/2\pi$. (Example 12.3 has already shown that wavelengths transform in ways unconnected with the Lorentz contraction.) Hence (13.7) entails

$$\sqrt{\frac{1 + v/c}{1 - v/c}} = \frac{\omega_R^{(R)}}{\omega_E^{(E)}} = \frac{\lambda^{(E)}}{\lambda^{(R)}} \equiv \rho = \frac{6.5}{5.6} = 1.161.$$

Squaring and rearranging we find

$$v/c = (\rho^2 - 1)/(\rho^2 + 1) = 0.148.$$

If your defence against a charge of knowingly running the light is accepted, you may then be charged with speeding. ∎

It is interesting to note the so-called nonrelativistic[9] (i.e. the Galilean) expressions obtained without the time-dilation factors $\sqrt{1 - v^2/c^2}$. In $S^{(R)}$, equation (13.7) without the square root would give

$$\left(\frac{\omega_R^{(R)}}{\omega_E^{(E)}}\right)_{NR,\, R\,\text{at rest}} = \frac{1}{1 - v/c}$$

$$= 1 + \left(\frac{v}{c}\right) + \left(\frac{v}{c}\right)^2 + \cdots,$$

$$\text{(collinear nonrelativistic case, } R \text{ at rest).} \tag{13.10}$$

For small v/c this differs from the true expression (13.8) only by the coefficient (1 instead of 1/2) of $(v/c)^2$. By contrast, in $S^{(E)}$, equation (13.9) without the square root would give

$$\left(\frac{\omega_R^{(R)}}{\omega_E^{(E)}}\right)_{G,\, E\,\text{at rest}} = 1 + \frac{v}{c} \qquad \text{(collinear nonrelativistic case } E \text{ at rest).} \tag{13.11}$$

To second order this differs both from the true (13.8), and also from the equally nonrelativistic (13.10) for R at rest.

Finally one should compare these nonrelativistic expressions with (2.50) for sound, of course after replacing c_s by c. (Compare also with Example 2.11.) Equation (2.50) describes the approach if we choose E to the left of R (upper signs) with $u_E \geq 0$ and $u_R \leq 0$; then it yields

$$\left(\frac{\omega_R^{(R)}}{\omega_E^{(E)}}\right)_{sound} = \frac{1 + |u_R|/c + w/c}{1 - |u_E|/c + w/c}. \tag{13.12}$$

If one sets the wind-speed $w = 0$, then (13.12) with $u_R = 0$ and $|u_E| = v$ reduces to (13.10), and with $u_E = 0$ and $u_R = v$ it reduces to (13.11). But why one should set $w = 0$ is unclear. Thus, in order to get from the true expression (13.8) to the nonrelativistic (13.10) "for R at rest", we must not only (i) disregard time-dilation, but must also (ii) assume that the medium is at rest relative to R; on the other hand, to get to the nonrelativistic (13.11) "for E at rest", we must (i) disregard time dilation, and assume (ii) that the medium is at rest relative to E. But these two situations are not linked by a Galilean transformation: if the medium is at rest relative to R, then it cannot be at rest relative to E, and vice versa. This is the same prima-facie paradox that was first observed by Euler, albeit in connection with aberration rather than with the Doppler effect (see Section 13.2.3).

Thus it is misleading (though common) to speak of (13.10) and (13.11) simply as *the* "nonrelativistic" versions of or approximations to (13.7): they depend not just on the generically Galilean disregard of time dilation, but also on very specific (and often quite arbitrary) assumptions about the relation of emitter and receiver to an aether (which does not exist, but is an essential element of nonrelativistic theories).

13.4 The Method of Invariants

Recall again the notation explained in Section 13.1, and equation (13.2) in particular.

The strategy is much the same as in Section 2.5.4 for sound: identify the pertinent invariants formed from the four-vector $\vec{K}$ and from the four-velocities $\vec{U}_R$ and $\vec{U}_E$ of emitter and receiver, evaluate them in both the frames $S^{(R,E)}$, and equate the results. But here the procedure is more streamlined, because all available invariants are obvious, namely the six dot products formed from the three four-vectors in the problem.

In fact $\vec{K} \cdot \vec{K} = 0$ and $\vec{U}_R \cdot \vec{U}_R = c^2 = \vec{U}_E \cdot \vec{U}_E$ are clearly irrelevant because they are universal constants that carry no information about the system. Further, $\vec{U}_R \cdot \vec{U}_E$ seems unpromising because it knows nothing about the wave, though it governs the relative speed of E and R (see Example 7.4). Hence we start with one of the two remaining candidates, say with $\vec{U}_E \cdot \vec{K}$.

In $S^{(E)}$ the components are given by $\vec{U}_E = (c, \mathbf{0})$ and $\vec{K} = (\omega^{(E)}/c, \mathbf{k}^{(E)})$, whence

$$\vec{U}_E \cdot \vec{K} = U_0^{(E)} K_0^{(E)} - \mathbf{U}^{(E)} \cdot \mathbf{K}^{(E)} = c(\omega^{(E)}/c) - 0 = \omega^{(E)} = \omega_E^{(E)}. \tag{13.13}$$

In $S^{(R)}$ the components of the same vectors are given by

$$\vec{U}_E = (c, \mathbf{u}_E^{(R)})/\sqrt{1 - (u_E^{(R)}/c)^2} = (c, -\mathbf{v})/\sqrt{1 - (v/c)^2} \tag{13.14}$$

and $\vec{K} = (\omega^{(R)}/c, \mathbf{k}^{(R)})$, whence

$$\vec{U}_E \cdot \vec{K} = U_0^{(R)} K_0^{(R)} - \mathbf{U}^{(R)} \cdot \mathbf{K}^{(R)} = (\omega^{(R)} + \mathbf{v} \cdot \mathbf{k}^{(R)})/\sqrt{1 - (v/c)^2}.$$

But we have $\mathbf{k}^{(R)} = \hat{\mathbf{k}}^{(R)}\omega^{(R)}/c = \hat{\mathbf{k}}^{(R)}\omega_R^{(R)}/c$, where the first step follows because $k = \omega/c$ in all frames and therefore also in $S^{(R)}$. Substitution yields

$$\vec{U}_E \cdot \vec{K} = \omega_R^{(R)}\frac{1 + \hat{\mathbf{k}}^{(R)}\cdot\mathbf{v}/c}{\sqrt{1-(v/c)^2}}. \tag{13.15}$$

Finally, equating (13.13) and (13.15), and rearranging, we obtain *the general Doppler formula*

$$\frac{\omega_R^{(R)}}{\omega_E^{(E)}} = \frac{\sqrt{1-(v/c)^2}}{1 + \hat{\mathbf{k}}^{(R)}\cdot\mathbf{v}/c} = \frac{\sqrt{1-(u_E^{(R)}/c)^2}}{1 - \hat{\mathbf{k}}^{(R)}\cdot\mathbf{u}_E^{(R)}/c}, \tag{13.16}$$

with the ratio expressed in terms of quantities observable in the receiver's rest frame $S^{(R)}$.

The converse is best derived by processing the other invariant $\vec{U}_R \cdot \vec{K}$ in the same way as we have just processed $\vec{U}_E \cdot \vec{K}$. One finds

$$\frac{\omega_R^{(R)}}{\omega_E^{(E)}} = \frac{1 - \hat{\mathbf{k}}^{(E)}\cdot\mathbf{v}/c}{\sqrt{1-(v/c)^2}} = \frac{1 + \hat{\mathbf{k}}^{(E)}\cdot\mathbf{u}_R^{(E)}/c}{\sqrt{1-(u_R^{(E)}/c)^2}}, \tag{13.17}$$

with the ratio now expressed in terms of quantites observable in the emitter's rest frame $S^{(E)}$. (It is left as an exercise to derive (13.17), and to show that it can be found equally well by using the standard rule to invert (13.16) regarded simply as a Lorentz transformation between $S^{(E)}$ and $S^{(R)}$.)

- Equation (13.16) is generally the more useful of the two, because the physicist is observer, i.e. receiver, more often than emitter.
- Note that $\hat{\mathbf{k}}^{(R,E)}\cdot\mathbf{v}$ are simply the projections $v_\parallel^{(R,E)}$ of $\mathbf{v}$ onto the directions of $\hat{\mathbf{k}}^{(R,E)}$ of the signal with respect to $S^{(R,E)}$, so that

$$\hat{\mathbf{k}}^{(R)}\cdot\mathbf{v} = v_\parallel^{(R)} = v\cos\phi^{(R)}, \qquad \hat{\mathbf{k}}^{(E)}\cdot\mathbf{v} = v_\parallel^{(E)} = v\cos\phi^{(E)},$$

where $\phi^{(R,E)}$ is the angle between $\mathbf{k}^{(R,E)}$ and $\mathbf{v}$. But one must bear in mind that $\phi^{(R)} \neq \phi^{(E)}$, whence the parallel projections thus defined are unequal: $v_\parallel^{(R)} \neq v_\parallel^{(E)}$.
- To obtain what are often called the analogous nonrelativistic expressions one drops the square-root factors. The somewhat dubious physics of this was discussed at the end of Section 13.3.

Example 13.2. A light source moving slowly is observed almost perpendicularly to the flight path. How far can the viewing angle be allowed to deviate from the perpendicular before the first-order contribution to the Doppler shift swamps the second-order term?

Solution. Evidently the question concerns the rest frame of the receiver, where the shift is given by (13.16). With $\hat{\mathbf{k}}^{(R)}\cdot\hat{\mathbf{v}} = -\hat{\mathbf{k}}^{(R)}\cdot\hat{\mathbf{u}}_E^{(R)} \equiv -\cos\phi^{(R)}$, expansion in powers of v/c yields

$$\frac{\omega_R^{(R)}}{\omega_E^{(E)}} = \frac{\sqrt{1 - (v/c)^2}}{1 - (v/c)\cos\phi^{(R)}}$$

$$= \left(1 - \frac{1}{2}\left(\frac{v}{c}\right)^2 + \cdots\right)\left(1 + \left(\frac{v}{c}\right)\cos\phi^{(R)} + \left(\frac{v}{c}\right)^2\cos^2\phi^{(R)} + \cdots\right)$$

$$= 1 + \left(\frac{v}{c}\right)\cos\phi^{(R)} + \left(\frac{v}{c}\right)^2\left[\cos^2\phi^{(R)} - \frac{1}{2}\right] + \cdots.$$

(In the collinear case $\phi^{(R)} = 0$, $\cos\phi^{(R)} = 1$, this reproduces (13.8).) When $\phi^{(R)} = \pi/2$ we have $\cos\phi^{(R)} = 0$, and the linear correction $(v/c)\cos\phi^{(R)}$ vanishes; whereas for large enough $\cos\phi^{(R)}$ it is obviously dominant. We require an angle $\phi^{(R)} \equiv \pi/2 - \delta\phi^{(R)}$ such that the linear and the quadratic terms are roughly equal in magnitude. But for small $\delta\phi^{(R)}$ one has $\cos\phi^{(R)} = \sin\delta\phi^{(R)} \approx \delta\phi^{(R)}$, while $|\cos^2\phi^{(R)} - 1/2| \approx 1/2$. Hence the condition reads

$$\left(\frac{v}{c}\right)\cos\phi^{(R)} \sim \left(\frac{v}{c}\right)^2\left|\cos^2\phi^{(R)} - \frac{1}{2}\right| \Rightarrow \delta\phi^{(R)} \sim v/2c. \quad \blacksquare \qquad (13.18)$$

Finally, by equating (13.16) and (13.17) we obtain

$$(1 + \hat{\mathbf{k}}^{(R)}\cdot\mathbf{v}/c)(1 - \hat{\mathbf{k}}^{(E)}\cdot\mathbf{v}/c) = 1 - (v/c)^2.$$

This rearranges into *the general aberration formula*

$$(\hat{\mathbf{k}}^{(R)}\cdot\hat{\mathbf{v}}) - (\hat{\mathbf{k}}^{(E)}\cdot\hat{\mathbf{v}}) = -(v/c)\{1 - (\hat{\mathbf{k}}^{(R)}\cdot\hat{\mathbf{v}})(\hat{\mathbf{k}}^{(E)}\cdot\hat{\mathbf{v}})\}, \qquad (13.19)$$

or, in terms of the angles $\phi^{(R,E)}$, into

$$\cos\phi^{(R)} - \cos\phi^{(E)} = -(v/c)\{1 - \cos\phi^{(R)}\cos\phi^{(E)}\}, \qquad (13.20)$$

which can be solved for either of $\cos\phi^{(R,E)}$ in terms of the other. For instance,

$$\cos\phi^{(R)} = \frac{\cos\phi^{(E)} - v/c}{1 - (v/c)\cos\phi^{(E)}}. \qquad (13.21)$$

Example 13.3. Section 13.2.2 derived Bradley's formula $\delta\phi \equiv \phi^{(R)} - \phi^{(E)} \simeq (v_\parallel/c)\sin\phi$ from first principles. Rederive it from (13.20).

Solution. We approximate to first order in v/c, which is assumed to be very small. The braces on the right of (13.20) are already prefaced by v/c, so that between them we need no longer distinguish $\phi^{(R)}$ from $\phi^{(E)}$, and can approximate this side as $-(v/c)(1 - \cos^2\phi^{(E)})$. On the left we can approximate

$$\cos\phi^{(R)} - \cos\phi^{(E)} = \cos(\phi^{(E)} + \delta\phi) - \cos\phi^{(E)} \approx \delta\phi(\mathrm{d}\cos\phi/\mathrm{d}\phi)$$

evaluated at $\phi = \phi^{(E)}$. Hence

$$-\delta\phi\sin\phi^{(E)} = -(v/c)\sin^2\phi^{(E)} \Rightarrow \delta\phi = (v/c)\sin\phi^{(E)}.$$

This method is notably quicker than starting with $\phi^{(R)} = \phi^{(E)} + \delta\phi$ in the explicit solution (13.21). (Try it and see!) ∎

13.5 Evidence: Doppler Shifts of Spectral Lines

13.5.1 First-Order Shift

When atoms drop from one energy state to another, lower in energy by ΔE, the (circular) frequency ω of the light they emit is close to $\omega_0 = \Delta E/\hbar$. In a low-pressure gas, there are three main reasons why ω differs from ω_0. (i) The interaction of the source with other atoms can shift the phase of the light, leading to frequency shifts governed mainly by the inverse of the mean time between collisions. Such *pressure broadening* becomes negligible at low enough pressures. (ii) Even for a completely isolated atom, there is an unavoidable frequency spread (called the natural width) $\Gamma_0 = 1/\tau_0$, with τ_0 comparable with the mean life of the upper state. For example, the first-excited to ground-state transition in hydrogen ($\omega_0 \approx 1.6 \times 10^{16}$ s^{-1}) has $\Gamma_0 \approx 6 \times 10^8$ s^{-1}. (iii) If the source is moving, then the frequency is Doppler shifted. As we shall see, under most conditions this effect swamps the natural width: given the Maxwell–Boltzmann velocity distribution of the atoms, the observed line shape can then serve to verify the Doppler effect. (Conversely, once we believe the theory of the Doppler effect, the line shape can reveal the velocity distribution.)

For brevity, write $\omega_E^{(E)} = \omega_0$, $\omega_R^{(R)} = \omega$, and u for the component of the source velocity towards the receiver, call it the x direction. Under normal laboratory conditions typical atomic velocities are so far below c that (except for specially designed experiments as in the next section) one can use the first-order Doppler formula

$$\omega = \omega_0\{1 + u/c\} \Rightarrow \xi \equiv \frac{\Delta\omega}{\omega_0} \equiv \frac{\omega - \omega_0}{\omega_0} = \frac{u}{c}. \tag{13.22}$$

The dimensionless measure ξ of the shift is introduced for convenience. Now consider a gas with molecules of mass M at temperature T. Fortunately the Maxwell–Boltzmann distribution[10] factorizes, in the sense that the fraction of all atoms in the range $\mathrm{d}u$ around u is proportional to $\exp(-Mu^2/2k_BT)\mathrm{d}u$, irrespective of the components of the velocity in the y and z directions. Because u is proportional to $\Delta\omega$, we see that the fraction of atoms subject to a Doppler shift in the range $\mathrm{d}\Delta\omega$ around $\Delta\omega$ corresponding to $\mathrm{d}\xi$ around ξ is

$$F\mathrm{d}\xi = a\sqrt{\pi}\exp(-\xi^2/a^2)\mathrm{d}\xi, \qquad a \equiv \sqrt{2k_BT/Mc^2}. \tag{13.23}$$

The prefactor $a\sqrt{\pi}$ has been chosen to satisfy the norming condition $\int_{-\infty}^{\infty} F\mathrm{d}\xi = 1$ (since every atom must have *some* value of ξ).

For rough orientation we compare the Doppler frequency spread $\Gamma_D = \omega_0 a$ with the natural width Γ_0 of the above-mentioned transition in hydrogen. Since $k_B \approx 8.6 \times 10^{-5}$ eV/K, while $Mc^2 \approx 940$ MeV (practically the same as the rest energy of the proton), one has

$$\Gamma_D \approx 1.6 \times 10^{16}\sqrt{2(8.6 \times 10^{-5})300/(940 \times 10^6)} \times \sqrt{T/300} = 3.7 \times 10^{10}\sqrt{T/300} \ \text{s}^{-1}.$$

But near room temperature this exceeds Γ_0 by a factor close to 100, so that the Doppler width is indeed dominant.

Equation (13.22) can be verified through (13.23) by measuring, as a function of frequency $f = \omega/2\pi$, either (i) the intensity radiated per unit frequency range by a low-pressure gas of excited atoms; or (ii) the energy absorbed per unit frequency range by the same gas of unexcited atoms, when illuminated by light uniformly distributed in frequency over the width of the line. At given f, the rates of the mutually inverse processes of emission and absorption are proportional to each other; and both are proportional to the number of atoms having velocities such that light with frequency f relative to the laboratory appears to them at their natural frequency f_0. Both curves should be proportional to F, i.e. Gaussians of width a; the absolute values of the quantity measured (and thus of the area under the curve) are not to the purpose.

Figure 13.4 records a test[11] by method (ii), using an absorption line with $\lambda_0 = c/f_0 = 780$ nm in ^{85}Rb atoms, at $T = 293$ K. It plots the intensity of transmitted light divided by its value well away from the spectral line, as a function of the detuning $(f - f_0)$. The absorption is weak (less than 10%) because the concentration is kept low in order to minimize pressure broadening. The total amount of gas is also kept low, in order to minimize complications from multiple interactions between the incident light and the target atoms. The dots are measured values. The curve is

$$1 - \exp[-(f - f_0)/f_0 a], \qquad f_0 a = f_0 \sqrt{2k_B T/Mc^2} = 306 \text{ MHz.,}$$

as dictated by (13.23). Evidently the agreement is satisfactory.

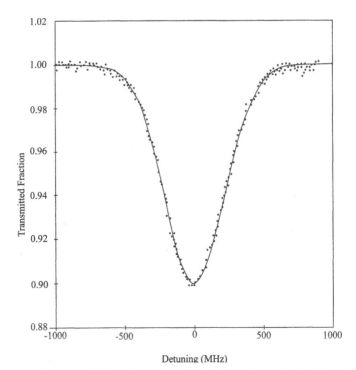

Figure 13.4   Gaussian spectral line, observed through absorption. For details see the text.

13.5.2 Second-Order Shift

It appears impossible to verify (13.16) to order $(v/c)^2$ by a direct measurement of $\Delta\omega$, because the terms of order v/c are so much larger,[12] and also because the velocity v itself cannot be determined accurately enough. Nevertheless verification is important because it is precisely the quadratic terms that depend on time dilation, and thereby on the specifically relativistic features of the theory. The design that makes spectroscopic tests possible goes back to Ives and Stilwell; here we outline a more recent experiment.[13]

The idea is to use as sources a beam of excited atoms; accept light emitted as near as makes no difference forwards and backwards; measure the corresponding wavelengths λ_F and λ_B very accurately with an interferometer; determine the source speed v indirectly from the difference $\lambda_B - \lambda_F$; and to use this v in the expression for the difference between the average $(\lambda_F + \lambda_B)/2$ and λ_0, which tests the theory.

Since the experiment measures wavelengths, we start by re-expressing the predicted shifts as

$$\frac{\lambda_{F,B}}{\lambda_0} = \frac{\omega_0}{\omega_{F,B}} = \frac{1 \mp (v/c)}{\sqrt{1 - (v/c)^2}} = 1 \mp (v/c) + \frac{1}{2}(v/c)^2 + \cdots, \tag{13.24}$$

where upper and lower signs refer respectively to Forwards ($\cos\phi = 1$) and Backwards ($\cos\phi = -1$). Note that in this ratio of wavelengths the quadratic and higher-order terms stem *wholly* from the square root, i.e. from time dilation. Evidently

$$\Delta\lambda \equiv \lambda_B - \lambda_F = 2\lambda_0(v/c). \tag{13.25}$$

Also

$$\delta\lambda \equiv (\lambda_F + \lambda_B)/2 - \lambda_0 = K\lambda_0(v/c)^2 = (K/4\lambda_0)(\Delta\lambda)^2, \tag{13.26}$$

where (v/c) is to be found from (13.25), and where the value of the constant K tests the theory:

$$\text{with time dilation: } K = 1/2; \qquad \text{without time dilation: } K = 0. \tag{13.27}$$

The layout is sketched in Figure 13.5. The collimated beam of excited hydrogen atoms is formed from "canal rays", by accelerating charged H_2^+ and H_3^+ ions out of a gas discharge through a known potential difference W (up to 76 kV); some of the molecular ions collide with background gas, and disintegrate in such a way that one of their constituents picks up a neutralizing electron and continues with very little change of velocity as an excited hydrogen atom. These atoms are the light sources. The beam passes though small holes in two primary mirrors, which reflect light emitted (nearly) backwards and (nearly) forwards away from the atomic beam. The light from both primary mirrors is then guided by other mirrors and lenses (not shown) into the entry slit of the spectrograph. The alignment of the primary mirrors (as of the entire optical system) is critical: they serve to define the effective observation angles $\phi_F = \phi$ and $\phi_B = \pi - \phi$, where ϕ must be accurately the same in both. In this experiment ϕ is so small that $\cos\phi \approx 0.9972$.

The beam contains hydrogen atoms with two different speeds v, depending on the mass $\mu = 2M$ or $\mu = 3M$ of the parent ion (where M is the mass of the proton):

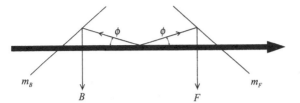

Figure 13.5 Crude sketch of arrangements in the experiment of Mandelberg and Witten. Heavy arrow: the atomic beam. The (near) forward and backward light rays B, F, coming from a central viewing region, are reflected away from the beam by the mirrors m_B and m_F. The mirrors are drilled to allow passage for the beam. The angles ϕ are shown much exaggerated.

evidently $\mu v^2/2 = |e|W$. Thus (check!) v/c is of order 10^{-2}. Observations were made on the H_α line, with $\lambda_0 = 6562\,\text{Å}$. (Recall $1\,\text{Å} = 10^{-10}\,$m.). Thus one expects the orders of magnitude $\Delta\lambda \sim (v/c)\lambda_0 \sim 100\,\text{Å}$ and $\delta\lambda \sim 1\,\text{Å}$. On the spectrograms each value of v yields two lines with wavelengths λ_F, λ_B; almost but not exactly halfway between them there is also a central line with wavelength λ_0, due to light from excited but stationary background gas. What is required is the wavelength difference $\delta\lambda$ between this central line and the centroid of the two flanking lines. One difficulty is that the widths of the individual lines are themselves of order $1\,\text{Å}$, i.e. comparable to $\delta\lambda$. Presumably this width is due to a velocity spread caused by the breakup of the parent ion.[14] Thus, in order to achieve the reported accuracy, it must have been necessary to determine the centre of each line with an uncertainty no greater than 1% of the line width, a notoriously difficult task, but in modern spectroscopy not necessarily impossible.

Figure 13.6 shows the observations of Mandelberg and Witten, together with some earlier ones confined to lower velocities. They quote their end-result as

$$K_{\text{exp}} = 0.498 \pm 0.025,$$

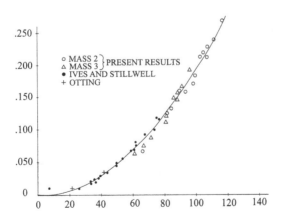

Figure 13.6 Experimental results on the second-order Doppler shift. Vertical axis: $\delta\lambda \equiv (\lambda_F + \lambda_B)/2 - \lambda_0$. Horizontal axis: $\Delta\lambda \equiv \lambda_B - \lambda_F$. Both scales in Å. The solid line is the theoretical prediction (13.26) with $K = 1/2$, namely $\delta\lambda = (\Delta\lambda)^2/8\lambda_0$. Reproduced from H. J. Mandelberg and L. Witten, "Experimental verification of the relativistic Doppler effect", *Journal of the Optical Society of America* (1962), volume **52**, by permission of the Optical Society of America.

which verifies the theory to within about 5%, and quite certainly rules out any possibility that $K = 0$, as suggested by the "nonrelativistic" theory.

13.6 Notes

1. Once they are familiar with the physics and embarked on calculations, most people adopt a less elaborate notation in order to save writing. For instance, one might well replace $\omega_E^{(E)} \to \omega_E$ and $\omega_R^{(R)} \to \omega_R$.

2. In such calculations one rarely meets any third frame S. If one does, with $\mathbf{u}_{R,E}$ referred to S, then one must remember that $\mathbf{v}$ is by no means the same as $\mathbf{u}_R - \mathbf{u}_E$. This is discussed in Example 7.4, where our present v would be called $u(R, E)$.

3. *A letter from the Reverend Mr.* James Bradley *Savilian Professor of Astronomy at* Oxford, *and F.R.S. to* Dr. Edmond Halley Astronom. Reg. &c. *giving an Account of a new discovered Motion of the Fix'd Stars. Philosophical Transactions of the Royal Society* (1729), volume **35**, page 637. Other details are given by A. B. Stewart, "The discovery of stellar aberration", *Scientific American* (March 1964), volume **210**, page 100.

4. This section identifies S with $S^{(E)}$ with S' with $S^{(R)}$.

5. In fact Bradley had set out to look for parallax, which would have yielded information about the distance from earth to star (also an important but a different question), rather than about the orbital speed of the earth; however, he found that any parallax was too small for him to measure. By today's goal-oriented criteria the work might well have been reckoned a failure, prejudicing his next research grant. Actually he appears to have paid for the telescope himself.

6. See D. Speiser, "L. Euler, the principle of relativity and the fundamentals of classical mechanics", *Nature* (1961), volume **190**, page 757.

7. See the remarks about closing speeds at the end of Section 4.2.4.

8. Recall that the time lapse between two events is least in the frame where they happen at the same place. Here this frame is $S^{(R)}$: hence the disposition of the square root.

9. This misleading but well-entrenched terminology was explained in Section 4.1.

10. In the nonrelativistic regime it entails the equipartition theorem, giving the mean-square value $M\langle u^2 \rangle/2 = k_B T/2 \Rightarrow \langle u^2/c^2 \rangle = k_B T/Mc^2$. Thus a^2 defined by (13.23) is comparable with the mean kinetic energy divided by the rest energy Mc^2.

11. The writer is grateful to Professor E. A. Hinds for so splendidly tailor-made an illustration, from the Sussex Centre for Optical and Atomic Physics. The temperature T is that of the Centre of an English summer afternoon.

12. Example 13.2 shows that attempts to measure the purely transverse shift fail through an inability to confine viewing angles close enough to $\pi/2$.

13. H. J. Mandelberg and L. Witten, "Experimental verification of the relativistic Doppler effect", *Journal of the Optical Society of America* (1962), volume **52**, page 529.

14. Call this velocity spread Dv (where D stands just for "difference", the usual symbols δ and Δ having been pre-empted for other kinds of differences). The order of magnitude of Dv can be estimated by noting that the breakup of the molecular ions and the subsequent electron capture are likely to involve energy transfers DE comparable with the binding energy of the electron in hydrogen, say up to $DE \sim 10\,\mathrm{eV}$. Hence we estimate $D(v^2/2M) = 2(Dv/v)(v^2/2M) \sim 10\,\mathrm{eV}$, whence $Dv/v \sim (1/2)(10\,\mathrm{eV})/(70\,\mathrm{keV}) \sim 10^{-4}$. This in turn suggests $D\lambda/\lambda \sim 10^{-4}$, so that $D\lambda \sim 10^{-4}\,(6000\,\mathring{A}) \sim 0.6\,\mathring{A}$.

Appendix A
Lorentz Transformations with Arbitrary Relative Velocity

Elsewhere in this book we use only the Lorentz transformation between two inertial frames S and S' in standard configuration. This configuration is simple in two respects. First, regardless of the velocity $\mathbf{v}$ of S' relative to S, the corresponding primed and unprimed axes are parallel. This means that the two frames become identical if $\mathbf{v} = \mathbf{0}$; in other words no rotation[1] is involved. Second, $\mathbf{v}$ points along the x (hence also along the x') axis; in other words $\mathbf{v} = v\hat{\mathbf{x}}$. We shall now find the transformation for the more general case where $\mathbf{v}$ can point in any direction; but we maintain the first simplification, i.e. corresponding axes are taken as parallel.

It proves convenient to introduce the unit vector $\hat{\mathbf{v}}$ along $\mathbf{v}$, and vector suffixes such that $r_\parallel$ is the component of $\mathbf{r}$ parallel to $\mathbf{v}$, while $\mathbf{r}_\perp$ embraces the two components of $\mathbf{r}$ perpendicular to $\mathbf{v}$; and similarly for primed vectors. In standard configuration for instance $r_\parallel = x$, while $\mathbf{r}_\perp = (y, z)$. Note that $\mathbf{r} = r_\parallel \hat{\mathbf{v}} + \mathbf{r}_\perp = \hat{\mathbf{v}}(\hat{\mathbf{v}} \cdot \mathbf{r}) + \mathbf{r}_\perp$, so that conversely $\mathbf{r}_\perp = \mathbf{r} - \hat{\mathbf{v}}(\hat{\mathbf{v}} \cdot \mathbf{r})$. With this notation the transformation reads

$$t' = \gamma(t - \mathbf{v} \cdot \mathbf{r}/c^2), \qquad r'_\parallel = \gamma(r_\parallel - vt), \qquad \mathbf{r}_\perp = \mathbf{r}'_\perp, \qquad (\text{A.1})$$

where $\gamma \equiv 1/\sqrt{1 - v^2/c^2}$. Equivalently,

$$t' = \gamma(t - \mathbf{v} \cdot \mathbf{r}/c^2), \qquad \mathbf{r}' = \mathbf{r} + \hat{\mathbf{v}}\{(\gamma - 1)\hat{\mathbf{v}} \cdot \mathbf{r} - \gamma vt\}. \qquad (\text{A.2})$$

The inverses are obtained as always by interchanging primed and unprimed and reversing the sign of $\mathbf{v}$ (and thereby of $\hat{\mathbf{v}}$).

Example A.1. Write down $(t', \mathbf{r}')$ when $\mathbf{v} = v(\cos\chi, \sin\chi, 0)$.

Solution. The velocity **v** is in the xy plane, at an angle χ to the x axis. Evidently $z' = z$. Also $\hat{\mathbf{v}} \cdot \mathbf{r} = x \cos \chi + y \sin \chi$, so that

$$t' = \gamma\{t - (v/c)(x \cos \chi + y \sin \chi)\},$$

$$\mathbf{r}' = \mathbf{r} + (\cos \chi, \sin \chi, 0)\{(\gamma - 1)(x \cos \chi + y \sin \chi) - \gamma v t\},$$

or more explicitly

$$x' = x\{\sin^2 \chi + \gamma \cos^2 \chi\} + y(\gamma - 1) \sin \chi \cos \chi - t \gamma v \cos \chi,$$

$$y' = x(\gamma - 1) \sin \chi \cos \chi + y\{\cos^2 \chi + \gamma \sin^2 \chi\} - t \gamma v \sin \chi. \quad \blacksquare$$

Similarly, (4.15) for velocities may be written

$$u'_\| = \frac{u_\| - v}{[1 - \mathbf{v} \cdot \mathbf{u}/c^2]}, \qquad \mathbf{u}'_\perp = \frac{\mathbf{u}_\perp}{\gamma[1 - \mathbf{v} \cdot \mathbf{u}/c^2]}, \qquad (A.3)$$

merging into

$$\mathbf{u}' = \frac{\mathbf{u} - \gamma \mathbf{v} + (\gamma - 1)\hat{\mathbf{v}}(\hat{\mathbf{v}} \cdot \mathbf{u})}{\gamma[1 - \mathbf{v} \cdot \mathbf{u}/c^2]}. \qquad (A.4)$$

The reader should verify that this reproduces (4.20) for u'^2.

Example A.2. Let **L** be the orbital angular momentum of a particle around the origin. From Chapter 9 it can be seen that

$$\boldsymbol{\lambda} \equiv \mathbf{L}/m = \mathbf{r} \times \mathbf{p}/m = \gamma(u)\boldsymbol{\chi}, \qquad \boldsymbol{\chi} \equiv \mathbf{r} \times \mathbf{u},$$

where the rest mass m of the particle is an invariant constant. Derive the Lorentz transformation for $\boldsymbol{\lambda}$.

Solution. Start by transforming $\boldsymbol{\chi}$. From (A.2) and (A.4),

$$\boldsymbol{\chi}' = \mathbf{r}' \times \mathbf{u}' = \frac{[\mathbf{r} + \hat{\mathbf{v}}\{(\gamma - 1)\hat{\mathbf{v}} \cdot \mathbf{r} - \gamma v t\}] \times [\mathbf{u} - \gamma \mathbf{v} + (\gamma - 1)\hat{\mathbf{v}}(\hat{\mathbf{v}} \cdot \mathbf{u})]}{\gamma[1 - \mathbf{v} \cdot \mathbf{u}/c^2]}$$

$$= \frac{\boldsymbol{\chi} + \gamma \mathbf{v} \times (\mathbf{r} - \mathbf{u}t) + (\gamma - 1)\hat{\mathbf{v}} \times \{(\hat{\mathbf{v}} \cdot \mathbf{r})\mathbf{u} - (\hat{\mathbf{v}} \cdot \mathbf{u})\mathbf{r}\}}{\gamma[1 - \mathbf{v} \cdot \mathbf{u}/c^2]},$$

where γ is shorthand for $\gamma(v)$. But standard vector identities entail

$$\hat{\mathbf{v}} \times \{(\hat{\mathbf{v}} \cdot \mathbf{r})\mathbf{u} - (\hat{\mathbf{v}} \cdot \mathbf{u})\mathbf{r}\} = \hat{\mathbf{v}} \times \{-\hat{\mathbf{v}} \times (\mathbf{r} \times \mathbf{u})\} = -\hat{\mathbf{v}} \times \{\hat{\mathbf{v}} \times \boldsymbol{\chi}\} = \boldsymbol{\chi} - \hat{\mathbf{v}}(\hat{\mathbf{v}} \cdot \boldsymbol{\chi}),$$

whence

$$\boldsymbol{\chi}' = \frac{\boldsymbol{\chi} - (1 - 1/\gamma(v))\hat{\mathbf{v}}(\hat{\mathbf{v}} \cdot \boldsymbol{\chi}) + \mathbf{v} \times (\mathbf{r} - \mathbf{u}t)}{[1 - \mathbf{v} \cdot \mathbf{u}/c^2]}$$

$$= \frac{\boldsymbol{\lambda}/\gamma(u) - (1 - 1/\gamma(v))\hat{\mathbf{v}}(\hat{\mathbf{v}} \cdot \boldsymbol{\lambda})/\gamma(u) + \mathbf{v} \times (\mathbf{r} - \mathbf{u}t)}{[1 - \mathbf{v} \cdot \mathbf{u}/c^2]}.$$

From (4.19) we have $\gamma(u') = \gamma(v)\gamma(u)[1 - \mathbf{v} \cdot \mathbf{u}/c^2]$, and substitution yields

$$\boldsymbol{\lambda}' = \gamma(u')\boldsymbol{\chi}' = \gamma(v)\{\boldsymbol{\lambda} - [1 - 1/\gamma(v)]\hat{\mathbf{v}} \cdot (\hat{\mathbf{v}} \cdot \boldsymbol{\lambda}) + \gamma(u)\mathbf{v} \times (\mathbf{r} - \mathbf{u}t)\}. \quad \blacksquare$$

Finally, and with some labour, (4.26) and (4.28) for accelerations may be re-written as

$$a'_\parallel = \frac{a_\parallel}{\gamma^3[1 - \mathbf{v}\cdot\mathbf{u}/c^2]^3}, \qquad a'_\perp = \frac{\mathbf{a}_\perp + [\mathbf{v}\times(\mathbf{u}\times\mathbf{a})]/c^2}{\gamma^2[1-\mathbf{v}\cdot\mathbf{u}/c^2]^3}, \tag{A.5}$$

eventually merging into

$$\mathbf{a}' = \frac{1}{\gamma^2[1-\mathbf{v}\cdot\mathbf{u}/c^2]^3}\left\{\mathbf{a} + \left(\frac{1}{\gamma}-1\right)\hat{\mathbf{v}}(\hat{\mathbf{v}}\cdot\mathbf{a}) + \frac{1}{c^2}\mathbf{v}\times(\mathbf{u}\times\mathbf{a})\right\}, \tag{A.6}$$

where $\gamma \equiv \gamma(v)$. The reader should check that this tallies with Problem 4.11 in the special case where $\mathbf{v}, \mathbf{u}, \mathbf{a}$ are coplanar.

Note

1. "Rotation" here means a *time-independent* change of orientation of the primed and unprimed axes relative to each other.

Appendix B
Vectors, Four-Vectors, and Transformation Matrices

Facility with four-vectors grows best from experience with ordinary vectors; but first one needs to look at ordinary vectors (just *vectors* from now on) in a way somewhat different from the usual elementary one, and more readily generalized. Sections B.1 and B.2 do just that. Section B.3 exploits this approach to show that as far as the formalities are concerned, four-vectors behave under Lorentz transformations very much like vectors behave under rotations in ordinary space. Moreover, the analogies between them suggest efficient methods for parametrizing and combining transformations; this is illustrated in section B.4.

B.1 What is a Vector? (Definition and Representations)

Vectors are usually introduced as quantities having direction as well as magnitude, in contrast to scalars, which have magnitude only. Although this definition goes a long way in practice, closer scrutiny shows that it lacks precision and is difficult to generalize.[1] To make headway one needs to re-define vectors through the way the values of their components change when we change the orientation of our coordinate axes, i.e. through *their transformation properties under rotations.*

The prototype of a vector is the *position vector* **r** of a point P relative to another point O chosen as the origin, or in other words the directed line segment OP from O to P. We regard it as a physical quantity in its own right, taken as fixed throughout this discussion. For simplicity we consider only two-dimensional space, i.e. a plane; and start by choosing x and y axes, arbitrarily. The Cartesian coordinates

$$r_1 \equiv x, \qquad r_2 \equiv y$$

of $\mathbf{r}$ are the distances shown in Figure B.1(a), from O to the perpendicular projections P_1, P_2 of P onto the axes. The figure also indicates the length r of the position vector, and the angle ϕ it makes with the x axis, i.e. its polar coordinates:

$$r \cos \phi = r_1, \qquad r \sin \phi = r_2, \qquad r = \sqrt{r_1^2 + r_2^2}. \tag{B.1}$$

With $\mathbf{r}$ we associate a 2×1 matrix r, (generally if confusingly called a *column vector*), and its matrix transpose r^T (also called a *row vector*):

$$\mathsf{r} \equiv \begin{bmatrix} r_1 \\ r_2 \end{bmatrix}, \qquad \mathsf{r}^\mathsf{T} \equiv [r_1, \ r_2]; \tag{B.2}$$

row and column matrices will be printed in sans serif, and the superfix T always identifies the transposed matrix. One says that the matrix r *represents* the position vector $\mathbf{r}$ with respect to the axes one has adopted. Because $\mathbf{r}$ and r carry the same information, the physicist tends not to distinguish between them unless he has to, referring to and treating them as if they were identical. Mathematically however they are distinct, and one needs to bear this in mind, if only because we are about to study situations where the values of $r_{1,2}$ change, so that r changes, while the position vector $\mathbf{r}$ remains the same.

Since it is the Cartesian coordinates that prove the most illuminating as regards four-vectors, we focus on them, and treat the polar coordinates as mere auxiliaries.

Obviously the values of the coordinates depend on the directions chosen for the axes. If we adopt new (primed) axes, turned through an angle χ relative to the unprimed as shown in Figure B.1(b), then with respect to the new axes the same $\mathbf{r}$ is represented by a new matrix, call it r$'$, i.e. by new coordinates $r_{1,2}'$, or equivalently by new (r', ϕ'). We determine them from the figure. Inspection shows directly that

$$r' = r, \qquad \phi' = \phi - \chi. \tag{B.3}$$

The first equality is fundamental, and is expressed by saying that *the lengths of position vectors are invariant*:

$$r = \sqrt{r_1^2 + r_2^2} = \sqrt{r_1'^2 + r_2'^2} = r'. \tag{B.4}$$

Accordingly we can dispense with the primes on r' unless they are wanted for emphasis.

The primed Cartesian coordinates are given by

$$r_1' = r' \cos \phi' = r' \cos(\phi - \chi) = r[\cos \phi \cos \chi + \sin \phi \sin \chi] = \cos(\chi) r_1 + \sin(\chi) r_2,$$

$$r_2' = r' \sin \phi' = r' \sin(\phi - \chi) = r[-\cos \phi \sin \chi + \sin \phi \cos \chi] = -\sin(\chi) r_1 + \cos(\chi) r_2.$$

In order to appreciate what is going on, it is not only convenient but essential to concentrate these relations into matrix form:

$$\begin{bmatrix} r_1' \\ r_2' \end{bmatrix} = \begin{bmatrix} \cos \chi, & \sin \chi \\ -\sin \chi & \cos \chi \end{bmatrix} \begin{bmatrix} r_1 \\ r_2 \end{bmatrix}. \tag{B.5}$$

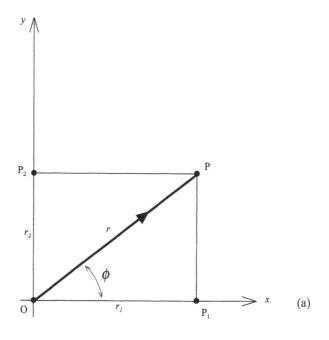

(a)

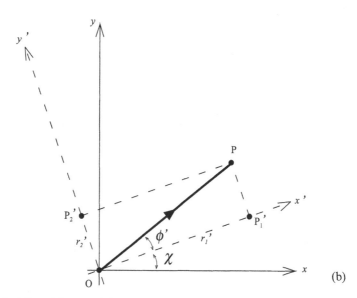

(b)

Figure B.1 The origin O and the point P remain fixed. The directed line segment from O to P, regarded as a physical object, is the position vector under discussion. Naturally it too remains fixed. (a) The directions of the x and y axes have been chosen arbitrarily. The point $P_1(P_2)$ is the foot of the perpendicular through P to the $x(y)$ axis. Note the polar angle ϕ, and the distances $OP = r$, $OP_1 = r_1$, $OP_2 = r_2$. (b) The primed axes have been rotated through an angle χ relative to the unprimed. The point $P_1'(P_2')$ is the foot of the perpendicular through P to the $x'(y')$ axis. Note the polar angle $\phi' = \phi - \chi$, and the distances $OP = r' = r$, $OP_1' = r_1'$, $OP_2' = r_2'$. The relation of $r_{1,2}'$ to $r_{1,2}$ is given in the text.

With Script capitals reserved for square matrices, we define the *rotation matrix*

$$\mathcal{R}(\chi) \equiv \begin{bmatrix} \cos \chi, & \sin \chi \\ -\sin \chi, & \cos \chi \end{bmatrix}, \tag{B.6}$$

and write (B.5) succinctly as

$$\mathbf{r}' = \mathcal{R}(\chi)\mathbf{r}. \tag{B.7}$$

Equation (B.7) shows how coordinates transform under rotations: meaning that it relates the coordinates found by two observers who measure the same **r**, but refer their results to the unprimed and to the primed axes respectively.

We are now in a position to give a practicable and generalizable definition of (two-dimensional) vectors generally:

Any pair of physical quantities v_1 and v_2 constitute a vector **v**, *represented by a column matrix* **v**, *if under rotations of the coordinate axes they transform in the same way as do the coordinates, i.e. if*

$$\begin{bmatrix} v_1' \\ v_2' \end{bmatrix} = \begin{bmatrix} \cos \chi, & \sin \chi \\ -\sin \chi & \cos \chi \end{bmatrix} \begin{bmatrix} v_1 \\ v_2 \end{bmatrix} \Rightarrow \mathbf{v}' = \mathcal{R}(\chi)\mathbf{v}. \tag{B.8}$$

Very properly, this definition leaves it up to the physicist to identify vectors: how observables transform depends on their physical properties, or on the properties of other physical quantities in terms of which they are defined.

B.2 Rotation Matrices

Suppose the axes are rotated first through an angle χ_1, transforming **v** into **v**$'$, and afterwards through χ_2, transforming **v**$'$ into **v**$''$. Then the transformation from **v** to **v**$''$ is evidently given by

$$\mathbf{v}'' = \mathcal{R}(\chi_2)\mathbf{v}' = \mathcal{R}(\chi_2)\mathcal{R}(\chi_1)\mathbf{v},$$

featuring the matrix product

$$\mathcal{R}(\chi_2)\mathcal{R}(\chi_1) = \begin{bmatrix} \cos \chi_2, & \sin \chi_2 \\ -\sin \chi_2, & \cos \chi_2 \end{bmatrix} \begin{bmatrix} \cos \chi_1, & \sin \chi_1 \\ -\sin \chi_1, & \cos \chi_1 \end{bmatrix}$$

$$= \begin{bmatrix} (\cos \chi_2 \cos \chi_1 - \sin \chi_2 \sin \chi_1), & (\cos \chi_2 \sin \chi_1 + \sin \chi_2 \cos \chi_1) \\ -(\sin \chi_2 \cos \chi_1 + \cos \chi_2 \sin \chi_1), & (\cos \chi_2 \cos \chi_1 - \sin \chi_2 \sin \chi_1) \end{bmatrix}$$

$$= \begin{bmatrix} \cos(\chi_2 + \chi_1), & \sin(\chi_2 + \chi_1) \\ -\sin(\chi_2 + \chi_1), & \cos(\chi_2 + \chi_1) \end{bmatrix} = \mathcal{R}(\chi_2 + \chi_1). \tag{B.9}$$

This is as it should be, because the two rotations carried out in successsion are identically the same[2] as a single rotation through the sum of the two angles.

Since a rotation through zero angle changes nothing, it is obvious that

$$\mathcal{R}(0) = 1 \equiv \begin{bmatrix} 1, & 0 \\ 0, & 1 \end{bmatrix},$$

where the "1" in the middle is shorthand for the unit matrix, also called the identity. Further, every rotation $\mathcal{R}(\chi)$ has an inverse,[3] namely the rotation through $-\chi$:

$$\mathcal{R}^{-1}(\chi) = \mathcal{R}(-\chi) = \begin{bmatrix} \cos\chi, & -\sin\chi \\ \sin\chi, & \cos\chi \end{bmatrix} = \mathcal{R}^{\mathrm{T}}(\chi), \qquad (B.10)$$

where

$$\mathcal{R}^{-1}(\chi)\mathcal{R}(\chi) = \mathcal{R}(\chi)\mathcal{R}^{-1}(\chi) = 1.$$

The relation

$$\mathcal{R}^{-1} = \mathcal{R}^{\mathrm{T}} \Rightarrow \mathcal{R}^{\mathrm{T}}\mathcal{R} = \mathcal{R}\mathcal{R}^{\mathrm{T}} = 1 \qquad (B.11)$$

is crucial, and reflects the invariance of r. To appreciate this, we first define the *scalar product* of any two vectors **a** and **b** as

$$\mathbf{a}\cdot\mathbf{b} \equiv \mathsf{a}^{\mathrm{T}}\mathsf{b} = \mathsf{b}^{\mathrm{T}}\mathsf{a} = a_1 b_1 + a_2 b_2, \qquad (B.12)$$

where the second expression is the ordinary matrix product

$$[a_1,\ a_2]\begin{bmatrix} b_1 \\ b_2 \end{bmatrix}.$$

Calling it a scalar implies that it is invariant under rotations, i.e. that

$$\mathsf{a}^{\mathrm{T}}\mathsf{b} = a_1 b_1 + a_2 b_2 = \mathsf{a}'^{\mathrm{T}}\mathsf{b}' = a_1' b_1' + a_2' b_2'. \qquad (B.13)$$

This follows directly from (B.11) in view of the transformation rule $\mathsf{v}' = \mathcal{R}\mathsf{v}$, which entails $\mathsf{v}'^{\mathrm{T}} = \mathsf{v}^{\mathrm{T}}\mathcal{R}^{\mathrm{T}}$:

$$\mathsf{a}'^{\mathrm{T}}\mathsf{b}' = \mathsf{a}^{\mathrm{T}}\mathcal{R}^{\mathrm{T}}\mathcal{R}\mathsf{b} = \mathsf{a}^{\mathrm{T}}1\mathsf{b} = \mathsf{a}^{\mathrm{T}}\mathsf{b}. \qquad (B.14)$$

In particular, we see that $\mathbf{v}\cdot\mathbf{v} = \mathsf{v}^{\mathrm{T}}\mathsf{v} = v_1^2 + v_2^2$ is invariant for any vector **v**. At first sight the invariance of the length r of the coordinate vector **r** appears to be just the special case with $\mathbf{a} = \mathbf{b} = \mathbf{r}$ and $\mathbf{a}\cdot\mathbf{b} = \mathbf{r}\cdot\mathbf{r} = r^2$. But in fact vectors are *defined* through the transformation rules (B.6), (B.7), which were derived by using the invariance of r to begin with.

B.3 Four-Vectors

Consider the coordinates $(X_0 \equiv ct, \mathbf{X} \equiv \mathbf{r})$ of an event, referred to an inertial frame S, and to some origin and axes chosen in S. With these we associate a 4×1 column matrix X and its transpose X^{T}:

$$\mathsf{X} \equiv \begin{bmatrix} X_0 \\ X_1 \\ X_2 \\ X_3 \end{bmatrix}, \qquad \mathsf{X}^{\mathrm{T}} = [X_0, X_1, X_2, X_3]. \qquad (B.15)$$

Lorentz transformations between frames in standard configuration read

$$\mathsf{X}' = \begin{bmatrix} X_0' \\ X_1' \\ X_2' \\ X_3' \end{bmatrix} = \begin{bmatrix} \gamma, & -\gamma\beta, & 0, & 0 \\ -\gamma\beta, & \gamma, & 0, & 0 \\ 0, & 0, & 1, & 0 \\ 0, & 0, & 0, & 1 \end{bmatrix} \begin{bmatrix} X_0 \\ X_1 \\ X_2 \\ X_3 \end{bmatrix} \equiv \mathcal{L}\mathsf{X}, \qquad (\text{B.16})$$

which defines the 4×4 *transformation matrix* $\mathcal{L}$. Since $X_{2,3}$ remain untouched, we ignore them for brevity whenever possible, and write

$$\mathsf{X} = \begin{bmatrix} X_0 \\ X_1 \end{bmatrix}, \qquad \mathsf{X}' = \mathcal{L}\mathsf{X}, \qquad \mathcal{L} = \begin{bmatrix} \gamma, & -\gamma\beta \\ -\gamma\beta, & \gamma \end{bmatrix}. \qquad (\text{B.17})$$

By analogy with (B.8) we adopt the definition *that any set of four physical quantities V_0, V_1, V_2, V_3 constitutes a four-vector $\vec{V}$, represented by a column matrix V, if they Lorentz-transform in the same way as event coordinates do, i.e. if*

$$\mathsf{V}' = \mathcal{L}\mathsf{V}. \qquad (\text{B.18})$$

Since Lorentz transformations leave $X_0^2 - \mathbf{X}^2$ invariant, we are led to look for a way of writing this combination in a convenient matrix form. To this end we define[4]

$$\mathcal{G} \equiv \begin{bmatrix} 1, & 0, & 0, & 0 \\ 0, & -1, & 0, & 0 \\ 0, & 0, & -1, & 0 \\ 0, & 0, & 0, & -1 \end{bmatrix}, \qquad \left(\text{or } \mathcal{G} = \begin{bmatrix} 1, & 0 \\ 0, & -1 \end{bmatrix} \text{ in the } 2 \times 2 \text{ version} \right). \quad (\text{B.19})$$

Notice that $\mathcal{G}$ is symmetric and that its square is the unit matrix:

$$\mathcal{G}^{\mathrm{T}} = \mathcal{G}, \quad \mathcal{G}^{\mathrm{T}}\mathcal{G} = \mathcal{G}^2 = 1 \Rightarrow \mathcal{G}^{\mathrm{T}} = \mathcal{G}^{-1}. \qquad (\text{B.20})$$

Now

$$X_0^2 - \mathbf{X}^2 = (X_0, \mathbf{X}) \begin{pmatrix} X_0 \\ -\mathbf{X} \end{pmatrix} = (X_0, \mathbf{X}) \begin{bmatrix} 1, & 0 \\ 0, & -1 \end{bmatrix} \begin{pmatrix} X_0 \\ \mathbf{X} \end{pmatrix} = \mathsf{X}^{\mathrm{T}}\mathcal{G}\mathsf{X}; \qquad (\text{B.21})$$

while $\mathsf{X}' = \mathcal{L}\mathsf{X}$ and $\mathsf{X}'^{\mathrm{T}} = \mathsf{X}^{\mathrm{T}}\mathcal{L}^{\mathrm{T}}$ lead to

$$X_0'^2 - \mathbf{X}'^2 = \mathsf{X}'^{\mathrm{T}}\mathcal{G}\mathsf{X}' = \mathsf{X}^{\mathrm{T}}\mathcal{L}^{\mathrm{T}}\mathcal{G}\mathcal{L}\mathsf{X}. \qquad (\text{B.22})$$

Thus the invariance property under study entails

$$X_0'^2 - \mathbf{X}'^2 = X_0^2 - \mathbf{X}^2 \Rightarrow \mathsf{X}^{\mathrm{T}}\mathcal{L}^{\mathrm{T}}\mathcal{G}\mathcal{L}\mathsf{X} = \mathsf{X}^{\mathrm{T}}\mathcal{G}\mathsf{X},$$

and because this is true for *any* X, it requires

$$\mathcal{L}^{\mathrm{T}}\mathcal{G}\mathcal{L} = \mathcal{G} \Rightarrow \mathcal{G}\mathcal{L}^{\mathrm{T}}\mathcal{G}\mathcal{L}\mathcal{L}^{-1} = \mathcal{G}\mathcal{G}\mathcal{L}^{-1} \Rightarrow \mathcal{G}\mathcal{L}^{\mathrm{T}}\mathcal{G} = \mathcal{L}^{-1}. \qquad (\text{B.23})$$

Accordingly, we can now define the four-scalar product of any two four-vectors $\vec{A}, \vec{B}$ as

$$\vec{A} \cdot \vec{B} \equiv \mathsf{A}^{\mathrm{T}}\mathcal{G}\mathsf{B} = \mathsf{B}^{\mathrm{T}}\mathcal{G}\mathsf{A} = A_0 B_0 - \mathbf{A} \cdot \mathbf{B} = A_0 B_0 - A_1 B_1 - A_2 B_2 - A_3 B_3, \quad (\text{B.24})$$

which is indeed a scalar, i.e. an invariant, since $\mathsf{A}'^{\mathrm{T}}\mathcal{G}\mathsf{B}' = \mathsf{A}^{\mathrm{T}}\mathcal{L}^{\mathrm{T}}\mathcal{G}\mathcal{L}\mathsf{B} = \mathsf{A}^{\mathrm{T}}\mathcal{G}\mathsf{B}$ in virtue of the first of equations (B.23).

Example B.1. Verify the general relation (B.23) explicitly for the special case of standard configuration.

Solution. Use[5] (B.19) and (B.20). On the right of (B.23),

$$\mathcal{L}^{-1}(\beta) = \mathcal{L}(-\beta) \qquad \text{(standard configuration)},$$

as is easily checked:

$$\mathcal{L}(-\beta)\mathcal{L}(\beta) = \begin{bmatrix} \gamma, & \gamma\beta \\ \gamma\beta, & \gamma \end{bmatrix} \begin{bmatrix} \gamma, & -\gamma\beta \\ -\gamma\beta, & \gamma \end{bmatrix}$$

$$= \begin{bmatrix} (\gamma^2 - \gamma^2\beta^2), & (-\gamma^2\beta + \gamma^2\beta) \\ (-\gamma^2\beta + \gamma^2\beta), & (\gamma^2 - \gamma^2\beta^2) \end{bmatrix} = \begin{bmatrix} 1, & 0 \\ 0, & 1 \end{bmatrix} \equiv 1.$$

But this tallies with the expression on the left of (B.23), where

$$\mathcal{G}\mathcal{L}^{\mathrm{T}}(\beta)\mathcal{G} = \begin{bmatrix} 1, & 0 \\ 0, & -1 \end{bmatrix} \begin{bmatrix} \gamma, & -\gamma\beta \\ -\gamma\beta, & \gamma \end{bmatrix} \begin{bmatrix} 1, & 0 \\ 0, & -1 \end{bmatrix}$$

$$= \begin{bmatrix} 1, & 0 \\ 0, & -1 \end{bmatrix} \begin{bmatrix} \gamma, & \gamma\beta \\ -\gamma\beta, & -\gamma \end{bmatrix} = \begin{bmatrix} \gamma, & \gamma\beta \\ \gamma\beta, & \gamma \end{bmatrix} = \mathcal{L}(-\beta). \quad \blacksquare$$

The analogies between ordinary vectors and four-vectors become even clearer if one rewrites $\mathcal{L}$ in terms of the *rapidity* ρ. This step is suggested by observing that the matrix elements γ and $\gamma\beta$ of $\mathcal{L}$ in standard configuration satisfy $\gamma^2 - (\gamma\beta)^2 = 1$, reminiscent of $\cosh^2\rho - \sinh^2\rho = 1$. Accordingly we define ρ by

$$\gamma = \cosh\rho, \quad \gamma\beta = \sinh\rho, \quad \beta = \tanh\rho, \tag{B.25}$$

and see that

$$\mathcal{L} = \begin{bmatrix} \cosh\rho, & -\sinh\rho \\ -\sinh\rho, & \cosh\rho \end{bmatrix}. \tag{B.26}$$

This is remarkably like (B.6), except for hyperbolic instead of trigonometric functions, and the same instead of opposite signs for the off-diagonal elements; the differences are dictated by the change of relative sign in the invariants, $X_0^2 - X_1^2$ instead of $X_1^2 + X_2^2$.

The rapidity is especially useful for combining successive collinear Lorentz transformations:

$$\mathcal{L}(\rho_2)\mathcal{L}(\rho_1) = \begin{bmatrix} \cosh\rho_2, & -\sinh\rho_2 \\ -\sinh\rho_2, & \cosh\rho_2 \end{bmatrix} \begin{bmatrix} \cosh\rho_1, & -\sinh\rho_1 \\ -\sinh\rho_1, & \cosh\rho_1 \end{bmatrix}$$

$$= \begin{bmatrix} (\cosh\rho_2\cosh\rho_1 + \sinh\rho_2\sinh\rho_1), & -(\cosh\rho_2\sinh\rho_1 + \sinh\rho_2\cosh\rho_1) \\ -(\sinh\rho_2\cosh\rho_1 + \cosh\rho_2\sinh\rho_1), & (\cosh\rho_2\cosh\rho_1 + \sinh\rho_2\sinh\rho_1) \end{bmatrix}$$

$$= \begin{bmatrix} \cosh(\rho_2 + \rho_1), & -\sinh(\rho_2 + \rho_1) \\ -\sinh(\rho_2 + \rho_1), & \cosh(\rho_2 + \rho_1) \end{bmatrix} = \mathcal{L}(\rho_2 + \rho_1), \tag{B.27}$$

in perfect analogy to (B.9). Incidentally we gain an elegant version of the collinear particle-velocity combination rule: velocity β_1 with respect to S' combines with velocity β_2 of S' with respect to S so as to yield with respect to S a resultant velocity β_{21} given by

$$\rho_{21} = \rho_2 + \rho_1 \;\Rightarrow\; \beta_{21} = \tanh(\tanh^{-1}\beta_2 + \tanh^{-1}\beta_1). \tag{B.28}$$

By virtue of the identity

$$\tanh(a+b) = (\tanh a + \tanh b)/(1 + \tanh a \tanh b),$$

(B.28) is indeed equivalent to the familiar but somewhat clumsier

$$\beta_{21} = (\beta_2 + \beta_1)/(1 + \beta_2\beta_1).$$

Example B.2. Use rapidities to re-derive the collinear Lorentz transformation (4.18) of γ.

Solution. The answer is immediate once it is realized that in the notation used above, the requisite "γ" is just γ_{21}:

$$1/\gamma_{21}^2 = 1/\cosh^2\rho_{21} = 1 - \tanh^2\rho_{21} = 1 - \tanh^2(\rho_2 + \rho_1)$$

$$= 1 - \left(\frac{\tanh\rho_2 + \tanh\rho_1}{1 + \tanh\rho_2\tanh\rho_1}\right)^2 = 1 - \left(\frac{\beta_2 + \beta_1}{1 + \beta_2\beta_1}\right)^2$$

$$= \frac{(1 - \beta_2^2)(1 - \beta_1^2)}{(1 + \beta_2\beta_1)^2}. \quad\blacksquare$$

B.4 Three Space-Dimensions

Both for rotations and for Lorentz transformations the physics becomes incomparably richer in more than the minimal number of dimensions, which are all that we have considered so far (namely two space dimensions and one time plus one space dimension respectively). Correspondingly the mathematics becomes more intricate, essentially because rotations around two different axes, or Lorentz transformations in two different directions, do not commute, meaning that their resultant depends on which is performed first. To study the consequences with any generality is far beyond our scope, and we settle for the simplest possible illustration in each case, using only transformations that differ very little from the identity.

B.4.1 Rotations

Consider two rotations, one through χ_1 around the x axis, i.e. in the yz plane, and the other through χ_2 around the y axis, i.e. in the zx plane. They and their inverses are represented by 3×3 matrices $\mathcal{R}_1, \mathcal{R}_1^{-1}$ and $\mathcal{R}_2, \mathcal{R}_2^{-1}$. Now ask what happens if we perform first $\mathcal{R}_1$, then $\mathcal{R}_2$, then $\mathcal{R}_1^{-1}$, and finally $\mathcal{R}_2^{-1}$, with the resultant $\mathcal{M} \equiv \mathcal{R}_2^{-1}\mathcal{R}_1^{-1}\mathcal{R}_2\mathcal{R}_1$. If the order of the operations were irrelevant, the effects of each rotation would be exactly annulled by its inverse, there would be no change at all, and $\mathcal{M}$ would equal the identity matrix 1. What actually happens is quite different. For simplicity we consider only very small angles, and work only to second order in

$\chi_{1,2}$. Then

$$\mathcal{R}_1 = \begin{bmatrix} 1, & 0, & 0 \\ 0, & \cos\chi_1, & \sin\chi_1 \\ 0, & -\sin\chi_1, & \cos\chi_1 \end{bmatrix} \approx 1 + \begin{bmatrix} 0, & 0, & 0 \\ 0, & -\chi_1^2/2, & \chi_1 \\ 0, & -\chi_1 & -\chi_1^2/2 \end{bmatrix},$$

$$\mathcal{R}_2 \approx 1 + \begin{bmatrix} -\chi_2^2/2 & 0, & \chi_2 \\ 0, & 0, & 0 \\ -\chi_2, & 0, & -\chi_2^2/2 \end{bmatrix}.$$

The inverses are obtained by changing the sign of $\chi_{1,2}$. It is straightforward to multiply the matrices explicitly, discarding all terms higher than second order,[6] and one finds

$$\mathcal{M} = \mathcal{R}_2^{-1}\mathcal{R}_1^{-1}\mathcal{R}_2\mathcal{R}_1 \approx 1 + \begin{bmatrix} 0, & \chi_2\chi_1, & 0 \\ -\chi_2\chi_1, & 0, & 0 \\ 0, & 0, & 0 \end{bmatrix} \approx \mathcal{R}_3(\chi_2\chi_1), \qquad \text{(B.29)}$$

where $\mathcal{R}_3$ is a rotation through the *second*-order-small angle $\chi_3 = \chi_2\chi_1$ around the z axis, i.e. in the xy plane. One can see this by comparison with (B.6) relabelled and rewritten three-dimensionally and approximated (only) to *first* order in χ_3:

$$\mathcal{R}_3(\chi_3) = \begin{bmatrix} \cos\chi_3, & \sin\chi_3, & 0 \\ -\sin\chi_3, & \cos\chi_3, & 0 \\ 0, & 0, & 1 \end{bmatrix} \approx 1 + \begin{bmatrix} 0, & \chi_3, & 0 \\ -\chi_3, & 0, & 0 \\ 0, & 0, & 1 \end{bmatrix}. \qquad \text{(B.30)}$$

In other words, one ends up not with the identity but with a slight twist.

B.4.2 Lorentz Transformations and Thomas Precessions

Now consider the analogous question where the $\mathcal{R}_{1,2}$ are replaced by Lorentz transformations $\mathcal{L}_{1,2}$ in the x, y directions, with small $\beta_{1,2}$. Working only to second order in the β's one has

$$\mathcal{L}_1 \approx 1 + \begin{bmatrix} \beta_1^2/2, & -\beta_1, & 0, & 0 \\ -\beta_1, & \beta_1^2/2, & 0, & 0 \\ 0, & 0, & 0, & 0 \\ 0, & 0, & 0, & 0 \end{bmatrix}, \quad \mathcal{L}_2 \approx 1 + \begin{bmatrix} \beta_2^2/2, & 0, & -\beta_2, & 0 \\ 0, & 0, & 0, & 0 \\ -\beta_2, & 0, & \beta_2^2/2, & 0 \\ 0, & 0, & 0, & 0 \end{bmatrix},$$

and explicit multiplication (or the same trick as for rotations) soon yields

$$\mathcal{L}_2^{-1}\mathcal{L}_1^{-1}\mathcal{L}_2\mathcal{L}_1 \approx 1 + \begin{bmatrix} 0, & 0, & 0, & 0 \\ 0, & 0, & -\beta_2\beta_1, & 0 \\ 0, & \beta_2\beta_1, & 0, & 0 \\ 0, & 0, & 0, & 0 \end{bmatrix} \approx \mathcal{R}_3(\beta_2\beta_1), \qquad \text{(B.31)}$$

where $\mathcal{R}_3(\beta_2\beta_1)$ is just the rotation through $\beta_2\beta_1$ around the z axis, i.e. in the xy plane. This is clear from comparison with (B.29), though we have stretched the notation by keeping the same symbol $\mathcal{R}_3$ for this rotation whether it occurs as a 3×3 matrix, or as part of a 4×4 matrix that affects neither velocities nor the time coordinate.

Thus a succession of Lorentz transformations, none of which involves a rotation, have combined into a resultant that is a pure rotation. Similar effects result from other successions of such Lorentz transformations; they are called *Thomas precessions*, some of them important to the fine structure of atomic spectra.

B.5 Notes

1. For instance, it offers little guidance regarding physical quantities whose dependence on direction (or directions) is more complicated, like those (called tensors) that connect the angular momentum of a body with its angular velocity, or the current density through a crystal with the electric field.

2. This applies only in the plane. In three dimensions, the resultant of rotations around different axes can depend on which is carried out first. One says that rotations *commute* in two dimensions, but not generally in three.

3. (i) Not all square matrices have inverses, and even if they have, it can happen that the "left-inverse" satisfying $\mathcal{R}_L^{-1}\mathcal{R} = 1$ differs from the "right-inverse" satisfying $\mathcal{R}\mathcal{R}_R^{-1} = 1$. But the cases we meet here are free of such complications. (ii) A set of matrices that includes the inverse of each and all products is said to form a group. Our set of $\mathcal{R}(\chi)$ for all χ forms the group of rotations around the z axis.

4. Commonly $\mathcal{G}$ is called the "metric tensor", and is written in lower case. We use a capital letter conformably with our script capitals for matrices generally. For three-vectors in ordinary space $\mathcal{G}$ reduces to the unit matrix, so that it needs and receives no mention.

5. For standard configuration, (B.16) shows that the $\mathcal{L}$ are symmetric, meaning that $\mathcal{L}^T = \mathcal{L}$. But we refrain from appealing to this relation, because it can fail for other configurations.

6. One can save labour by using $\mathcal{C} \equiv [\mathcal{R}_1^{-1}\mathcal{R}_2 - \mathcal{R}_2\mathcal{R}_1^{-1}]$; to write

$$\mathcal{M} = \mathcal{R}_2^{-1}\{\mathcal{R}_2\mathcal{R}_1^{-1} + \mathcal{C}\}\mathcal{R}_1 = 1 + \mathcal{R}_2^{-1}\mathcal{C}\mathcal{R}_1.$$

Now set $\mathcal{R}_{1,2} \equiv 1 + \delta\mathcal{R}_{1,2}$, and similarly for the inverses; and observe that the $\delta\mathcal{R}$ are at best first-order small. But it is readily seen that (i) $\mathcal{C} = [\delta\mathcal{R}_1^{-1}\delta\mathcal{R}_2 - \delta\mathcal{R}_2\delta\mathcal{R}_1^{-1}]$; (ii) this expression is already second-order small, whence to second order $\mathcal{M} \approx 1 + \mathcal{C}$; and (iii) $\mathcal{C}$ to second order may be evaluated using the $\delta\mathcal{R}$ to first order only; which is easy.

Appendix C
Motion under Given Forces

C.1 Newton's Second Law

The pre-Einstein version of Newton's second law reads

$$m\frac{d^2\mathbf{r}}{dt^2} = m\frac{d\mathbf{u}}{dt} = \frac{d\mathbf{p}}{dt} = \mathbf{F}, \tag{C.1}$$

where $m, \mathbf{r}, \mathbf{u} \equiv d\mathbf{r}/dt$, and $\mathbf{p} \equiv m\mathbf{u}$ are the (constant) mass, position, velocity, and momentum of a particle acted on by a force $\mathbf{F}$. In this form the law is primarily a way of suggesting that it is the acceleration of the particle that needs explanation, rather than $\mathbf{u}$ or some other time derivative of its position: clearly (C.1) cannot determine actual motions until one knows $\mathbf{F}$, and is likely to prove useful only if most of the prescriptions specifying forces (as functions of position or of time or of both) are reasonably simple. In particular the law is not subject to full experimental test except jointly with some force law.

Essentially the same reservation besets any attempt to adapt the law to the Einsteinian relativity principle: the result must be covariant, but one cannot verify whether it is except in conjunction with a covariant theory for the forces. The only such theory well established in physics is Maxwell's electromagnetism, and to demonstrate its covariance one must first study the Lorentz transformations of the electromagnetic fields. Since that is far beyond our remit, this appendix will merely state, without proof, the equation of motion for a charged particle referred to some given inertial frame, where the $\mathbf{E}$ and $\mathbf{B}$ fields are known from the outset; but we will not consider at all how to transform this equation to other frames. (Nor do we allow for radiative reaction, i.e. for the fact that accelerated particles radiate, and that in principle the radiated fields should be added to those given in advance.)

It seems plausible to look for the requisite generalization in the form

$$d\mathbf{p}/dt = \mathbf{F} = ?,$$

where $\mathbf{p} = m\mathbf{u}/\sqrt{1 - u^2/c^2}$ is the true rather than the Newtonian momentum. In fact, for a particle of mass m and charge q the correct equation turns out to read

$$\frac{d\mathbf{p}}{dt} = \frac{d}{dt}\left\{\frac{m\mathbf{u}}{\sqrt{1 - u^2/c^2}}\right\} = q(\mathbf{E} + \mathbf{u} \times \mathbf{B}) \equiv \mathbf{F}, \qquad (C.2)$$

so that $\mathbf{F}$ is just the familiar Lorentz force.

The derivative on the left acts both on the numerator and on the denominator:

$$m\left\{\frac{1}{(1 - u^2/c^2)^{1/2}}\frac{d\mathbf{u}}{dt} + \mathbf{u}\frac{(\mathbf{u} \cdot d\mathbf{u}/dt)/c^2}{(1 - u^2/c^2)^{3/2}}\right\} = q(\mathbf{E} + \mathbf{u} \times \mathbf{B}) = \mathbf{F}. \qquad (C.3)$$

Remarkably, the force need not be parallel to the acceleration $d\mathbf{u}/dt$, but may have a component along $\mathbf{u}$ as well. Taking the scalar product with $\mathbf{u}$ one finds

$$m\left\{\frac{(\mathbf{u} \cdot d\mathbf{u}/dt)}{(1 - u^2/c^2)^{3/2}}\right\} = q\mathbf{u} \cdot \mathbf{E} = \mathbf{u} \cdot \mathbf{F}, \qquad (C.4)$$

since $\mathbf{u} \cdot (\mathbf{u} \times \mathbf{B})$ vanishes. The expression on the left we recogize as $d\varepsilon/dt$:

$$m\left\{\frac{(\mathbf{u} \cdot d\mathbf{u}/dt)}{(1 - u^2/c^2)^{3/2}}\right\} = \frac{d}{dt}\left\{\frac{mc^2}{\sqrt{1 - u^2/c^2}}\right\} = \frac{d\varepsilon}{dt} = q\mathbf{u} \cdot \mathbf{E} = \mathbf{u} \cdot \mathbf{F}. \qquad (C.5)$$

Thus the rate of change of ε equals the rate at which the force does work on the particle, exactly as in Newtonian physics; only the expression for ε is different. Note that $d\varepsilon/dt$ depends only on $\mathbf{E}$: magnetic fields can do no work on charged particles.

Collecting these expressions and using $dt = d\tau/\sqrt{1 - u^2/c^2}$ we see that

$$\frac{d\vec{p}}{d\tau} = \frac{d(\varepsilon/c, \mathbf{p})}{d\tau} = q(\mathbf{u} \cdot \mathbf{E}/c, \mathbf{E} + \mathbf{u} \times \mathbf{B})/\sqrt{1 - u^2/c^2}.$$

Since the expression on the left is a four-vector, the expression on the right must be a four-vector too, though one cannot verify this directly until one learns how to Lorentz-transform the fields.

Equation (C.3) is awkward to use, and can be simplified somewhat by moving the second term on the left over to the other side, and then substituting in it from (C.5). This yields

$$\frac{1}{\sqrt{1 - u^2/c^2}}\frac{d\mathbf{u}}{dt} = \frac{1}{m}\left\{\mathbf{F} - \frac{1}{c^2}\mathbf{u}(\mathbf{u} \cdot \mathbf{F})\right\} = \frac{1}{m}\left\{\left(1 - \frac{u^2}{c^2}\right)\mathbf{F} + \frac{1}{c^2}\mathbf{u} \times (\mathbf{u} \times \mathbf{F})\right\}, \quad (C.6)$$

a clean expression for the acceleration in terms of the force (and, inevitably, of the velocity).

C.2 Examples

C.2.1 Motion Parallel to a Constant Homogeneous E Field

If $\mathbf{B}$ is zero and $\mathbf{E}$ is constant, and if the particle moves only in the direction of $\mathbf{E}$, then (C.5) reduces to

$$m(1 - u^2/c^2)^{-3/2}u\,du/dt = uqE \implies du/dt = (qE/m)(1 - u^2/c^2)^{3/2}. \qquad (C.7)$$

Inspection shows that this is precisely the expression (8.13) for a particle experiencing *constant proper acceleration* qE/m. In our present elementary approach we can only observe this coincidence as a fact: to foresee it one would need to be able to Lorentz-transform fields between the laboratory and the instantaneously co-moving frame.

C.2.2 Motion in a Constant Homogeneous B Field

Equation (C.5) shows that ε and therefore the speed u are now constants. Since $\mathbf{F} = q\mathbf{u} \times \mathbf{B}$ is perpendicular to $\mathbf{B}$ and $\mathbf{u} \cdot \mathbf{F} = 0$, the first equality in (C.6) shows that the component of $\mathbf{u}$ parallel to $\mathbf{B}$, call it $u_{\parallel}$, is also constant; only the components in the plane perpendicular to $\mathbf{B}$, call them $\mathbf{u}_{\perp}$, change with time.

For simplicity we consider only the case where $u_{\parallel} = 0$, so that $u \equiv |\mathbf{u}| = |\mathbf{u}_{\perp}|$. Then (C.6) prescribes

$$\frac{d\mathbf{u}_{\perp}}{dt} = \left(\frac{q\sqrt{1 - u^2/c^2}}{m}\right)\mathbf{u}_{\perp} \times \mathbf{B} = \left(\frac{qc^2}{\varepsilon}\right)\mathbf{u}_{\perp} \times \mathbf{B}, \tag{C.8}$$

where (qc^2/ε) is constant. This has the same structure as the nonrelativistic equation, except that m there is now replaced by $\varepsilon/c^2 = m/\sqrt{1 - u^2/c^2}$. Accordingly,[1] $\mathbf{u}_{\perp}$ precesses uniformly around $\mathbf{B}$, with angular frequency

$$\omega = qc^2 B/\varepsilon = (qB/m)\sqrt{1 - u^2/c^2}. \tag{C.9}$$

We choose the z axis parallel to $\mathbf{B}$; then it is easily verified by substitution that (C.8) is satisfied by

$$\begin{pmatrix} u_x \\ u_y \end{pmatrix} = u\begin{pmatrix} \cos(\omega t + \phi) \\ -\sin(\omega t + \phi) \end{pmatrix}. \tag{C.10}$$

Correspondingly, parallel to the xy plane the particle moves with angular velocity ω around a circle:

$$\begin{pmatrix} x - x_0 \\ y - y_0 \end{pmatrix} = r\begin{pmatrix} \sin(\omega t + \phi) \\ \cos(\omega t + \phi) \end{pmatrix}, \qquad r = u/\omega. \tag{C.11}$$

The speed u, the phase angle ϕ, the radius r, and the coordinates (x_0, y_0) of the centre of the circle are constants whose values depend on the initial conditions. One checks that on differentiation (C.11) reduces to (C.10).

Two features need stressing. (i) Substituting for ω from (C.9) into (C.11) one finds

$$r = \frac{u\varepsilon}{c^2} \cdot \frac{1}{qB} = \frac{p}{qB}, \tag{C.12}$$

where $p = u\varepsilon/c^2$ is the magnitude of the momentum. Therefore the radius of the track is a direct measure of the momentum; in fact the relation between radius and momentum is exactly the same as nonrelativistically. (ii) By contrast, (C.9) shows that for given $\mathbf{B}$ the angular frequency ω decreases inversely with the energy, whereas nonrelativistically it was a constant dependent only on the charge-to-mass ratio.

The weightiest *evidence* for these relations is that they enter critically into the design of modern particle accelerators and storage rings, which function as intended. A more direct if less accurate check was made with a proton synchrotron at the Carnegie

Institute of Technology,[2] using protons with kinetic energy 385 MeV, which corresponds to $u/c \simeq 0.7$. The test compares two different theoretical expressions for q/ε. On the one hand, equation (C.9) predicts $(q/\varepsilon)_1 = \omega/Bc^2$, whose left-hand side was found, at a predetermined value of the orbit radius r, from the known value of B and from the measured circulation frequency ω of the protons. On the other hand, expressing ε directly in terms of $u = \omega r$, one has $(q/\varepsilon)_2 = q\sqrt{1 - u^2/c^2}/Mc^2 = q\sqrt{1 - \omega^2 r^2/c^2}/Mc^2$, where $M = 938.3\,\text{MeV}/c^2$ is the proton mass. The experimental result is quoted as

$$[(qc^2/\varepsilon)_1 - (qc^2/\varepsilon)_2]/(qc^2/\varepsilon)_2 = -0.0006 \pm 0.001,$$

verifying the predictions within the experimental error of 0.1%.

It is left as an exercise for the reader to generalize the theory in this section to particles that have a constant but nonzero momentum parallel to **B**.

C.2.3 Motion in a Coulomb Field

If **B** is zero and $\mathbf{E} = -\nabla\phi$ derives from the electrostatic potential $\phi = q'/4\pi\varepsilon_0 r$ of a point charge q' at the origin, then the force experienced by another point charge q of opposite sign is

$$\mathbf{F} = -\hat{\mathbf{r}}A/r^2 = -\nabla V, \qquad V = -A/r, \qquad A = |qq'|/4\pi\varepsilon_0. \tag{C.13}$$

Equation (C.6) with this force is the relativistic Kepler problem. In the hydrogen atom, $A = e^2/4\pi\varepsilon_0$; according to Newton's law of gravitation (i.e. disregarding its modification by the general theory), the motion of the planets around the sun (mass M) is governed by the same equation but with $A = GMm$.

The chief novelty is that relativistically the planetary orbits are in general not closed, except for the special and as we shall see quite unrepresentative case of those that happen to be circular (of which more below). In other words, if $r(\theta)$ is the radial distance as a function of the polar angle θ, then as a rule $r(2\pi n)$ differs from $r(0)$ for any integer $n = 1, 2, \ldots$: there is no whole number of revolutions after which the motion would repeat exactly.[3] Figure C.1 illustrates what happens. This is in sharp contrast to the nonrelativistic case ($c \to \infty$), where all possible planetary orbits are ellipses, which close after one revolution (i.e. $r(2\pi) = r(0)$), and where the differences between circular and nearly circular orbits are minor and merely quantitative.

Although we shall not go through the argument, the standard method for dealing with nonrelativistic orbits works just as easily in the relativistic case; in terms of the total angular momentum $L \equiv |\mathbf{r} \times \mathbf{p}|$ and of the total energy

$$E = \varepsilon + V = mc^2/\sqrt{1 - u^2/c^2} - A/r \tag{C.14}$$

it yields the general orbit through

$$1/r = P + Q\cos(\lambda\theta), \tag{C.15}$$

where

$$\lambda = \sqrt{1 - A^2/c^2 L^2}, \quad P = EA/c^2 L^2 \lambda^2, \quad Q = [(E/c\lambda)^2 - (mc)^2]^{1/2}/\lambda L, \tag{C.16}$$

and where we have chosen axes so that $\theta = 0$ corresponds to a perihelion.[4]

From this one can determine the precession of the perihelion, namely the angular displacement $\Delta\theta$ between successive minima of r, as shown in the figure. With the nth perihelion ($\cos\theta = 1$) at θ and the next at $(\theta + 2\pi + \Delta\theta)$, equation (C.15) entails

$$\lambda\theta = 2n\pi, \quad \lambda(\theta + 2\pi + \Delta\theta) = 2(n+1)\pi \Rightarrow \Delta\theta = 2\pi(1/\lambda - 1). \tag{C.17}$$

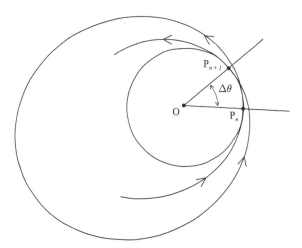

Figure C.1   The relativistic precession of the perihelion in a Coulomb potential. The centre of attraction is at O. The arrows indicate the direction of motion along the orbit; P_n and P_{n+1} are two successive perihelia, spaced by the angle $\Delta\theta$ defined in the text. The circle is not part of the orbit: it has been drawn with radius equal to the minimum distance, so that the perihelia are points where the orbit touches the circle. (The scale of the figure is unrealistic: its equation is $r = \{2 + \cos(2\theta/\sqrt{5})\}^{-1}$).

Because λ is less than unity, $\Delta\theta$ is positive: the perihelia advance, in the sense that the planet needs to turn through more than a full revolution to get from one to the next. For nearly nonrelativistic motion $\lambda \approx 1 - A^2/2c^2L^2$ is very nearly one, and we can approximate

$$\Delta\theta \approx \pi(A/cL)^2. \tag{C.18}$$

Presently we shall estimate $\Delta\theta$ for some interesting orbits that are near-circular as well as near-nonrelativistic.

By contrast to the general case described by (C.15) and (C.16), circular orbits are easy to describe from first principles. For circular motion r, u, and the magnitude p of the momentum are constant, and the angular momentum and Newton's second law read

$$L = pr = \varepsilon ur/c^2, \qquad u^2/r = Ac^2/\varepsilon r^2. \tag{C.19}$$

From this one finds

$$r = Lc^2/\varepsilon u \;\Rightarrow\; V = -A\varepsilon u/Lc^2, \tag{C.20}$$

and, eliminating ε,

$$u = A/L \;\Rightarrow\; \lambda = \sqrt{1 - u^2/c^2}. \tag{C.21}$$

These yield a neat expression for the total energy:

$$E = \varepsilon + V = \varepsilon - A\varepsilon u/Lc^2 = \varepsilon(1 - Au/Lc^2) = \varepsilon(1 - u^2/c^2),$$
$$E = mc^2\sqrt{1 - u^2/c^2} = mc^2\sqrt{1 - (A/Lc)^2} \qquad \text{(circular orbit)}. \tag{C.22}$$

Reassuringly, the same relation follows from the general orbit equation (C.15) if one makes r constant by setting $Q = 0$.

In (C.18) we can now approximate $\Delta\theta$ for nearly circular orbits[5] by using (C.21):

$$\Delta\theta \approx \pi u^2/c^2. \tag{C.23}$$

For the planet Mercury (fastest because nearest the sun), the mean orbital speed is 48 km/s, whence $\Delta\theta \approx 0.0166''$ (seconds of arc). Astronomers like to express this as an advance, call it Θ, of so many seconds of arc per (earth) century; since the orbital period is 88 (earth) days, one has $\Theta \approx (100 \times 365/88)\Delta\theta = 6.9''$/century. After currently accepted corrections for other much larger effects (mostly from perturbations by other planets), observation yields $43''$/century; this is in good agreement with the general theory of gravitation, which includes special relativity and much other physics besides.

Finally, it is interesting to consider the Bohr model of the hydrogen atom, which views the electron as a classical particle orbiting a fixed proton ($A = e^2/4\pi\varepsilon_0$), but with its angular momentum constrained to values $L = l\hbar$, $l = 1, 2, 3, \ldots$. In that case (C.22) gives the energy levels as

$$E_l = mc^2\sqrt{1 - (\alpha/l)^2}, \qquad \alpha \equiv e^2/4\pi\varepsilon_0\hbar c \approx 1/137, \qquad \text{(circular Bohr orbits)}, \qquad \text{(C.24)}$$

where α is called the fine-structure constant (for reasons cited below). Since $\alpha \ll 1$, this entails

$$E_l = mc^2\left\{1 - \frac{1}{2}\left(\frac{\alpha}{l}\right)^2 - \frac{1}{8}\left(\frac{\alpha}{l}\right)^4 + \cdots\right\},$$

or

$$E_l - mc^2 = -\frac{m(e^2/4\pi\varepsilon_0\hbar)^2}{2l^2}\left\{1 + \frac{\alpha^2}{4l^2} + \cdots\right\}, \qquad \text{circular Bohr orbits).} \qquad \text{(C.25)}$$

The first term on the right, which does not feature c, is Bohr's original expression for the strictly nonrelativistic system. By pure coincidence it also happens (with $l = 1, 2, 3, \ldots$) to reproduce the values of all the true quantum-mechanical levels governed by the (equally non-relativistic) Schrödinger equation, albeit without revealing the true relation between energy and angular momentum.

The full importance of the relativistic effects (in particular of the second term on the right of (C.25)) becomes apparent only on quantizing the general case of not necessarily circular orbits.[6] The result, which we merely quote, reads

$$E_{n,l} = mc^2\left[1\left(\frac{\alpha}{n' + \sqrt{l^2 - \alpha^2}}\right)^2\right]^{-1/2}, \qquad n \equiv n' + l, \qquad n' = 0, 1, 2, 3, \ldots. \qquad \text{(C.26)}$$

Circular orbits correspond to $n' = 0$, in which case (C.26) reduces to (C.24). But now expansion in powers of α yields

$$E_{n,l} - mc^2 = -\frac{m(e^2/4\pi\varepsilon_0\hbar)^2}{2n^2}\left\{1 + \frac{\alpha^2}{n^2}\left(\frac{n}{l} - \frac{3}{4}\right) + \cdots\right\}. \qquad \text{(C.27)}$$

To leading order, orbits with the same n but different l are predicted to have the same energy ("are degenerate"); but the relativistic correction, namely the second term on the right, splits the degeneracy. This splitting of the energy levels[7] is the fine structure to which α owes its name.

C.3 Notes

1. The trick is to differentiate again: $d^2\mathbf{u}_\perp/dt^2 = (qc^2/\varepsilon)(d\mathbf{u}_\perp/dt) \times \mathbf{B} = (qc^2/\varepsilon)^2(\mathbf{u}_\perp \times \mathbf{B}) \times \mathbf{B} = -(qc^2B/\varepsilon)^2\mathbf{u}_\perp$. This shows that both Cartesian components of $\mathbf{u}_\perp$ oscillate harmonically with the same circular frequency qc^2B/ε. Substitution back into (C.8) then links the two components, yielding (C.10).

2. D. J. Grove and J. G. Fox, "e/m for 385-Mev protons", *Physical Review* (1953), volume **90**, page 378.

3. Unless the parameter λ defined below is, coincidentally, a rational number N_1/N_2, in which case the motion does repeat after a number N_2 of revolutions.
4. A perihelion is a point of nearest approach to the sun (or to the centre of attraction more generally).
5. In the limit of exactly circular orbits, $\Delta\theta$ of course becomes unobservable, but mathematically speaking it does not vanish; rather it assumes the value (C.23).
6. The most accessible treatment is the classic by A. Sommerfeld, *Atomic structure and spectral lines*, third edition, Methuen, 1930.
7. By yet another coincidence, (C.26), (C.27) yield identically the same energy levels as does the Dirac equation for electrons, although with a different interpretation of the quantum number l. In this loose sense, the errors from the merely semi-classical Bohr-type quantization happen to compensate the errors from ignoring the intrinsic spin of the electron.

Appendix D
Wave Equations

Elsewhere we deal only with plane waves. This can be done simply, and compatibly with any reasonable invariant dispersion relation. But the fundamental physics (including the dispersion relation) of particular types of waves resides in the wave equations that they satisfy, of which plane waves are merely special though important solutions. Here we consider some important wave equations, written as partial differential equations, and show how they are constrained by requiring them to be form-invariant, conformably to the relativity principle.

Since such equations involve the partial derivatives $\partial/\partial t$ and ∇, one must start by asking how these behave under Lorentz transformations. We shall show presently that

$$\vec{\Box} \equiv \left(\frac{\partial}{c\partial t}, -\mathbf{\nabla}\right) = \left(\frac{\partial}{\partial X_0}, -\frac{\partial}{\partial \mathbf{X}}\right) \tag{D.1}$$

transforms as a four-vector, where the minus sign is to be noted.[1] Accordingly *the wave operator*

$$\Box^2 \equiv \vec{\Box} \cdot \vec{\Box} = \frac{1}{c^2}\frac{\partial^2}{\partial t^2} - \nabla^2 \tag{D.2}$$

is invariant, and this is the combination that occurs in the equations. Observe that the speed c enters simply to ensure invariance, regardless of the speed of the waves to be described eventually.

We restrict ourselves to waves that (i) propagate through empty space; (ii) can cross freely, so that it is possible to set up arbitrary linear combinations (superpositions) of any waves that are realizable individually; and (iii) obey an equation of no higher than second order. Condition (i) ensures that no velocity (like that of a material medium) can enter the equation as a parameter, nor any other vectors such as might characterize anisotropic materials. Condition (ii) ensures that the equation is linear in whatever function, call it ψ, describes the disturbance. But the most general form-invariant equation subject to these restrictions reads

$$\{\Box^2 + \kappa^2\}\psi(t, \mathbf{r}) = 0, \tag{D.3}$$

where κ is a constant with the dimensions of inverse length. (There can be no terms linear in the components of $\vec{\Box}$, because by assumption (ii) no other four-vector is available for forming four-scalar products.) One verifies by substitution that (D.3) has plane-wave solutions, with arbitrary amplitudes A, of the form

$$\psi = A \exp\{-i\vec{K} \cdot \vec{X}\} = \exp\{-i(\omega t - \mathbf{k} \cdot \mathbf{r})\}, \tag{D.4}$$

subject only to the dispersion relation

$$\omega^2/c^2 - k^2 = \vec{K} \cdot \vec{K} \equiv \kappa^2. \tag{D.5}$$

The simplest possibility is $\kappa^2 = 0$; this obtains for the electromagnetic field, with ψ standing for any Cartesian component of $\mathbf{E}$ or of $\mathbf{B}$. Indeed

$$\Box^2 \psi = 0 \tag{D.6}$$

is universally referred to as *the* wave equation. We have studied its plane-wave solutions in Chapters 12 and 13, and recall only that $\omega^2/c^2 - k^2 = 0$ ensures that all such waves travel at the invariant speed c.

The other possibility is that κ^2 is a nonzero constant, which Example 12.2 shows must be positive. Then (D.3) is called *the Klein–Gordon equation*. It has many applications; for instance to waveguides, as in Problems 12.4 and 12.5. Most important, with $\kappa = mc/\hbar$ it governs the quantum wavefunctions of particles having mass m; the plane-wave solutions (D.4), (D.5) for this case have been discussed in Section 12.6.

It remains only to establish the transformation properties of $\Box$. We do this by applying the familiar chain rule for partial derivatives to the event coordinates $(X_0, \mathbf{X})$ and $(X_0', \mathbf{X}')$, connected in standard configuration by

$$X_0' = \gamma(X_0 - \beta X_1), \quad X_1' = \gamma(X_1 - \beta X_0), \quad X_{2,3}' = X_{2,3}.$$

Evidently $\partial/\partial X_{2,3} = \partial/\partial X_{2,3}'$, and we need only link $\partial/\partial X_{0,1}$ with $\partial/\partial X_{0,1}'$. One calculates

$$\frac{\partial}{\partial X_0} = \frac{\partial X_0'}{\partial X_0}\frac{\partial}{\partial X_0'} + \frac{\partial X_1'}{\partial X_0}\frac{\partial}{\partial X_1'} = \gamma\frac{\partial}{\partial X_0'} - \gamma\beta\frac{\partial}{\partial X_1'} = \gamma\left(\frac{\partial}{\partial X_0'} - \beta\frac{\partial}{\partial X_1'}\right),$$

$$\frac{\partial}{\partial X_1} = \frac{\partial X_0'}{\partial X_1}\frac{\partial}{\partial X_0'} + \frac{\partial X_1'}{\partial X_1}\frac{\partial}{\partial X_1'} = -\gamma\beta\frac{\partial}{\partial X_0'} + \gamma\frac{\partial}{\partial X_1'} = \gamma\left(\frac{\partial}{\partial X_1'} - \beta\frac{\partial}{\partial X_0'}\right).$$

This is the transformation from primed to unprimed, to be compared with the general rule

$$A_0 = \gamma(A_0' + \beta A_1'), \quad A_1 = \gamma(A_1' + \beta A_0')$$

for any four-vector $\vec{A}$. But the comparison shows that it is indeed $(\partial/\partial X_0, -\partial/\partial X_1)$ that transform by the same rules as (A_0, A_1), which is what was claimed in (D.1).

At the end of Section 12.6 we pointed out that quantum wavefunctions are not directly observable, which re-opens the question whether the phase $\vec{K} \cdot \vec{X}$ of quantum plane waves need be invariant, and whether $\vec{K}$ need transform like a four-vector (as in fact it does, and as the notation already suggests). To close this gap in the reasoning one appeals to the underlying Klein–Gordon equation. As we have just seen, (D.4) solves (D.3) only if $\vec{K}$ satisfies (D.5), where κ^2 is an invariant constant. But the arguments in Section 7.1 and Appendix B.3 readily extend to show, purely as a matter of algebra,[2] that if $K_0^2 - \mathbf{K}^2$ is invariant, then $(K_0, \mathbf{K})$ does transform like a four-vector; which is what we wanted.

Notes

1. To help remember the sign, think about the change δf in some function $f(\vec{X})$, resulting from a small change $\delta \vec{X} = (c\delta t, \delta \mathbf{r})$. Evidently

$$\delta f = \{\delta X_0 \partial/\partial X_0 + \partial \mathbf{X} \cdot \mathbf{V}\} f = \{\delta X_0 \Box_0 - \delta \mathbf{X}(\Box) f = \{\delta \vec{X} \cdot \vec{\Box}\} f.$$

Thus the differential operator between the braces is an invariant, whereas without the minus sign from (D.1) it would read $\{\delta X_0 \Box_0 + \delta \mathbf{X} \cdot \Box\} = \{\delta X_0 \partial/\partial X_0 - \delta \mathbf{X} \cdot \mathbf{V}\}$, which is not.

2. To be precise, we do need to assume, in the language of Appendix B, that the column vector K formed from the components of $\vec{K}$ Lorentz-transforms linearly into $K' = \mathcal{L}K$, with *some* matrix $\mathcal{L}$ guaranteeing that $K'^T K' = K^T K$. Then the argument in Section B.3 shows at once that L is indeed the standard matrix transforming four-vectors.

Appendix E
Black-Body Radiation: The Lorentz Transformation of Planck's Law

E.1 The Problem and its Solution

The universe is permeated by black-body radiation (BBR), thought to be a legacy from its youth. At any given point, this radiation is isotropic with respect to a special reference frame S, which one might plausibly claim is at rest in an absolute sense. In S the temperature[1] of the BBR is 2.726 ± 0.010 K. With respect to other frames the BBR is anisotropic (different intensities in different directions), and satellite measurements of the anisotropy determine our velocity relative to S (see Example E.1). To analyse such measurements we need the Lorentz transform of the wave-vector distribution of BBR. This transformation allows for aberration and Doppler shifts: the direction and frequency of any given photon are different in different frames. Unfortunately, the transformation requires other physics as well, as will appear presently.

We streamline the problem by treating the radiation not explicitly as electromagnetic waves, but rather as photons, i.e. particles of zero rest mass:

$$\mathbf{p} = \hbar \mathbf{k}, \qquad \varepsilon = cp = \hbar \omega. \tag{E.1}$$

According to Planck and Einstein, in thermal equilibrium at temperature T relative to S the average[2] number of photons in a given state is

$$f(\mathbf{p}) = \frac{1}{\exp[\varepsilon / k_B T] - 1}. \tag{E.2}$$

The number of states in the range $d^3 r$ around $\mathbf{r}$ and with the wave-vector in a range $d^3 k$ around $\mathbf{k}$ is $2 d^3 k \, d^3 r / (2\pi)^3 = 2 d^3 p \, d^3 r / (2\pi\hbar)^3$, where the factor 2 allows for the two polarization states for given **k**. Hence the number of photons in these ranges is

$$dN = f(\mathbf{p}) \frac{2d^3p\, d^3r}{(2\pi\hbar)^3}. \tag{E.3}$$

It needs saying that dN is the number of photons in these ranges *at a given time t with respect to S*. Correspondingly, f is the distribution function with respect to S. It does not depend on t because it describes a photon gas in equilibrium. It does not depend on **r** because in equilibrium the photon gas is homogeneous.

Our problem is to determine the corresponding distribution function, call it f', which describes the same photon gas but with respect to another frame S'. This function is defined so that

$$dN' = f'(\mathbf{p}') \frac{2d^3p'\, d^3r'}{(2\pi\hbar)^3} \tag{E.4}$$

is the number of photons in the ranges d^3p' around $\mathbf{p}'$ and d^3r' around $\mathbf{r}'$, *at a given time t' with respect to S'*. The primes on $\mathbf{p}'$ and $\mathbf{r}'$ on the right are merely for notational convenience later on: (E.4) is a *definition*, where the components of $\mathbf{p}'$ are simply independent variables, unrelated at this stage to the components of $\mathbf{p}$ in (E.3). Similarly, at this stage $d^3p'\, d^3r'$ is unrelated to $d^3p\, d^3r$. The distribution f' is independent of t' and of $\mathbf{r}'$ for the same reasons that were cited for f. The distributions f and f' are different functions of their respective arguments: our task is to relate them. The answer turns out to be remarkably simple: *f and f' are equal when **p**' is related to **p** by the Lorentz transformation from S to S'*. One says that the distribution is *a four-scalar or equivalently an invariant function*. Section E.2 proves this: it is important to realize that the conclusion is far from self-evid nt. To explicate what it means, we take S and S' to be in standard configuration but ith their relative motion along the z axis (not along the x axis as elsewhere in this book); then, with $\beta \equiv v/c$ and $\gamma \equiv 1/\sqrt{1-\beta^2}$,

$$\{p'_z = \gamma(p_z - \beta p), \quad p'_{x,y} = p_{x,y}\} \Rightarrow f'(\mathbf{p}') = f(\mathbf{p}). \tag{E.5}$$

Equation (E.5) simplifies because $f(\mathbf{p})$ depends nly on ε, for which a Lorentz transformation yields

$$\varepsilon = cp = \gamma(\varepsilon' + \beta p'_z) = cp'\gamma(1 \quad \beta\cos\theta');$$

we write the spherical-polar components of $\mathbf{p}'$ as ($',\theta',\phi'$). (Recall $0 \le \phi' < 2\pi$, $0 \le \theta' \le \pi$, $-1 \le \cos\theta' \le 1$.) Then (E.5) reads

$$f(\mathbf{p}) = \frac{1}{\exp[cp/k_BT] - 1} = f'(\mathbf{p}') = \frac{1}{\exp[cp'\gamma(1 \quad \beta\cos\theta')/k_BT] - 1}. \tag{E.6}$$

Somewhat more elegantly, once one knows that t e distribution is invariant, one reasons that it must be a function, call it $F(\vec{V} \cdot \vec{p})$, f the only invariant $\vec{V} \cdot \vec{p}$ that involves the photon four-momentum $\vec{p}$ and the four-v locity $\vec{V}$ of the BBR rest frame. Thus we have

$$f(\mathbf{p}) = F(\vec{V} \cdot \vec{p}) = \frac{1}{\exp[\vec{V} \cdot \vec{p}/k\ T] - 1}; \tag{E.7}$$

the rightmost expression is evident from (E.2) in the frame S, where the components of $\vec{p}$ and of $\vec{V}$ are $(p, \mathbf{0})$ and $(c, \mathbf{0})$, so that $\varepsilon = cp = V_0 p_0 - \mathbf{V} \cdot \mathbf{p} = \vec{V} \cdot \vec{p}$. On the other hand, the components with respect to S' are $(p', \mathbf{p}')$ and $\gamma(c, -c\boldsymbol{\beta})$, so that

$$\vec{V} \cdot \vec{p} = (V_0' p_0' - \mathbf{V}' \cdot \mathbf{p}') = \gamma(cp' + c\boldsymbol{\beta} \cdot \mathbf{p}') = cp'\gamma(1 + \cos\theta'),$$

whence $f(\vec{V} \cdot \vec{p})$ expressed in primed components reproduces (E.6).

To write out (E.4) explicitly, we abbreviate $d^3 r' = dV'$, so that $dN'/dV' \equiv dn'$ becomes the *number density* of photons with momenta in $d^3 p'$. Moreover, it is convenient to consider ranges

$$d^3 p' = p'^2 dp' d\Omega', \qquad d\Omega' = d\phi' \sin\theta' d\theta' = d\phi' |d\cos\theta'|,$$

where $d\Omega'$ is the element of solid angle spanned by $(d\phi', d\theta')$. Thus

$$dn' \equiv \frac{dN'}{dV'} = \frac{2p'^2 dp' d\Omega'}{(2\pi\hbar)^3} f'(\mathbf{p}') = \frac{2p'^2 dp' d\Omega'}{(2\pi\hbar)^3} \cdot \frac{1}{\exp[cp'\gamma(1 + \beta\cos\theta')/k_B T] - 1}.$$

$$(E.8)$$

Purely for convenience, one might describe this as a Planckian distribution with a direction-dependent temperature T':

$$f'(\mathbf{p}') = \frac{1}{\exp[cp'/k_B T'] - 1}, \qquad T' \equiv T/\gamma(1 + \beta\cos\theta'). \qquad (E.9)$$

Finally we revert from particles to waves, and from photon numbers to energies. The joint energy of the photons counted by dn' is $cp'dn'$; their (circular) frequencies are $\omega' = cp'/\hbar$; accordingly *the energy density of the BBR per unit range of frequencies per unit solid angle is*

$$\rho' \equiv \frac{dE'}{dV' d\omega' d\Omega'} = \frac{2}{(2\pi)^3} \cdot \frac{(\hbar\omega'^3/c^3)}{\exp[\hbar\omega'\gamma(1 + \beta\cos\theta')/k_B T] - 1}. \qquad (E.10)$$

Observe from Section 13.4 that the combination $\omega'\gamma(1 + \beta\cos\theta')$ is just what one would get by Doppler shifting ω' through a relative velocity $-c\boldsymbol{\beta}$.

Since their speed is c, the *energy flux* due to the photons in $d^3 p'$ is

$$d\boldsymbol{\Phi}' = \hat{\mathbf{p}}' c(cp' dn') = \hat{\mathbf{p}}' c\rho' d\omega' d\Omega'. \qquad (E.11)$$

Example E.1. As a function of $\mathbf{p}'$, the observed intensity of the cosmic background radiation can be fitted quite accurately to the formula (E.9), but with $T' = T_0' + T_1' \hat{\mathbf{p}}' \cdot \hat{\mathbf{u}}'$, where $T_{0,1}'$ and the unit vector $\hat{\mathbf{u}}'$ are parameters of the fit. T_1' is called the dipole anisotropy. Measured values in the rest frame of the earth are $T_0' \approx 2.7$ K and $T_1' \approx 3.3 \times 10^{-3}$ K. What is our speed $c\beta$ relative to the cosmos?

Solution. Write our velocity as $c\boldsymbol{\beta} = \hat{\boldsymbol{\beta}} c\beta$. On the assumption, which we shall check presently, that $\beta \ll 1$, equation (E.9) yields

$$T' = T/\gamma(1 + \beta\cos\theta') = T - T\beta\cos\theta' + \cdots = T_0' + T_1' \hat{\mathbf{p}}' \cdot \hat{\mathbf{u}}'.$$

The second step drops terms of order β^2 or higher; the last step equates theory to observation. Since θ' is the angle between $\boldsymbol{\beta}$ and the photon direction $\mathbf{p}'$, we see that

$\hat{\beta} = -\hat{u}$. Equating coefficients one finds

$$\beta = T_1'/T_0' = (3.3 \times 10^{-3})/2.7 = 1.2 \times 10^- \quad \Rightarrow \quad c\beta = 370 \text{ km/s}.$$

Evidently β is indeed small enough to justify neglect'ng β^2. ∎

E.2 The Proof that $f'(p') = f(p)$

The proof of (E.5) is delicate because it must find r nges $d^3p\,d^3r$ and $d^3p'\,d^3r'$ such that dN and dN' are equal because they count the sa e set of photons, even though the dN occupy $d^3p\,d^3r$ simultaneously with respect t S (but not with respect to S'), and the dN' occupy $d^3p'\,d^3r'$ simultaneously with res ect to S' (but not with respect to S). Also one must realize that the Lorentz ransformation connecting the *argument* $\mathbf{p}$ of f with the *argument* $\mathbf{p}'$ of f', thoug it is of course relevant to the connection between the *ranges* d^3p and d^3p', connect the ranges in a way that is far from obvious, and makes it fatal to jump to conclusi ns. These points are easily and often missed, resulting in proofs that by virtue of c mpensating errors look simple mathematically but physically are nonsense, and cre te serious confusion when they collapse under questioning. An elementary proof foll ws. It has been made as simple as possible by restricting it³ to particles that are in quilibrium (an assumption we have used already), and propagate freely, i.e. without collisions with walls, with other particles, or with each other.

For brevity we disregard the x, y components of four-vectors, because they do not affect the argument; and we write $p_z \equiv p$.

We ask: what ranges dp' and dz' with respect to S are occupied at $t' = 0$ by those particles that with respect to S occupy given ranges d and dz at some common time, say at $t = 0$? (The values zero for both t and t' are c osen for definiteness: any other fixed values, not necessarily the same for both, woul serve just as well.)

It is comparatively easy to find dp, because the part'cles are free by assumption and their momenta therefore constant in time (with res ect to either frame). Thus, by virtue of the Lorentz transformation

$$p' = \gamma[p - \beta\varepsilon/c], \tag{E.12}$$

a small change dp induces a change dp' given (for a times t') by

$$dp' = d[\gamma(p - \beta\varepsilon/c)] = \gamma[1 - (\beta/c)d\varepsilon/dp]dp = \gamma[1 - \beta u/c]dp, \tag{E.13}$$

where u is the velocity of the particle:

$$d\varepsilon/dp = (d/dp)\sqrt{c^2p^2 + m^2c^4} = c^2p/\varepsilon = u.$$

It is less easy to find dz', because (unlike the mome ta) the positions vary with time. Consider a typical particle in the range, say particle umber i. Its world line referred to S and to S' is

$$z_i(t_i) = z_i(t_i = 0) + u_i t_i, \qquad z_i'(t_i') = {}_i'(t_i' = 0) + u_i' t_i', \tag{E.14}$$

where the velocities u_i, u_i' are constants because the p rticle is free. The suffix i is just a device allowing one to refer easily to events with coo dinates (t_i, z_i) and correspond-ing (t_i', z_i') on this particular trajectory.

We must express $z_i'(t_i' = 0)$ in terms of $z_i(t_i = 0)$; this seems to be the most slippery step in the proof. Lorentz transformation yields

$$z_i = \gamma[z_i' + \beta ct_i'], \qquad t_i = \gamma[t_i' + \beta z_i'/c];$$

for events on the trajectory, z_i and z_i' are functions respectively of t_i and t_i', given by (E.14). Hence

$$z_i(t_i = 0) + u_i t_i = z_i(t_i = 0) + u_i \gamma[t_i' + \beta z_i'(t_i')/c] = \gamma[z_i'(t_i' = 0) + u_i' t_i' + \beta ct_i'].$$

On setting $t_i' = 0$ the second equality becomes

$$z_i(t_i = 0) + u_i \gamma \beta z_i'(t_i' = 0)/c = \gamma z_i'(t_i' = 0) \;\Rightarrow\; z_i'(t_i' = 0) = \frac{z_i(t_i = 0)}{\gamma(1 - \beta u_i/c)}.$$

At this point it is safe to drop the auxiliary suffix i: we see that particles in a small range dz at $t = 0$ occupy at $t' = 0$ the small range

$$dz' = \frac{dz}{\gamma[1 - \beta u/c]}. \tag{E.15}$$

This is the requisite counterpart to (E.13).

The end-result now follows at once. Since the same set of particles occupies the primed and unprimed ranges, and since the primed and unprimed momenta are related by (E.12), we see that

$$dN = f(p)dpdz = f'(p')dp'dz' = f'(p')[dp\gamma(1 - \beta u/c)]\left[\frac{dz}{\gamma(1 - \beta u/c)}\right]$$

$$= f'(p')dpdz; \tag{E.16}$$

on equating the second and the last expressions and cancelling $dpdz$ this establishes (E.5), namely $f(p) = f'(p')$.

E.3 Applications

What one most wants to know in S' is how the energy density U' and of the momentum density $\mathbf{P}'$ depend on β. Both are given by integrals over $\mathbf{p}'$, needing the same two tricks to evaluate them.

The energy density is

$$U' = \int dn'\varepsilon' = \int \frac{2d^3p'}{(2\pi\hbar)^3} f(\mathbf{p}')\varepsilon'$$

$$= \frac{2}{(2\pi\hbar)^3} \int d\Omega' \int_0^\infty dp' p'^2 \frac{cp'}{\exp[cp'\gamma(1 + \beta\cos\theta')/k_BT] - 1}. \tag{E.17}$$

Here we have already implemented the first trick, which is to integrate over p' before Ω'. The second trick is to choose the exponent as our new integration variable ξ. Then

$$p' \equiv \frac{k_BT}{c\gamma(1 + \beta\cos\theta')}\xi \tag{E.18}$$

leads to

$$U' = \frac{2c(k_BT/c\gamma)^4}{(2\pi\hbar)^3} \int d\Omega' \frac{1}{(1 + \beta\cos\theta')^4} \int_0^\infty d\xi \frac{\xi^3}{\exp[\xi] - 1}. \tag{E.19}$$

The rightmost integral is a pure number:

$$\int_0^\infty d\xi \xi^3/(e^\xi - 1) = \pi^4/1 \ .$$

Next, $\int_0^{2\pi} d\phi'$ yields just a factor 2π, because the integran does not depend on ϕ'. Thus

$$\int d\Omega' \frac{1}{(1+\beta\cos\theta')^4} = 2\pi \int_{-1}^1 d\cos\theta' \frac{1}{(1+\beta\cos\theta')^4} = -\frac{2}{3} \cdot \frac{1}{(1+\beta\cos\theta')^3}\bigg|_{\cos\theta'=-1}^{\cos\theta'=+1}$$

$$= \frac{2\pi}{3\beta}\left[\frac{1}{(1-\beta)^3} - \frac{1}{(1+\beta)^3}\right] = \frac{2\pi}{3\beta} \cdot \frac{2(3+\beta^3)}{(1-\beta^2)^3} = 4\pi\gamma^6(1+\beta^2/3). \quad \text{(E.20)}$$

Substitution into (E.19) yields

$$U'(\beta) = \gamma^2(1+\beta^2/3)\frac{\pi^2(k_BT)^4}{15(\hbar c)^3} = \gamma^2(1+\beta^2/3)U, \quad (E.21)$$

where $U \equiv U'(\beta = 0) = \pi^2(k_BT)^4/15(\hbar c)^3$ is Planck's res lt.

The momentum density $\mathbf{P}'$ is

$$\mathbf{P}' = \int dn'\mathbf{p}' = \frac{2}{(2\pi\hbar)^3}\int d\Omega' \int_0^\infty dp' p'^2 \frac{\mathbf{p}'}{\exp[cp'\gamma 1+\beta\cos\theta')/k_BT] - 1}.$$

One can readily convince oneself that $P'_{x,y}$ vanish by sy metry. The z component, i.e. the component along $\boldsymbol{\beta}$, is obtained from the expression for ' by replacing the factor $\varepsilon' = cp'$ in its integrand by $p'_z = p'\cos\theta'$. Thus

$$P'_z = \frac{2(k_BT/c\gamma)^4}{(2\pi\hbar)^3}\int d\Omega' \frac{\cos\theta'}{(1+\beta\cos\theta')^4}\int d\xi \frac{\xi^3}{\exp[\xi] - 1}.$$

Here we require

$$\int d\Omega' \frac{\cos\theta'}{(1+\beta\cos\theta')^4} = 2\pi\int_{-1}^1 d\cos\theta' \frac{\cos\theta'}{(1+\beta \text{ os}\theta')^4} = -16\pi\beta\gamma^6/3.$$

Substitution then yields

$$\mathbf{P}' = -\boldsymbol{\beta}(4\gamma^2/3)U/c. \quad (E.22)$$

Example E.2. A flat plate of area A is moving with sp ed $c\beta$ along its normal through isotropic black-body radiation at temperature T. Both fa es reflect perfectly. Calculate the force on the plate in its own rest frame S'.

Solution. The force is found by summing the momentum c anges of the photons that impinge on the plate in unit time. Consider the leading face first. Ch ose the polar axis along the motion. From elementary kinetic theory, the number of photons (all with speed c!) falling per unit time on unit area at angles in the range $d\Omega'$ around Ω', nd in the range dp' around p', is $dn' \cdot c|\cos\theta'|$, with dn' from (E.8). Reflection reverses the normal component of the momentum, so that each such photon imparts an impulse $2p'|\cos\theta|$ to the plate, opposing the motion. Summing these impulses over unit area and unit time yiel s the pressure $\mathcal{P}'_+$:

$$\mathcal{P}'_+ = \frac{2}{(2\pi\hbar)^3}2\pi\int_0^1 d\cos\theta'\int_0^\infty dp' p'^2 \frac{2cp'\cos^2\theta'}{\exp[cp'\gamma(-\beta\cos\theta')/k_BT] - 1};$$

the minus sign in the exponent allows for the fact that hotons falling on the leading face necessarily move in the opposite direction. The integral run only over $0 \le \theta' \le \pi/2$ rather than over $0 \le \theta' \le \pi$, because $\mathcal{P}'_+$ stems from photons coming from only one side. On changing variables as for U' this reduces to

$$\mathcal{P}'_+ = U \frac{1}{\gamma^4} \int_0^1 d\cos\theta' \frac{\cos^2\theta'}{(1 - \beta\cos\theta')^4} = \mathcal{P}\frac{1}{\gamma^4(1-\beta)^3},$$

where $\mathcal{P} \equiv U/3$ is the pressure of the radiation in its own rest frame. The pressure $\mathcal{P}'_-$ on the trailing face is given by reversing the sign of β. Hence the total force opposing the motion is

$$F' = A[\mathcal{P}'_+ - \mathcal{P}'_-] = A\mathcal{P}\frac{1}{\gamma^4}\left[\frac{1}{(1-\beta)^3} - \frac{1}{(1+\beta)^3}\right] = 2\beta(3 + \beta^2)\gamma^2 A\mathcal{P}. \quad \blacksquare \qquad (E.23)$$

E.4 Notes

1. See, for example, M. White, D. Scott, and J. Silk, "Anisotropies in the cosmic microwave background", *Annual Reviews of Astronomy and Astrophysics* (1994), volume **32**, page 319.
2. For brevity we shall drop the "average". All the results below are average values; the fluctuations around them can be studied separately.
3. In fact the result is perfectly general: with all the arguments of the functions connected by a Lorentz transformation one has $f'(\mathbf{p}', \mathbf{r}', t') = f(\mathbf{p}, \mathbf{r}, t)$ even if the distribution varies with position and time, and even if the particles are not free. Our proof is a poor man's version of that given by N. G. van Kampen, "Lorentz-invariance of the distribution in phase space", *Physica* (1969), volume **43**, page 244. However, our argument is at least independent of the mass of the particles and of the form of $f(\mathbf{p})$; thus it applies to any initial distribution independent of $\mathbf{r}$, since any such distribution of free particles will remain constant in time, whether or not it happens to describe thermal equilibrium. For a different and simpler but somewhat artificial proof, specific to free photons, see G. R. Henry *et al.* "Distribution of blackbody cavity radiation in a moving frame of reference", *Physical Review* (1968), volume **176**, page 1451, who also give several applications.

Problems

Chapter 2

2.1 $\mathbf{r}$ is the position of an isolated particle, and $\mathbf{u}$ $d\mathbf{r}/dt$, while $\mathbf{A}$, $\mathbf{B}$, $\mathbf{Q}$, $\mathbf{U}$, $\boldsymbol{\Omega}$ are given constant vectors. In various reference fra   es (a) to (e) observations of the particle lead to the following conclusions:

(a) $\mathbf{u} = \mathbf{U}$;
(b) $\mathbf{r} = \mathbf{U}t$;
(c) $\mathbf{r} = \mathbf{B}\sqrt{|t|} + \mathbf{Q}$;
(d) $d\mathbf{u}/dt = \mathbf{A}$;
(e) $d\mathbf{u}/dt = \boldsymbol{\Omega} \times \mathbf{u}$.

Which of the frames (a) to (e) are inertial? Giv   reasons briefly.

2.2 A river is flowing at 2 miles/h. An oarsman c n make 3 miles/h through the water. Rowing upstream, he hears a splash, an   realizes 15 minutes later that it was made by his (half-empty) whisky bottle fal ing overboard. He turns round and retrieves it as fast as he can. How long be   re he does so?
   Hint: choose the most convenient reference fr   e. Not all information is useful information.

2.3 Two trains of lengths L_1, L_2 travel on parallel   racks in opposite directions, at speeds u_1, u_2. Consider four events: A, the fro ts pass; B, the front of train 1 passes the tail of train 2; C, the front of 2 passes the tail of 1; and D, the tails pass. Calculate the time intervals and the dista ces separating all pairs of these events (a) relative to train 1; (b) relative to trai   2; (c) relative to the track.

2.4 A particle hitting a fixed flat buffer with speed   at an angle θ to the normal is reflected specularly (like a light ray from a fla   mirror), so that the tangential component of its velocity is unchanged while th   normal component is reversed.

(a) Determine its speed u_r and its angle θ_r to t e normal after reflection, if the buffer is constrained to recede with speed $w < u\cos\theta$ (unaffected by the collision).
(b) Evaluate your result in the special case wh re $u/w = 10$ and $\theta = 45°$.
Answers: $u_r/w = 8.70$ and $\theta_r = 54.4°$.

2.5 A radioactive source moving with speed v relative to the laboratory emits alpha particles. In the rest frame S' of the source their speed is u', and their angular distribution isotropic (i.e. equal numbers are emitted into equal elements of solid angle with respect to S'). With respect to the laboratory, the proportions of alphas emitted forward and backward are f and $b = 1 - f$. The forward–backward asymmetry is $\sigma \equiv (f - b)$. Calculate σ and sketch it as a function of $\xi \equiv v/u'$.

2.6 When calculating decay rates in particle physics, one often needs to determine the most general Lorentz-invariant function of the velocities or momenta of the products, subject to various restrictions. Some such restrictions might arise, for instance, if the energy release is small compared with the rest masses. The following is an artificially simplified example of such a problem.

Given three arbitrary velocities $\mathbf{u}_1, \mathbf{u}_2, \mathbf{u}_3$, the most general scalar combination quadratic in their components is

$$F = a_1 u_1^2 + a_2 u_2^2 + a_3 u_3^2 + b_{12}\mathbf{u}_1 \cdot \mathbf{u}_2 + b_{23}\mathbf{u}_2 \cdot \mathbf{u}_3 + b_{31}\mathbf{u}_3 \cdot \mathbf{u}_1,$$

where the coefficients a_i and b_{ij} are constants. If F is invariant under Galilean transformations, show that

$$a_1 + a_2 + a_3 + b_{12} + b_{23} + b_{31} = 0.$$

Does invariance impose any stronger constraints on the coefficients?

2.7 For a one-dimensional ideal gas, the Maxwell–Boltzmann velocity distribution with respect to the laboratory S reads

$$f(u)du = N \exp(-mu^2/2kT)du,$$

where $f(u)du$ is the fraction of molecules having velocities between u and $u + du$, and N is a norming constant such that $\int_{-\infty}^{\infty} du f(u) = 1$. Recall that the mean-square speed is given by

$$\int_{-\infty}^{\infty} du f(u)u^2 = kT/m.$$

(a) Find the velocity distribution $f'(u')$ with respect to an observer S' whose own velocity with respect to the laboratory is v. Check the normalization, i.e. that $\int_{-\infty}^{\infty} du' f'(u') = 1$.
   Hint: the main difficulty is, probably, to understand the meaning of the prime on the function $f'(u')$. With u' and du' the appropriate Galilean transforms of u and du, the products $f'(u')du'$ and $f(u)du$ count the same molecules (albeit labelled differently, because by reference to S' and to S respectively). Hence $f(u')du' = f(u)du$.
(b) Determine the mean velocity and the mean-square velocity with respect to S'. Can you find them without detailed calculations?

2.8 (a) Relative to a railcar moving at $5\,\text{m/s}$ along a straight, a mass $m_1 = 0.1\,\text{kg}$ moving forwards with speed $u_1 = 1\,\text{m/s}$ collides head on with a mass $m_2 = 0.05\,\text{kg}$ moving in the opposite direction with speed $u_2 = 5\,\text{m/s}$. If

m_2 is at rest after the collision, what is the velocity w_1 of m_1 in magnitude and direction? How much kinetic energy has been lost?

(b) Now describe the collision relative to the track. Is momentum conserved? How much kinetic energy has been lost?

2.9 The angular momentum of a particle around the origin is defined as $\mathbf{L} \equiv \mathbf{r} \times \mathbf{p}$. Show that its Galilean transform is $\mathbf{L}' = \mathbf{L} + \mathbf{v} \times (m\mathbf{r} - \mathbf{p}t)$.

2.10 A shell is fired with velocity having horizontal and vertical components u_x, u_y. At the highest point of its path the shell explodes into two fragments having masses m_1 and m_2, which separate in a horizontal direction. The explosion produces extra kinetic energy E. Show that the fragments hit the ground separated by a distance

$$\frac{u_y}{g} \left\{ 2E \left[\frac{m_1 + m_2}{m_1 m_2} \right] \right\}^{1/2}.$$

Hint: choose the most convenient reference frame.

2.11 Given momentum conservation formulated as in equation (2.20), prove mass conservation as formulated in (2.16).

2.12 Liquid helium is confined to a very long straight tube at temperature T. Thermodynamics shows that, with the tube at rest, the state of thermal equilibrium is that which minimizes the free energy $F = E - TS$, where E, S are the internal energy and entropy of the helium. Under Galilean transformations, T and S are invariant. To study superfluidity one needs to identify the equilibrium state when the tube is moving parallel to its axis with velocity u. Show that this state minimizes not F but $F - uP$, where P is the total momentum of the fluid.

2.13 A train 200 m long travels along a straight at 200 km/h. A whistle mounted outside at the front sounds A (440 Hz). What frequencies are heard by receivers mounted (a) outside on the rear of the train; (b) on the track ahead; (c) on the track behind?

2.14 As in problem 2.13 (a) to (c), but in a headwind blowing at 100 km/h.

2.15 On a windless day, two cars start simultaneously from the origin, with equal speeds u. Sounding a siren with frequency f_A, car A goes due north; car B goes due east. Working only to first order in u/c_s, determine

(a) from what direction the sound of A reaches B;
(b) the frequency f_B heard by B.
Hint: start by calculating the time $\tilde{t}$ when A emitted the sound that reaches B at time t, and thence the position of A at time $\tilde{t}$.

2.16 In still air, sound of frequency f falls normally on a stationary baffle, which reflects it without changing its frequency. Now consider sound of the same frequency incident on the same baffle receding from the sound wave at speed $w < c_s$. What is the frequency of the sound after reflection?
Hint: transform to the rest frame of the baffle, solve the problem there, and transform back again. Remember that for the baffle to recede without creating a

vacuum, the air must follow it: in other words there must be a wind blowing with speed w, parallel to the incident sound wave.
Answer: $f_{\text{refl}}/f_{\text{inc}} = (1 - w/c_s)/(1 + w/c_s)$.

Chapter 4

The inertial frames S and S' are in standard configuration, with relative speed v.

4.1 A metre-stick pointing in the x direction moves along the x axis with speed $0.8c$, its mid-point passing the origin at $t = 0$. Where are its end-points at that time?

4.2 Two events happen at the same place and 4 s apart with respect to S. (a) What is their spatial separation with respect to S' where they happen 6 s apart? (b) What is the speed of S' relative to S?

4.3 Two events happen on the x axis at the same time and 1 km apart with respect to S. (a) What is the time interval between them with respect to S', where they happen 2 km apart? (b) What is the speed of S' relative to S?

4.4 The distance to a star is 20 light years measured relative to earth-fixed axes. (Disregard the rotation and the orbital motion of the earth.) A probe travels from earth to star and returns immediately, moving at constant speed u relative to earth on each leg of the journey. The probe's clock is robust enough to remain unaffected by take-off and landing. On return, it has advanced by 30 years. (a) Determine u. (b) How much have earth-fixed clocks advanced between the probe's departure and return?
  [You will find that the probe's clock has advanced by less than have clocks on earth: in common parlance the probe's clock has run slow. This used to be presented as the so-called clock (or twin) "paradox", on the grounds that the probe might with equal right think of the earth as the traveller, and therefore expect the earth clock to go slow. In fact there is no paradox because there is no symmetry between the two: the probe has accelerated (as it can tell from the acceleration stresses), while the earth has not. We know how to account for both clocks relative to the earth's frame because it is inertial, but we are not tooled up to give a full account of the earth clock relative to the probe's rest frame, which is not. The writer believes that such an account requires the equivalence principle, which is beyond the scope of this book: see, for example, R. D. Sard, *Relativistic mechanics*, chapter 6, Benjamin, New York, 1970; or V. Fock, *The theory of space time and gravitation*, sections 61 & 62, Pergamon, London, 1959. Attempts at a full description without the equivalence principle are doomed to obscure rather than clarify.]

4.5 A two-stage rocket is so designed that the firing of each stage increases the velocity of the payload by the same amount **u** relative to the frame where the rocket was at rest before the firing. The final speed of the payload relative to base is $c/2$. Determine u/c. (Gravitational fields should be ignored.)

4.6 Relative to the laboratory a rod of rest length l_0 moves in its own line with velocity u. A particle moves in the same line with equal and opposite velocity. How long does it take for the particle to pass the rod

(a) relative to the rest frame of the rod?

(b) relative to the rest frame of the particle?

(c) relative to the laboratory?

Hint: what are the pertinent distances and speeds in each case?

Answers: $l_0(1 + u^2/c^2)/2u$, $l_0(1 - u^2/c^2)/2u$, $l_0\sqrt{(1 - u^2/c^2)}/2u$.

4.7 In its rest frame, a right-angled triangle has sides of length 1, 1, $\sqrt{2}$, in metres. Determine the angles of the triangle observed from a frame relative to which it is moving with speed $c/2$ in a direction parallel to the longest side.

4.8 A galaxy is receding from earth at speed u. In its own rest frame its diameter is L. Its distance from us is D. From the farthest and the nearest points of the galaxy, light signals are emitted that reach us simultaneously. What is the time interval between their emission as measured relative to the rest frame of the galaxy? *Hint: not all information is useful information.*

4.9 **Relative speeds.** With respect to the laboratory, two particles A and B have velocities **a** and **b**. In Newtonian physics their relative speed $u(A, B)$ is $|\mathbf{a} - \mathbf{b}|$. In Einsteinian physics it is defined as the speed of either particle in the rest frame of the other, say of B in the rest frame of A. By using a brute-force Lorentz transformation from the laboratory to A's rest frame, show that

$$u^2(A, B) = \frac{(\mathbf{a} - \mathbf{b})^2 - a^2b^2/c^2 + (\mathbf{a} \cdot \mathbf{b})^2/c^2}{(1 - \mathbf{a} \cdot \mathbf{b}/c^2)^2} = \frac{(\mathbf{a} - \mathbf{b})^2 - (\mathbf{a} \times \mathbf{b})^2}{(1 - \mathbf{a} \cdot \mathbf{b}/c^2)^2}.$$

(From its definition, $u(A, B)$ is an invariant; this will open an easier route to it in Chapter 7).

4.10 Neutral π mesons decay into pairs of photons, which are emitted isotropically in the rest frame of the mesons. A beam of such mesons travels with velocity **v** relative to the laboratory. The proportions of photons emitted forward and backward are f and $b = 1 - f$. The forward–backward asymmetry is $\sigma \equiv (f - b)$. Treating the photons as if they were particles having speed c, calculate and sketch σ as a function of $\beta = v/c$. Compare with the nonrelativistic problem 2.4.

4.11 Relative to S', both the velocity $\mathbf{u}'$ and acceleration $\mathbf{a}'$ of a particle lie in the $x'y'$ plane, at angles ϕ' and χ' to the x' axis. Relative to S the acceleration **a** is at an angle χ to the x axis. Show that

$$\tan\chi = \gamma\left\{\tan\chi' + \frac{u'v}{c^2} \cdot \frac{\sin(\chi' - \phi')}{\cos\chi'}\right\}.$$

Chapter 5

5.1 (a,b) Recalculate the answers to Problems 4.2 and 4.3 by appeal to the invariance properties of Δs^2 and of Δt, but without using the Lorentz transformations explicitly.

(c) Briefly and clearly, explain which method you prefer, and why.

5.2 The (t, x) coordinates of three events A, B, C are (1m/c, 1m), (2m/c, 3m), (3m/c, 2m).

(a) For each of the three intervals AB, BC, CA state whether it is timelike, spacelike, or lightlike.

(b) What are the proper time differences or proper distances between each pair?

(c) Which events could have caused which others?

(d) Illustrate all this by entering the three events on the usual space–time diagram with axes x and ct, and by drawing the light cones for all three. Briefly and clearly, explain how the answers to parts (a), (b), (c) could have been deduced from the diagram alone.

5.3 One pair of events is separated by the intervals $(\Delta t_{12}, \Delta \mathbf{r}_{12})$, and another pair by $(\Delta t_{34}, \Delta \mathbf{r}_{34})$. Show that $(c^2 \Delta t_{12} \Delta t_{34} - \Delta \mathbf{r}_{12} \cdot \Delta \mathbf{r}_{34})$ is an invariant.

5.4 For the encounter described in Problem 4.6, with $u = c/2$, draw properly scaled space-time diagrams in the rest frames of the rod, of the particle, and of the laboratory. Each diagram should show the world lines of the particle and of both ends of the rod.

5.5 (a) On the space–time diagram with axes x, ct, draw the light cone through O, and several contours of constant positive and negative values of s^2, including those with $s^2 = \pm 1\,\mathrm{m}^2$. Which (if any) of these contours are possible trajectories of particles?

(b) On the same diagram draw the x' and ct' axes. Derive the scale factor $\sqrt{(1 + \beta^2)/(1 - \beta^2)}$ along these axes by considering their intersections with the contours for $s^2 = \pm 1\,\mathrm{m}^2$. (See the small-print paragraph at the end of Chapter 5.)

5.6 **Keeping in touch, I.** Consider the same return trip as in Problem 4.4 and Example 5.2. Between launch and return, brief radar pulses are sent at intervals of θ s (earth time) from earth to probe.

On a space–time diagram referred to earth, draw the trajectories of several of these pulses.

On its way out and back, the probe receives, respectively, $n_{1,2}$ pulses at time intervals of $\Delta \tau_{1,2}$ (probe time). Determine all these quantities. Check that $(n_1 \Delta \tau_1 + n_2 \Delta \tau_2) = $ (trip duration by probe clock).
Answers: $n_{1,2} = (L/u\theta)(1 \mp u/c)$; $\Delta \tau_{1,2} = \theta \sqrt{(1 \pm u/c)/(1 \mp u/c)}$.

5.7 **Keeping in touch, II.** Consider the same return trip as in the preceding question. Between launch and return, brief radar pulses are sent at intervals of θ s (probe time) from probe to earth.

Draw whatever space–time diagram(s) you think most helpful for answering the following questions.

Earth receives $n_{3,4}$ such pulses sent, respectively, on the probe's way out and back, at time intervals $\Delta t_{3,4}$ (earth time). Determine all these quantities. Check that $(n_3 \Delta t_3 + n_4 \Delta t_4) = $ (trip duration by earth clock).
Answers: $n_3 = n_4 = (L/u\theta)\sqrt{1 - u^2/c^2}$; $\Delta t_{3,4} = \theta \sqrt{(1 \pm u/c)/(1 \mp u/c)}$.

Chapter 6

6.1 Two science-fiction space probes synchronize their clocks, and go to a star 10 light-years away. SF1 leaves first, travelling at speed $c/4$. SF2 leaves later, travelling at speed $c/2$, and the two probes arrive simultaneously. Draw the world lines of both probes. What is the difference between their clock readings on arrival? (Assume takeoff and landing introduce no appreciable difference.)

6.2 At time $t = 0$ you are at the origin. You must attend a ceremony at $\mathbf{r} = (L, 0, 0)$ scheduled for time $t = T > L/c$. You wish to attend (a) as young as possible; or (b) as old as possible. Explain as fully as you can what trajectories could serve, assuming that you can move (i) only along the x axis, or (ii) anywhere in space.

6.3 The acceleration of a particle confined to the x axis has constant magnitude a, and is always directed towards the origin. The particle oscillates with amplitude A and period T, and with maximum speed far below c. Calculate the amount $\Delta\tau$ by which its proper time falls behind laboratory time over one period, and show that $\Delta\tau/T = 2aA/3c^2$.

6.4 The bob of a grandfather clock oscillates with an amplitude of 0.1 m. Determine how many years it takes for the proper time of the bob to lag 1 s behind the time indicated by the clock.

6.5 Two clocks are synchronized and then flown once around the earth at the same height and at constant ground speed in opposite directions along the equator. Both flight times are 45 h. Calculate the difference between the clock readings after the trips. [This problem is inspired by measurements reported by J. C. Hafele and R. E. Keating, "Around-the-world atomic clocks: Predicted relativistic time gains", and "Observed relativistic time gains," *Science* (1972), volume **177**, pages 166 and 168. Under similar (but not identical) conditions they found differences of the order of 3×10^{-7} s].

6.6 A thing moves along the x axis with velocity $u(t) = c/\cosh(\omega t)$. Its own clock is set so that its proper time τ reads zero when it passes the origin, which it does when $t = 0$. (This is an idealized trajectory, which no particle can follow exactly. Why not?)

(a) Express t and x as functions of τ.
(b) Evidently τ lags behind t. Determine the total lag Δ accumulated between $t \to -\infty$ and $t \to \infty$. (It can be defined formally by

$$\Delta \equiv \left\{ \lim_{t \to \infty} (t - \tau) - \lim_{t \to \infty} (t - \tau) \right\},$$

or equivalently by the same expression but with limits $\tau \to \pm\infty$.)
Answer: $\Delta = (1/\omega) \log 4$.

6.7 A space station circles the earth 300 miles above the surface. An astronaut due to spend 90 days there decides to brighten the decor with a grandfather clock. On arrival at midnight he winds it up, fixes it with its foot towards the earth, adjusts it to 12 o'clock, and then forgets about it. What time does it show when he leaves at 6 am on day 90? (Pretend the station does not tumble.)

6.8 Grossly simplified, an optical clock consists of a standing light wave between two parallel mirrors fixed to an optical bench, a distance L apart. There are N half-waves, so that $\lambda = 2L/N$, and the frequency of the light is $f = c/\lambda = cN/2L$. Since N is an integer, it cannot change continuously, and small disturbances leave it unaltered: this is what stabilizes the clock. Time is measured by allowing some of the light to leak out, and counting its oscillations.

Discuss and estimate the fractional error in times measured by the clock, due to likely accelerations parallel to the bench.

Chapter 7

7.1 (a) Two particles move towards each other with equal speeds u along the x axis. What is their relative speed w? If $u/c \ll 1$, one may approximate $w \approx u(a + bu^2/c^2 + \cdots)$. Determine a and b.

(b) Two particles move towards the origin, each with speed u, one along the x axis and the other along the y axis. What is their relative speed w? If $u/c \ll 1$, one may approximate $w \approx u(a + bu^2/c^2 + \cdots)$. Determine a and b.

Answers: (a) $a = 2$, $b = -2$; (b) $a = \sqrt{2}$, $b = -1/2\sqrt{2}$.

7.2 The component A_μ of a four-vector $\vec{A}$ is zero with respect to *all* inertial frames. (It does not matter which component: you may take μ as 0, *or* 1, *or* 2, *or* 3.) Show that *all* components of $\vec{A}$ are zero with respect to *all* inertial frames, i.e. that $\vec{A}$ is the *null vector*: $\vec{A} = \vec{0} = (0, \mathbf{0})$.

7.3 Verify that Newton's first law may be expressed by saying that for any free particle $\vec{X}(\tau) = \vec{U}\tau$, where $\vec{U}$ is a constant vector with $\vec{U} \cdot \vec{U} = c^2$ and $U_0 > 0$.

7.4 The coordinate four-vector of a particle is given by $\vec{X} = \vec{W}f(\tau)$, where τ is the proper time. We are told only that $\vec{W}$ is a *constant* timelike vector with $W_0 > 0$. What can you conclude (a) about $f(\tau)$, and (b) about the trajectory of the particle with respect to an inertial frame S?

Hint: you may need to use $\vec{U} \cdot \vec{U} = c^2$, where $\vec{U} \equiv d\vec{X}/d\tau$.

Answer: (a) $f(\tau) = c\tau/\sqrt{\vec{W} \cdot \vec{W}}$; (b) *free particle with* $\mathbf{r} = \mathbf{u}t$, *where* $\mathbf{u} = \mathbf{W}/W_0$.

7.5 Two four-vectors are orthogonal: $\vec{A} \cdot \vec{B} = 0$.

(a) Show that they cannot both be timelike.

(b) What can you say about them if both are lightlike?

(c) Suppose $\vec{A}$ is timelike and $\vec{B}$ spacelike. What can you say about $(B_0, \mathbf{B})$ in the frame where $\mathbf{A} = \mathbf{0}$? What can you say about $(A_0, \mathbf{A})$ in the frame where $B_0 = 0$?

(d) Suppose both are spacelike. What can you say about $(B_0, \mathbf{B})$ in the frame where $A_0 = 0$?

7.6 With respect to an inertial frame S, a particle moves anticlockwise in the xy plane, with constant speed u around a circle of radius R, centred on the origin.

(a) Determine its coordinate four-vector $(X_0, \mathbf{X})$ and its four-velocity $(U_0, \mathbf{U})$ as functions first of t and then as functions of the particle's proper time τ.

(b) Frame S' is in standard configuration with S, with relative velocity v. Determine $(X_0', \mathbf{X}')$ and $(U_0', \mathbf{U}')$ as functions of τ.

Chapter 8

8.1 Two inertial frames S, S' are in standard configuration, with $v = c/\sqrt{2}$. Relative to S, a particle moves in the xy plane, at constant speed $u = c/\sqrt{2}$ counterclockwise in a circle of radius R around the origin O; hence its acceleration has constant magnitude u^2/R, and is always directed towards O. (a) Determine its acceleration $\mathbf{a}' = (a_1', a_2')$ relative to S' when relative to S the particle is at $(x, y) = (R, 0), (0, R), (-R, 0)$, and $(0, -R)$. (b) Are these accelerations $\mathbf{a}'$ directed towards O'?

8.2 With respect to an inertial frame S, a particle starts from rest at proper time τ_1 and moves along the x axis, with proper acceleration $\alpha(\tau)$. As in Example 8.2, its four-velocity is governed by

$$dU_0/d\tau = \alpha(\tau)U_1/c, \qquad dU_1/d\tau = \alpha(\tau)U_0/c.$$

(a) Show that

$$U_0(\tau) = c\cosh\left(\int_{\tau_1}^{\tau} d\tau\,\alpha(\tau)/c\right), \qquad U_1(\tau) - c\sinh\left(\int_{\tau_1}^{\tau} d\tau\,\alpha(\tau)/c\right).$$

Hint: either (i) write down and solve the differential equations for $U_\pm \equiv (U_0 \pm U_1)$; or (ii) change the independent variable from τ to θ defined by $d\theta = \alpha(\tau)d\tau$.
(b) The particle starts from rest at time $-\infty$, and $\alpha(\tau) = (\alpha_0/\tau)/(1 + \tau^2/T^2)$, where α_0 and T are constants. What is the final velocity?
Answer: $u(\infty) = c\tanh(\alpha_0 T)$.

8.3 Constant proper acceleration. For such motion, Example 8.2 derived

$$(A_0, A_1) = (\alpha/c)(U_1, U_0) = \alpha(\sinh(\alpha\tau/c), \cosh(\alpha\tau/c)). \tag{*}$$

(a) Hence, or otherwise, show that $\vec{U}$ and $\vec{A}$ obey the equations

$$d^2\vec{U}/d\tau^2 = \lambda^2\vec{U}, \qquad d^2\vec{A}/d\tau^2 = \lambda^2\vec{A}, \qquad \lambda^2 \equiv (\alpha/c)^2. \tag{**}$$

(b) Now suppose you are given the first of equations (**). With appropriate initial conditions at $\tau = 0$, solve it so as to *derive* (*) for $\vec{U}(\tau)$. (c) The same question for $\vec{A}$.
Hint: the most general solution of the second-order equation for $\vec{U}$, say, can obviously be written as

$$\vec{U} = \vec{P}\exp(\lambda\tau) + \vec{Q}\exp(-\lambda\tau),$$

where $\vec{P}, \vec{Q}$ are constant four-vectors (i.e. independent of τ), which one must determine. You may want to use the fact that $\vec{U} \cdot \vec{U} = c^2$ (for all τ), and results from Problem 7.5. For simplicity, choose directions so that $\mathbf{U}$ and $\mathbf{A}$ lie along the x axis.

8.4 Show that the equation $d^2 \vec{U}/d\tau^2 = -\lambda^2 \vec{U}$ has no physically acceptable solutions.
Hint: as for Problem 8.3.

8.5 With respect to the laboratory frame S, a system executes simple-harmonic motion according to $\beta(t) = \beta_0 \sin(\omega t)$, where $\beta \equiv u/c < 1$. The system will be damaged if its proper acceleration exceeds a critical value α^*. What is the maximum tolerable value β_0^* of β_0?
Answer: if $\alpha^* < \omega c/\sqrt{3}$, then $\beta_0^* = \alpha^*/\omega c$. If $\alpha^* > \omega c/\sqrt{3}$, then $\beta_0^* = \sqrt{1 - 2\omega c/\sqrt{27}\alpha^*}$.

8.6 Two frames S, S' are in standard configuration. With respect to S, a light pulse circulates (at speed c) around a narrow ring in the xy plane, of radius R. What are the maximum and minimum values of the magnitude $|\mathbf{a}'|$ of its acceleration with respect to S'?
Answer: $(c^2/R)(1 \pm v/c)/(1 \mp v/c)$.

8.7 **Relative acceleration?** Consider two particles neither of which need move uniformly. It might seem sensible to define their relative acceleration as the acceleration of one in the instantaneous rest frame of the other. Explain whether variables so defined supply useful information.

Chapter 9

Remember that in this book "mass" always means rest mass, as explained in Section 9.1.

9.1 For a particle of mass m and speed βc, plot

(a) ε/mc^2 as a function of β and of p/mc;
(b) β as a function of ε/mc^2 and of p/mc.

9.2 A particle observed from an inertial frame S has energy 5 GeV and momentum 3 GeV/c. (1 GeV = 1000 MeV.)

(a) What is its speed (as a multiple of c)?
(b) What is its mass (expressed in terms of GeV/c^2)?
(c) A frame S' moves relative to S in the same direction as the particle. With respect to S', if the momentum of the particle is 4 GeV/c, what is its energy?
(d) What is the velocity of S' relative to S?
Answer to (d): $-0.186c$.

9.3 A photon of energy ε is absorbed by an initially stationary particle of mass m, forming a new particle.

(a) What are the energy and momentum of the new particle?
(b) What is its velocity?
(c) What is its mass?

9.4 A π meson at rest decays thus: $\pi \rightarrow \mu + \nu$. What is the total energy of the μ meson?
The masses in units of MeV/c^2 are $m(\pi) = 140$, $m(\mu) = 105$, $m(\nu) = 0$.
Answer: 109 MeV.

9.5 A neutrino of energy ε is scattered through $90°$ by an initially stationary electron (mass m). Calculate the energy of the scattered neutrino, and evaluate your result when $\varepsilon = mc^2$.
Answer: $\varepsilon/2$.

9.6 *K* mesons (mass M) in flight at speed $c\beta$ decay into pairs of π mesons (mass m). What are the maximum and minimum energies of the daughters?
Hint: start by convincing yourself that maximum and minimum are realized by π mesons moving parallel and antiparallel to the K meson. Check your algebra by verifying that for $m = M/2$ both daughters have velocity $c\beta$.

9.7 What is the minimum energy needed to split a ^{4}He nucleus (an α particle) into two ^{2}H nuclei (deuterons)?
  The masses of the *atoms*, in modern amu, are 4.002600 and 2.0141014. Express your answer in MeV.
Recall note 8 to Section 9.3.2.

9.8 A heavy radioactive nucleus at rest emits an α particle, whose mass is 4 amu. The masses of the parent and of the daughter nuclei differ by 6 amu, and are so much larger than 4 amu that the daughter can be considered as remaining at rest. Determine the speed and the kinetic energy of the α particle.

9.9 Consider neutral π mesons decaying in flight as described in Example 9.2. In the meson rest frame S' the angular distribution of the photons is isotropic, i.e. the fraction of photons emitted into an element $d\Omega'$ of solid angle is just $d\Omega'/4\pi$.

(a) Find the angular distribution of the photons with respect to the laboratory frame S, i.e. the function $f(\theta, \phi)$ such that on average a fraction $f d\Omega$ of the photons are emitted into an element of solid angle $d\Omega \equiv d\phi d\cos\theta$. Here θ is the angle between the photon momentum and the beam direction, and ϕ the usual azimuthal angle. Check that your result satisfies the obvious requirement $\int_0^{2\pi} d\phi \int_{-1}^{1} d\cos\theta f = 1$. Plot f against θ (or if you prefer against $\cos\theta$) for $\beta = v/c = 0.1$ to 0.9 in steps of 0.1.
Hint: the fraction of decays populating the solid angle $d\Omega \equiv d\phi d\cos\theta$ with respect to S is identically the same as the fraction populating $d\Omega' \equiv d\phi' d\cos\theta'$ with respect to S'.
Answer: $f = 1/4\pi\gamma^2(1 - \beta\cos\theta)^2$, where $\gamma \equiv 1/\sqrt{1 - \beta^2}$.

(b) Find the energy distribution of the photons, i.e. the function $g(k_0)$ such that on average a fraction $g dk_0$ of the photons are emitted into an energy range $c dk_0$ around ck_0. Check that your result satisfies the obvious requirement $\int_{k_0(\text{min})}^{k_0(\text{max})} dk_0 g = 1$.
Answer: $g = 1/\beta\gamma m$.

9.10 Relativistic equipartition? At temperature T, the probability that a free particle has momentum with magnitude in the range dp around p is $dp f(p) = dp p^2 \exp(-\beta\varepsilon)/Z$, where $\varepsilon(p)$ is the particle energy, $\beta \equiv 1/kT$, and $Z \equiv \int_0^\infty dp p^2 \exp(-\beta\varepsilon)$ is a norming constant which ensures that

$\int_0^\infty dp f(p) = 1$. Therefore the average energy is

$$\langle \varepsilon \rangle = \int_0^\infty dp f(p) \varepsilon = \frac{\int_0^\infty dp\, p^2 \exp(-\beta \varepsilon) \varepsilon}{\int_0^\infty dp\, p^2 \exp(-\beta \varepsilon)}$$

$$= \frac{-(d/d\beta) \int_0^\infty dp\, p^2 \exp(-\beta \varepsilon)}{\int_0^\infty dp\, p^2 \exp(-\beta \varepsilon)} = -\frac{dZ/d\beta}{Z} = -\frac{d}{d\beta} \log Z.$$

All this is true regardless of the expression used for $\varepsilon(p)$.

(a) At low temperatures ($kT \ll mc^2$) it is overwhelmingly probable that ε is very close to mc^2. Then one may use the approximation (9.20) $\varepsilon \simeq mc^2 + p^2/2m - p^4/8mc^3$, and approximate further by

$$\exp(-\beta[mc^2 + p^2/2m - p^4/8mc^3]) \simeq \exp(-\beta[mc^2 + p^2/2m])(1 + \beta p^4/8mc^3).$$

Use this approximation to show that $\langle \varepsilon \rangle \simeq mc^2 + 3kT/2 + (15/8)(kT)^2/mc^2$. *Hint: when you meet* $\log(1 + \beta p^4/8mc^3)$, *approximate it as* $\beta p^4/8mc^3$. *You may quote the integrals* $\int_0^\infty dx \exp(-x^2) x^{2n} = 1.3 \ldots (2n-1)\sqrt{\pi}/2^{n+1}$, *for* $n = 1, 2, 3, \ldots$.

(b) At high temperatures ($kT \gg mc^2$) it is overwhelmingly probable that ε is much greater than mc^2. Then one may use the approximation (9.21) $\varepsilon \simeq cp + m^2c^3/2p$, and approximate further by

$$\exp(-\beta[cp + m^2c^3/2p]) \simeq \exp(-\beta cp)(1 - \beta m^2c^3/2p).$$

Use this approximation to show that $\langle \varepsilon \rangle \simeq 3kT + m^2c^4/2kT$. *Hint: when you meet* $\log(1 - \beta m^2c^3/2p)$, *approximate it as* $-\beta m^2c^3/2p$. *You may quote the integrals* $\int_0^\infty dx \exp(-x) x^n = n!$, *for* $n = 0, 1, 2, \ldots$.

(c) The average value of p^2c^2/ε is

$$\left\langle \frac{p^2c^2}{\varepsilon} \right\rangle = \frac{\int_0^\infty dp\, p^2 \exp(-\beta \varepsilon) p^2 c^2/\varepsilon}{\int_0^\infty dp\, p^2 \exp(-\beta \varepsilon)}, \qquad \varepsilon = \sqrt{m^2c^4 + p^2c^2}.$$

Show, by an appropriate integration by parts in the numerator, that $\langle p^2c^2/\varepsilon \rangle = 3kT$ (exactly).

9.11 For the **Cockroft–Walton reaction** decribed in Section 9.3.2, with $K_p = 0.5\,\text{MeV}$, calculate the kinetic energy K_α of an α particle emitted (a) at right angles, (b) parallel, and (c) antiparallel to the incident proton. Given their accuracy, discuss for each case whether one needs to distinguish between K_α and $Q/2$. *Hint: Start by showing that it is perfectly safe to approximate all the kinetic energies nonrelativistically by* $p^2/2m$.

9.12 **The time energy indeterminacy relation** in quantum mechanics shows that if a state decays with a mean life T, then its energy ε can be determined only within $\Delta \varepsilon$, where

$$T\Delta \varepsilon \geq \hbar/2. \tag{i}$$

For a long-lived unstable particle of mass m (rest energy mc^2) and mean life T_0 in its rest frame (i.e. in a state with three-momentum $\mathbf{p} = 0$), this entails

$$T_0 \Delta mc^2 \geq \hbar/2. \tag{ii}$$

(a) By considering the small change $\Delta\varepsilon$ of energy that results from a small change Δm of mass *at constant* $\mathbf{p}$, and by linking T and T_0 through time dilation, show that if (ii) is true in the rest frame of the particle, then (i) is true in any frame.

(b) Show that, by contrast, (i) would fail if $\Delta\varepsilon$ were taken to be the change in ε resulting from a small change Δm of mass *at constant velocity*.

(If you are puzzled why (a) works while (b) fails, consult your teacher of quantum mechanics.)

9.13 Newtonian conservation ruled out another way. In an inertial frame S', consider an elastic collinear collision along the x' axis between two particles with masses m_1, m_2 and initial (final) velocities $u_1', u_2', (w_1', w_2')$. Start by assuming the Newtonian conservation laws

$$m_1 u_1' + m_2 u_2' = m_1 w_1' + m_2 w_2', \quad m_1 u_1'^2 + m_2 u_2'^2 = m_1 w_1'^2 + m_2 w_2'^2. \tag{i), (ii}$$

(a) Use (i), (ii) to derive

$$u_1' - u_2' = -(w_1' - w_2'). \tag{iii}$$

By the relativity principle we can drop the primes from (i), (ii), (iii), obtaining relations of the same form with respect to another inertial frame S in standard configuration with S', with relative velocity $-v$.

(b) Lorentz-transform (iii) to S, and then use the unprimed version of (iii) to derive

$$1 + v(u_1 + u_2)/c^2 + v^2 u_1 u_2/c^4 = 1 + v(w_1 + w_2)/c^2 + v^2 w_1 w_2/c^4. \tag{iv}$$

(c) Because (iv) must hold for arbitrary values of v, one is entitled to equate separately the coefficients of v and of v^2 on the two sides. Hence show that $u_1 + u_2 = w_1 + w_2$ and $u_1^2 + u_2^2 = w_1^2 + w_2^2$. But (unless the masses are equal) these equations are incompatible with the unprimed versions of (i) and (ii), which shows that (i) and (ii) are not form-invariant, and therefore that they cannot be true.

Chapter 10

10.1 (a) What are the dimensions of $e^2/4\pi\varepsilon_0$ if one uses natural units ($c = 1$)?

(b) Evaluate $e^2/4\pi\varepsilon_0$ in such units, taking the unit of $[T]$ as 1 s, and the unit of $[M]$ as 1 MeV.

Answer: 4.80×10^{-24} MeV $\cdot$ s.

10.2 The size of a hydrogen atom in its ground state is of the order of the Bohr radius $a_0 = (4\pi\varepsilon_0/e^2)\hbar/m$, where m is the electron mass. Calculate the ratio of a_0 to the Compton wavelength of the electron.

10.3 The classical radius r_e of the electron is defined so that a charge e distributed uniformly over the surface of a sphere of radius r_e has energy equal to the electron mass m. Calculate r_e, its ratio to the Compton wavelength, and its ratio to a_0.

10.4 The deuteron (the nucleus of heavy hydrogen) consists of a proton and a neutron, with a binding energy $E \approx 2\,\text{MeV}$, which is small by nuclear standards. The masses of a neutron and a proton are almost equal, $M \sim 1000\,\text{MeV}$.

(a) The structure of the deuteron is usually considered on the basis of the Schrödinger equation, which takes kinetic energies to be (momentum)^2/ 2(mass). Estimate the accuracy of this approximation.

(b) According to the meson theory of nuclear forces, the forces responsible for the binding have a range of order $r_e \sim 1/\mu$, where $\mu \sim 140\,\text{MeV}$ is the mass of the π meson. According to quantum mechanics, the radius of the deuteron is $R \sim \hbar/\sqrt{ME}$.

(c) Evaluate the ratio r_e/R.

(d) Evaluate R and r_e in fermis (1 fermi $= 10^{-15}$ m), and then as multiples of the π meson Compton wavelength.

Chapter 11

11.1 K mesons are to be produced by shooting π mesons at effectively stationary protons, through the reaction

$$\pi + p \rightarrow K + \Lambda.$$

Determine the minimum total energy needed by the π mesons, given the masses (in MeV)

π	K	p	Λ
140	494	938	1116

Answer: 902 MeV.

11.2 The proton (mass 938 MeV) has an unstable excited state called N^* (mass 1232 MeV). If an N^* is to be made by letting a stationary proton absorb a photon, what is the energy that the photon must have?
Answer: 340 MeV.

11.3 **Multiparticle decays.** A K meson having mass m decays from rest into three π mesons having identical masses μ.

(a) In the configuration where one of the daughters is produced at rest, calculate the energies and momenta of the other two.
Answers: 177 MeV, 108 MeV/c.

(b) In the configuration that produces one daughter (say number 3) with the greatest possible energy $\varepsilon_3 = \varepsilon_{\max}$, the other two are emitted in the opposite direction with equal energies (i.e. $\varepsilon_1 = \varepsilon_2$). Calculate $\varepsilon_{\max}$.

Answer: 187 MeV.

(c) Prove the assertion made in part (b) about the way to maximize ε_3.

Hint: for the proof, think of the process as if there were only two daughters, one of them daughter 3 of mass μ, and the other daughters 1 and 2 jointly, of total mass M (along the lines of Section 11.2); and notice that ε_3 is maximized by minimizing M. The same argument tells you how to maximize the energy of any one daughter when there are arbitrarily many, with arbitrary masses.

11.4 K mesons can also decay into two π mesons. For such decays in flight at speed $0.9c$, determine the maximum angle to the flight path at which a daughter can emerge.

Answer: $44.8°$.

11.5 For Compton scattering, calculate the angle χ between the momentum of the recoiling electron and the incident photon.

Answer: $\tan \chi = \cot(\theta/2)/(1 + q_i/mc)$

11.6 **Photons with nonzero mass?** Consider the Compton effect, supposing that photons have mass μ, much smaller than the electron mass and q_i/c, where q_i is the momentum of the incident photon.

(a) Neglecting terms higher than quadratic in μ, show that Compton's formula then becomes

$$\lambda_f - \lambda_i = \lambda_c(1 - \cos\theta) + \frac{1}{2}\left(\frac{\mu}{m}\right)^2 \left\{ \frac{\lambda_i^2}{\lambda_c}(1 - \cos\theta) + (\lambda_i + \lambda_c)(1 - \cos\theta)^2 \right\}$$

$$+ \cdots.$$

(b) Find rough upper limits on μ (in eV) using Gingrich's data quoted in Section 11.4.2.

Hints: photon momentum and wavelength are related by $\lambda = q/h$ regardless of the value of μ. Use the approximation from Section 9.2.1 appropriate to $p/\mu c \gg 1$.

11.7 For elastic collisions between equal-mass particles, plot the incident speed in S^L against the speeds in S^*. Make sure the slopes are correct at both ends.

11.8 The elastic scattering process shown in Figure 11.4 is often described in terms of the **Mandelstam variables**

$$s \equiv (\vec{p}_1 + \vec{p}_2) \cdot (\vec{p}_1 + \vec{p}_2), \quad t \equiv (\vec{p}_1 - \vec{p}_3) \cdot (\vec{p}_1 - \vec{p}_3), \quad u \equiv (\vec{p}_1 - \vec{p}_4) \cdot (\vec{p}_1 - \vec{p}_4).$$

(a) Show that $s + t + u = 2m_1^2 + 2m_2^2$.

(b) Show that the four-momentum transfer $\vec{p}_1 - \vec{p}_3$ is spacelike, i.e. that t is negative. Express t in terms of s and of the scattering angle θ^* in the centre-of-momentum frame.

(c) Express u similarly. Is $(\vec{p}_1 - \vec{p}_4)$ timelike or spacelike or can it be either?

11.9 With respect either to the centre of momentum or to the laboratory frame it is obvious that all the three-momenta in elastic scattering lie in the same plane, called **the scattering plane**. Is there such a plane with respect to *any* inertial frame?

Hint: the question can be answered only by a general proof or by a counter-example.

11.10 Gathering dust. A body moves through a stationary dust cloud of density ρ. At time $t = 0$ the mass and the velocity of the body are m_0 and u_0. All the dust particles hit by the body adhere to it, but without changing its cross-sectional area A.

(a) Show that Newtonian mechanics gives the velocity of the body as

$$u(t)/u_0 = \{1 + t/T\}^{-1/2}, \tag{i}$$

$$T = m_0/2\rho A u_0. \tag{ii}$$

(b) Show that relativistically (i) still applies, but now with

$$T = m_0/2\rho A u_0 \sqrt{1 - (u_0/c)^2}. \tag{iii}$$

11.11 What rate of burn with respect to proper time, i.e. what function $\dot{m} \equiv dm/d\tau$, will produce constant acceleration α of a rocket with respect to its instantaneous rest frame?

11.12 Photon rocket. A sci-fi rocket operating in gravity-free space has initial mass m_i, a constant burn rate $\dot{m}$ with respect to its proper time τ (so that $m = m_i - \dot{m}\tau$), and an exhaust speed equal to c.

(a) Relative to its initial rest frame S, how long does it take to reach speed $u = c/2$?
(b) What is the corresponding value of τ?
Answers: (a) 0.4413 $m_i/\dot{m}$, (b) 0.4226 $m_i/\dot{m}$.
Hint: you are unlikely to be able to determine $u(t)$, but it is not too hard to find $t(u)$, which is given by

$$\frac{u/c}{2(1 + u/c)} + \frac{1}{4}\log\left(\frac{1 + u/c}{1 - u/c}\right) = \frac{\dot{m}}{m_i}t.$$

11.13 Photon rocket 2. Suppose that the rocket reduces its mass instantaneously. Then the burn is equivalent to the decay from rest of a particle of mass m_i into another of mass m_f plus a single photon. Calculate the final velocity of the daughter particle, compare it with the velocity given by the rocket formula with $w = c$, and comment briefly.

11.14 Problem 4.5 imagined a two-stage rocket, such that each burn increases the speed by the same amount with respect to the frame where the rocket was at rest before the burn. The final speed relative to the initial rest frame is $c/2$. If the exhaust speed is $w = c/2$, what fraction of the *initial* mass is needed as propellant (a) for the first burn? (b) for the second burn?
Answers: 0.4226, 0.2440.

11.15 Pandora's box. A rigid one-dimensional box (rest frame S) contains a particle of mass m that bounces elastically, with energy $\varepsilon = \sqrt{m^2 + p^2}$ and momentum $\pm p$, between opposite faces. Thus the time averages relative to S are $\langle\text{energy}\rangle = \varepsilon$ and $\langle\text{momentum}\rangle = 0$. The frame S' is in standard configuration with S. Calculate the time averages (relative to S') of the particle's energy and momentum.

Caution: the results are not what one might have expected. Thus $\langle\text{momentum}\rangle' \neq \gamma\beta\varepsilon$, and $\langle\text{energy}\rangle' \neq \gamma\varepsilon$. (The differences stem from the fact that under Lorentz transformations the inevitable stresses in the box itself relative to S contribute to the total energy and momentum relative to S'. The details are beyond our scope here.)

11.16 Thermodynamic equilibrium. In Example 11.3, it proves convenient to define the average temperature $\bar{\theta} \equiv (n_1\theta_1 + n_2\theta_2)/n$, and the four-scalar $\Delta \equiv (\vec{U}_1 \cdot \vec{U}_2/c^2 - 1)$. If $\mathbf{u}_1 = \mathbf{u}_2$, then Δ vanishes, and $\theta^* = \bar{\theta}$. If (relative to the frame S), $u_{1,2}/c \ll 1$, then $\Delta = (\mathbf{u}_1 - \mathbf{u}_2)^2/2c^2 + (\text{terms of order } 1/c^4)$.

(a) Show that the equation for θ^* may be written as

$$(1 + s\theta^*/c^2)^2 = (1 + s\bar{\theta}/c^2)^2 + \frac{2n_1n_2}{n^2}(1 + s\theta_1/c^2)(1 + s\theta_2/c^2)\Delta.$$

(b) For small Δ one may expand in powers of $1/c^2$:

$$\theta^* = \bar{\theta} + A_0 + A_2/c^2 + \cdots,$$

where A_0, A_2 are functions of Δ, and thereby of $\mathbf{u}_{1,2}$. Verify that the nonrelativistic correction is

$$A_0 = \frac{n_1n_2}{n^2} \cdot \frac{(\mathbf{u}_1 - \mathbf{u}_2)^2}{2s},$$

and determine the first relativistic correction, i.e. A_2.

Chapter 12

12.1 S and S' are in standard configuration. A source at rest with respect to S' emits light in the $x'y'$ plane, at an angle ϕ' to the positive x' axis.

(a) Given $\phi' = \pi/4$ and $v = c/2$, what are the velocity components of this light in the x and y directions with respect to S?

(b) Given $v/c = 0.99$, tabulate and plot ϕ for values of ϕ' varying between 0 and 180° at intervals chosen to give what you consider to be reasonable accuracy.

12.2 Relative to the laboratory frame S, a simple-harmonic light wave with frequency ω propagates through empty space in the positive x direction. The frequency is ω' relative to another frame S' in standard configuration with S. Tabulate and plot v/c for values of ω'/ω from 0 to 2 at intervals of 0.1.

12.3   A laser source is mounted at the edge of a turntable of radius R. When the turntable is at rest, the light beam from the source is directed radially outwards, along the x axis. At time $t = 0$ the table starts to rotate with constant angular velocity $\Omega < c/R$. Determine the angle ϕ between the beam and the x axis at any later time.
Hint: assume that ϕ depends only on the instantaneous velocity of the source.
Answer: $\phi = \Omega t + \cot^{-1}\sqrt{(c/\Omega R)^2 - 1}$.

12.4   A uniform hollow waveguide at rest relative to a frame S is parallel to the x axis and has square cross-section with sides of length a. We pretend that inside there is a vacuum and that the walls are perfect conductors. Then the simplest type of electromagnetic wave propagating along the guide (so that $k_x = k$) obeys the dispersion relation $\omega/c = \sqrt{k^2 + \pi^2/a^2}$.

(a) Is $\vec{K} = (\omega/c, k, 0, 0)$ timelike, spacelike, or lightlike?
(b) Show that $u_p u_g = c^2$. Sketch ω, u_p/c, and u_g/c as functions of ka for $-\infty < ka < \infty$.
(c) Evaluate u_p/c and u_g/c for $ka = 1$. Determine u_p'/c, and u_g'/c relative to another frame S' in standard configuration with S.
(d) Can a Lorentz transformation (with v in the x direction) reverse the sign of k_x, of u_p, or of u_g?

12.5   Suppose the waveguide in the preceding problem has its closed sides parallel to the y and to the z axes. Then, in its rest frame, for one of the modes with the indicated frequency the electric field inside varies as

$$\mathbf{E} = \hat{\mathbf{y}}\cos(\omega t - kx)\sin(\pi z/a), \qquad 0 \le y, z \le a.$$

How does the frequency change under Lorentz transformations with $\mathbf{v}$ parallel to the y or to the z axis?
Caution: think carefully. You do not need to know anything about the Lorentz transformations of electromagnetic fields.

12.6   Suppose that in the experiment described in Section 12.5, the two beams are slightly separated, and that one traverses a stationary disk, while the other traverses an identical disk spinning so that the light path is everywhere at right angles to the velocity $\mathbf{v}$ of the material. Calculate the beat frequency for this case, to leading order in v/c. Estimate its maximum value, in Hz. Is it likely to be observable?

12.7   **Light in a plasma.** Electrodynamics shows that to a reasonable approximation transversely polarized waves in a cold electron plasma obey the dispersion relation

$$\omega^{(0)2} = \omega_p^2 + c^2 k^{(0)2}$$

relative to a frame $S^{(0)}$ where the undisturbed plasma is at rest. Here ω_p is the so-called plasma frequency, given by $\omega_p^2 = N^{(0)}e^2/\varepsilon_0 m$, with e, m the electron charge and mass, and $N^{(0)}$ the number of electrons per unit volume. The plasma is said to be cold if with respect to $S^{(0)}$ the total energy $\varepsilon^{(0)}$ of each electron differs negligibly from mc^2. (Do not confuse this energy $\varepsilon^{(0)}$ with the vacuum permittivity ε_0.)

(a) Using the Lorentz-transformation rules for ω and $\mathbf{k}$, determine the dispersion relation $\omega(\mathbf{k})$ relative to another frame S having velocity $\mathbf{v}$ with respect to $S^{(0)}$. Verify that under Lorentz transformation the plasma remains optically isotropic, and the plasma frequency unchanged.

(b) In $S^{(0)}$ we could equally well have defined the plasma frequency by $\omega_p^2 \equiv N^{(0)}(e^2/\varepsilon_0)c/P_0^{(0)}$, where $P_0^{(0)} = \varepsilon^{(0)}/c$ is the time component of the energy-momentum four-vector $\vec{P}$ for an electron. By Lorentz-transforming $\vec{P}$ and the number density N from $S^{(0)}$ to S, show directly from this definition that the plasma frequency is indeed an invariant.

Note that such a plasma is isotropic not only in its rest frame but in all frames. This is quite unlike the behaviour of familiar materials like glass, which become anisotropic when they move, as illustrated in Example 12.1.

12.8 **Sound in a plasma.** The plasma featured in Problem 12.7 can also support compressional (i.e. longitudinally polarized) waves. In $S^{(0)}$ these obey the dispersion relation

$$\omega^{(0)2} = \omega_p^2 + s^2 k^{(0)2},$$

where $s \ll c$ is a constant which would be equal to the speed of sound if the electrons were neutral.

(a) Under what conditions are the four-vectors $\vec{K}$ for such waves timelike, spacelike, or lightlike?

(b) Determine the group and the phase velocities with respect to the rest frame $S^{(0)}$, and verify that $u_g^{(0)} u_p^{(0)} = s^2$.

(c) Following the method of Example 12.1, or otherwise, show that the invariant dispersion relation reads

$$\vec{K} \cdot \vec{K} = \omega_p^2/s^2 - (1/s^2 - 1/c^2)(\vec{K} \cdot \vec{W})^2.$$

(d) Express this appropriately to a plasma having velocity $\mathbf{w}$. Assuming $w/s \ll 1$ as well as $w/c \ll 1$ (but working to all orders in s/c) linearize in $\mathbf{w}$ and show that the dispersion relation becomes

$$\omega \approx (1 - s^2/c^2)\mathbf{w} \cdot \mathbf{k} + \sqrt{\omega_p^2 + s^2 k^2}.$$

12.9 With respect to an inertial frame S, the standard nonrelativistic Schrödinger wave-function for a free particle of mass m reads

$$\psi \sim \exp\{-i\phi\}, \qquad \phi = (\varepsilon t - \mathbf{p} \cdot \mathbf{r})/\hbar,$$

where $\varepsilon \equiv p^2/2m$. By the same prescription the phase with respect to another inertial frame S' is $\phi' = (\varepsilon' t' - \mathbf{p}' \cdot \mathbf{r}')/\hbar$.

Use the Galilean transformations from chapter 2 to show that

$$\phi' = \phi - m(v^2 t/2 - \mathbf{v} \cdot \mathbf{r})/\hbar \neq \phi.$$

*Thus the phase of the Schrödinger wave function is not invariant under Galilean transformations. (What saves the situation eventually is the fact that $\phi' - \phi$ does not depend on the state of the particle, i.e. on **p**.)*

12.10 In its rest frame $S^{(0)}$ a glass rod has the frequency-independent refractive index $n^{(0)}$. Electromagnetic theory tells us that a light beam traversing the rod has the same frequency inside and out, $\omega_{out}^{(0)} = \omega_{in}^{(0)}$, and wave-vectors $k_{out}^{(0)} = c\omega_{out}^{(0)}$, $k_{in}^{(0)} = n^{(0)}k_{in}^{(0)}$. Reflections should be disregarded.

With respect to the laboratory frame S, the rod moves with velocity w in the same direction as the light. By Lorentz-transforming the outside and the inside waves between $S^{(0)}$ and S, show that the laboratory frequencies are related by

$$\frac{\omega_{in}}{\omega_{out}} = 1 + \frac{(n^{(0)} - 1)}{1 + c/w}.$$

12.11 A glass cylinder which in its own rest frame has length a and refractive index n moves with speed $c\beta$ along an evacuated tube of length L. Calculate the time T needed by a light pulse to traverse the tube in the same direction as the cylinder. *Answer:* $T = L/c + (a/c)(n - 1)\sqrt{(1 - \beta)/(1 + \beta)}$.

Chapter 13

13.1 From Figures 13.2, identify the dates on which the earth was travelling directly towards, and directly away from, the projection of each star onto the ecliptic.

13.2 For stars $45°$ above the ecliptic, at what distance does maximum apparent displacement due to parallax equal the maximum aberration?

13.3 If **v** and the signal direction $\hat{\mathbf{k}}$ are collinear (parallel or antiparallel) in one of the frames $S^{(E,R)}$, then they are collinear in the other frame too: thus collinearity can be prescribed to the same effect in either. But perpendicularity cannot. Determine the frequency ratios when $\hat{\mathbf{k}}$ is perpendicular to **v** in (a) $S^{(R)}$; and (b) $S^{(E)}$.
Answers: (a) $\omega_R^{(R)}/\omega_E^{(E)} = \sqrt{1 - (v/c)^2}$; (b) $\omega_R^{(R)}/\omega_E^{(E)} = 1/\sqrt{1 - (v/c)^2}$.

13.4 Consider green light ($\lambda \simeq 6.5 \times 10^{-7}$ m) received on earth from points diametrically opposite on the sun's equator. What is the wavelength difference between such rays?
(The sun's radius is 109 earth radii, and its rotation period is 25.4 days.)

13.5 A roadside speed trap works by measuring the beat frequency Ω between a radar signal (3000 MHz) and its reflection from cars approaching effectively head-on. The police wish to prosecute drivers exceeding 130 km/h. What is the minimum value of Ω leading to prosecution?
Hint: at such low speeds the reflected signal can be treated as if it came from the geometrical-optics image of the radar gun mirrored by the car, the image moving at twice the speed of the car.

13.6 Astronomers observe that on a cosmic scale light sources recede from earth with velocity proportional to their distance: this is the main evidence for the expansion of the universe. Such velocities are determined through the Doppler shifts they produce, called recessional red-shifts. The largest measured to date is of the order $\lambda^{(R)}/\lambda^{(E)} \sim 6$. (a) What speed does this indicate? (b) What speed would one infer without the factor stemming from time dilation?

13.7 Two monochromatic light sources approach each other head-on with equal speeds βc relative to the laboratory. When they pass, the frequency of the light each receives from the other is halved. Find β.
Answer: $\beta = 3 - \sqrt{8} \approx 0.1716$. *You might enjoy the contrast with Example 2.10.*

13.8 For the received to exceed the emitted frequency, what conditions must be imposed (a) on the angle, call it $\phi^{(R)}$, between $\mathbf{k}^{(R)}$ and $\mathbf{u}_E^{(R)}$; (b) on the angle, call it $\phi^{(E)}$, between $\mathbf{k}^{(E)}$ and $\mathbf{u}_R^{(E)}$? The relative speed v is given, and is the same in both cases.

13.9 A mirror parallel to the yz plane is receding with velocity βc along its normal, i.e. along the x axis. A plane monochromatic light wave propagating in the positive x direction catches up with and is reflected by the mirror.

(a) Determine the frequency of the reflected beam.
Hint: transform the incident and the reflected beams, separately, from the laboratory to the mirror rest frame, and then back again. (A mirror at rest reflects without changing the frequency.) Answer: $\omega_{\text{refl}}/\omega_{\text{inc}} = (1 - \beta)/(1 + \beta)$.
(b) Compare the result with the Doppler shift one would predict by taking the reflected beam to come from the geometrical-optics image of the source, given that such an image moves with twice the speed of the mirror. Show that there is (approximate) agreement only for $\beta \ll 1$.

13.10 As part (a) of the preceding question, but with light incident at an angle ϕ_{inc} to the x axis. Show that

$$\tan(\phi_{\text{refl}}/2) = \frac{1 + \beta}{1 - \beta} \tan(\phi_{\text{inc}}/2), \qquad \frac{\omega_{\text{refl}}}{\omega_{\text{inc}}} = \frac{1 - 2\beta \cos(\phi_{\text{inc}}) + \beta^2}{1 - \beta^2}.$$

Hints: as for 13.9(a); and see Example 9.3.

13.11 The navigator of a spaceship observes a radio signal from a beacon known to be monochromatic (though with an unknown frequency) in its own rest frame, and known also to be moving (relative to her) along a trajectory $\mathbf{r}(t) = ((t - t_0)v, -h, 0)$, where t_0, v, h are constants. She measures the frequency $\omega(t)$ of the signal, and the angle $\phi(t)$ between its direction and her x axis, both as functions of ship time t.

From these measurements it is possible to determine the values of t_0, v, and h. Write a set of *explicit* instructions that will allow her to do so *as conveniently as possible*.
Hint: as a preliminary, determine the time of emission $\tilde{t}$ of the light arriving at time t, governed by the equation $|\mathbf{r}(\tilde{t})| = c(t - \tilde{t})$.

Index

Bold face indicates a chapter, section, or
  appendix

Aberration
  nonrelativistic, 28, 31
  relativistic, **13.2**; 161, 169–70, 198,
    224
Acceleration
  four-acceleration, **Chapter 8**
  Galilean transformation, 15
  Lorentz transformation, **4.2.5**; 177
  proper, 94–97, 213, 220
  relative, 214
  stresses, **6.3**; 74, 75
Active transformations, **2.3.4**; 75
Aether models (or theories) of light, 38, 146,
    164, 167
  evidence, **3.4.4**
Amplitude and phase, 26–27, 145
Angular momentum
  Galilean transformation, 207
  in hydrogen atom, 193
  Lorentz transformation, 176–77
Atomic mass units: *see* Units
Atomic units: *see* Units

Ballistic models (or theories) of light, **3.4.2**
  evidence, **3.4.3**
Black-body radiation
  cosmic background radiation, 8, 77, 200,
    204
  energy and momentum densities, 202–03
  force on body moving through, 203–04
  Lorentz transformation of Planck's law,
    Appendix E

Boltzmann: *see* Distribution functions;
  Equipartition
Bradley (stellar aberration), **13.2.2**, **13.2.3**;
  174, 224

Causality, 70
Centre-of-momentum and laboratory frames,
  23–24, 125, 127, 133–34, 140
Charged particles, **Appendix C**
  motion in constant B field, 190–91;
  motion in Coulomb field, 191–93
  motion in constant E field, 189–90;
"Clock-paradox", 208, 211
Clocks: *see* also Light–clocks
  clock hypothesis, 74;
  evidence, **6.4**
  gravitational effects on, 8, 82
  ideal or robust, **6.3**, **6.4**; 74
Cockroft–Walton reaction, **9.3.2**; 216
Colliding beams, 128–29
Co-moving frame, 77, 94, 137
Common sense, 6, 40
Compton effect
  evidence, **11.4.2**
  theory, **11.4.1**; 219
Compton wavelength, 113, 129, 158, 217
Conservation laws for Einsteinian energy and
    momentum, **9.1**; **Chapter 11**
  evidence, **9.3**
  and the relativity principle, 103–05
Conservation laws, Newtonian
  energy, 25
  failure of, 101–02, 217
  mass, **2.4.1**, **9.3.4**
  momentum, **2.4.2**
  and the relativity principle, **2.4.5**; 207

Continuity conditions for light waves, 155, 159
Coordinates: *see also* Galilean transformations; Lorentz transformations
transformation under rotations, **B.1, B.2, B.4.1**; 18
transformation under translations, 18
Cosmic background radiation: *see* Black-body radiation
Coulomb potential, **C.2.3**
Covariance: *see* Form-invariance

Decay, two-body, **11.1**; 106, 107, 139–40, 219
multiparticle, 218
de Broglie: *see* Quantum mechanics
Dimensions and dimensional analysis, 78–79, 118
Dispersion relations: *see* also Fizeau effect; Refractive index
2.5.3; 27, 145, 147–48, 150, 158, 223
Distribution functions: *see* also Equipartition
Lorentz transformation, **E.2**; 199
Maxwell-Boltzmann, 170, 206
Doppler effect
for black-body radiation, 161, 198, 200
for binary X-ray source, 45–6
nonrelativistic, **2.5.4**
collinear, 32
general, 33–34
in reflection, 207–08
relativistic, **13.3–13.5**
collinear, **13.3**; 224, 225
evidence, first order, 170–71
evidence, second order, 172–74
general, **13.4**
in reflection, 225
and time-dilation, 172
Drag coefficient: *see* Fresnel

Einstein relation for photons: *see* Quantum mechanics (Planck–Einstein relation)
Einsteinian vs Newtonian physics, definition, 10
Elastic scattering, **11.5**; 219, 220
centre-of-momentum and laboratory frames, 133–34
laboratory opening angle between equal-mass particles, 134–35
Elasticity (relativistic): *see* Rigidity
Electromagnetism, appendix C; vii, 3, 38, 39
Emission theory: *see* Ballistic models
Energy: *see also* Conservation laws; Four-momentum; Galilean transformations; Lorentz transformations; Mass and energy
Einsteinian, 104–05, 105–06

internal or rest, 104
kinetic, 105
release, 105, 114
and speed, **9.3.1**; 104
Newtonian (kinetic and potential), **2.4.4**
Energy density: *see* Black-body radiation
Energy-momentum four-vector: *see* Four-momentum
Equipartition, 174, 215–16
Equivalence principle, 208
Euler's paradox
for aberration, 164
for the Doppler effect, 34–35
Event, 11, 13, 53, 55, 67–69, 70, 208
Experimental evidence (how assessed), 6–7, 64–5
Extreme-relativistic limit or regime, 54, 106

Fine-structure constant, 54, 115, 193
Fine structure of the hydrogen atom, 193
FitzGerald: *see also* Lorentz contraction; 66
Fizeau effect, **12.5**; 159
dispersive correction to, 155–56
Force: *see also* Newton's second law; **2.4.3**, **Appendix C**; 101, 188–89
Form-invariance: *see also* Relativity principle
7.4, 9.1, 13.4; 20–21, 33–34, 102–03, 122–23, 147–48, 167–68, 199–200
test for, **2.3.3**
Four-acceleration: *see* Acceleration
Four-momentum: *see also*: Conservation laws; Energy (Einsteinian); Lorentz transformations
definition and properties, **9.1, 9.2**
for zero-mass particles, **9.2.2**
Four-scalar product, 85, 183
Four-vectors, **Chapter 7**
definition, 83–84
lightlike, spacelike, and timelike, 85
matrix representations, **B.3**; 84
Four-velocity, *see* Velocity
Frame hopping, 71
Fresnel drag coefficient, 153
Frequency and wave-vector, 27
frequency four-vector, 145, 147
Galilean transformation, 27–28
Lorentz transformation, **12.2**

Galilean transformations, **Chapter 2**
acceleration, 15
angles, 16
angular momentum, 207
derivation, **4.3**
dispersion relation, **2.5.3**
energy, **2.4.4**
force, **2.4.3**

frequency and wave-vector, **2.5.2**
inverse, 14
momentum, **2.4.2**
position coordinate, 13–14
  differences, 15
time, **2.2.1**; 14, 38
velocity
  combination rule, 15, 29
  differences, 16
  particle, 14–15
  group, 28, 29
  phase, 28–30
  wave-vector, **2.5.2**
General theory of relativity: *see* Gravitation
Gravitation,
  effect on clocks, 8, 18, 82
  general theory of, 6, 8
  and inertial frames, 82
Group velocity: *see* Velocity

History, 7
Homogeneity of space, 18, 63
Hydrogen atom, **C.2.3**; 54, 118, 193, 217

Indeterminacy relation for time and energy,
    140, 216–17
Inertia: *see* Newton's first law
Inertial frames: *see also* Centre-of-momentum
    and laboratory frames; Co-moving
    frame **2.3.1, 2.3.2**; 3, 10, 11, 12, 37, 39
  and gravitational fields, 8, 18, 82
Inertial observer: *see* Inertial frames
Instantaneous rest frame: *see* Co–moving
    frame
Intervals: *see* Space-time intervals
Invariance: *see* Form-invariance; Relativity
    principle
Invariant, definition of, 12
Invariant speed: *see* Light speed in vacuo;
    Speed
Inverse transformations: *see* Lorentz
    transformations; Galilean
    Transformations
Isotropy of space, 18
Ives and Stilwell experiment, 172

Kennedy–Thorndyke experiment, **3.4.4**; 50
Kinematics (definition), 12
Klein–Gordon equation, 196

Laboratory frame: *see* Centre-of-momentum
    and laboratory frames
Left-right symmetry, 63
Light-clocks
  longitudinal, **3.3.2**; 80
  transverse, **3.3.1**

Light cone: *see* Space-time diagrams
Light speed in vacuo: *see also* Light-clocks
  independent of apparatus velocity, **3.4.4**
  independent of source velocity, **3.4.2,
    3.4.3**
  invariance of, **3.1**; 3, 58, 89
Lightlike: *see* Four-vectors; Space-time
    intervals
Lorentz contraction, **3.3.2, 4.2.3**; 65, 75
  causes, 57
  derived from light-clock, **3.3.2**
  derived from Lorentz transformation,
    4.2.3;
  and visual appearances, 57, 66
Lorentz force, 189
Lorentz transformations (arbitrary relative
    velocity), **Appendix A**
  coordinates, 175; acceleration, 177; angular
    momentum (orbital), 176
Lorentz transformations (standard
    configuration)
  see also: Black-body radiation; Lorentz
    contraction; Time dilation;
    Matrix representation
  acceleration, **4.2.5**
  angles and directions, 56, 59–60, 109, 134,
    152, 161–63, 169, 209, 215, 221–12
  derivation, **4.3**
  in $1 + 1$ dimensions, **7.3**
  energy and momentum, **9.2.3**
  event coordinates (time and position), **4.1**
  four-vectors, **7.1**
  frequency and wave-vector, **12.2**; 145
  γ, 58–59, 88–89, 152
  gradient, **Appendix D**
  intervals: *see* Space-time intervals
  inverse, 14, 54, 63, 88, 152
  momentum: *see* energy and momentum
  position: *see* event coordinates
  time: *see* event coordinates
  velocity
    combination rule, 184–85
    group, **12.3**
    particle, **4.2.4**; 87
    phase **12.4**
  wavelength, 151
  wave operator, 195–96
  wave-vector: *see* frequency and wave-
    vector

Mach's principle, 9
Mandelstam variables, 219
Mass: *see also* Rockets; Thermodynamics;
    Zero mass
  continuously variable, **11.6.2**; 220
  discrete, **11.6.1**

and energy, 104, 123
  evidence, **9.3.1–9.3.3**
  invariance of, **2.4.1**; 103
  Newtonian, conservation of, 22, 115
  "rest mass", 103
  total, 122, 125–26
  and production thresholds, 11.3
Matrix representations of
  four-vectors, **B.3**; 84
  Lorentz transformations **B.3**; in 3D,
    B.4.2
  three-vectors, **B.1**
  rotations in 2D, **B.2**; in 3D, **B.4.1**

Maxwell–Boltzmann: *see* Distribution
  functions; Equipartition
Maxwell's equations, *see* Electromagnetism
Media, propagation of light in: *see*:
  Dispersion relations; Fizeau effect;
  Refractive index
Metric tensor, 183, 187
Minimal reading, 7
Momentum: *see* Conservation; Four-
  momentum; Galilean transformations;
  Lorentz transformations
Momentum density: *see* Black-body radiation

Natural units: *see* Units
Newton's first law, 3, 11–12, 13, 212
Newton's second law, 13, 24, 101, 188–89,
  192
Newton's third law, 13
Newton's rule for inelastic collisions, 19–20
Newtonian physics (definition), 10
Noninertial frames, *see also* Acceleration
  stresses; 12, 74–75
Nonuniform motion, **6.1**
"Nonrelativistic" limit or regime, *see also*
  Euler's paradox 10, 54, 89, 94, 104,
  106, 135, 147, 153, 155, 159, 165,
  166–67, 168

Oblique axes on space-time diagrams, 72–73
Observer: *see* Inertial frames
Orbits, **C.2.3**
Orthogonality between four-vectors, 93, 212

Parallax, 174, 224
Passive transformation, **2.3.4**; 75
Perihelion, 191–93, 194
Phase: *see also* Amplitude and phase;
  Quantum mechanics invariance of, in
  classical physics, 28, 36, 146–47
  invariance of, for relativistic quantum
  wave-functions, 196–97

Phase velocity: *see* Velocity
Photons: *see also* Black-body radiation;
  Compton effect, Quantum mechanics
  (Planck-Einstein relation)
  with nonzero mass, 37–38, 219
  photon rocket, 139–40, 220–21
  as zero-mass particles, **9.2.2**; 146, 198
Planck's law: *see* Black-body radiation
Planck–Einstein relation: *see* Quantum
  mechanics
Plane waves: *see* Waves
Plasma, intergalactic and interstellar, 46
  waves, 223
Point particles (definition), 35
Position coordinates: *see* Coordinates
Positron annihilation, **9.3.3**
Production thresholds, **11.3**
  centre-of-momentum and laboratory
    frames, 127, 128
  colliding beams, **11.3.2**
  stationary targets, 127–28, 140
Proper acceleration, 94, 213
Proper distance, 68
Proper time, **6.2**; 8, 40, 68, 94–97
  undefined at light speed, 89, 107

Quantum mechanics
  de Broglie relation and wavelength, 132,
    157, 158
  discrete versus continuously variable
    masses, 136
  hydrogen atom, 54, 118–19, 193, 217
  indeterminacy relation for time and energy,
    140, 216–17
  phases of Schroedinger wavefunctions, 158,
    159, 224
  of relativistic wavefunctions, **12.6**; 145,
    196–97
  Planck–Einstein relation (photon energy
    and frequency), 113, 129, 157, 198

Rapidity, 184–85
Reference frames: *see also* Inertial frames;
  Noninertial frames; Centre-of-mass
  and laboratory frames and
  transformation rules, **2.1**; 3, 10
Reflection invariance, 63
Refractive index, **12.4.3**; 147–48, 153,
  155–56, 156–67
Relative velocity: *see* Velocity
Relativity principle
  applicability, 8, 17–18
  and the conservation laws, **2.4.5**; 104
  Einstein's, **3.2**
  Galilean, **2.3.2**

"Rest mass" and Rest energy: *see* Mass (and energy)
Representations: *see* Matrix representations
Rigidity, relativistic limits on, 80–81, 82
Rockets, **11.6.3**; 220, 221
Rotating frames: *see* Noninertial frames
Rotation of coordinate axes: **B.1**, **B.2**, **B.4.1**

Scattering: *see* Compton effect; Elastic scattering; Mandelstam variables
"Segregate and square", 122, 124–25, 131
Simultaneity, 4, 13, 16, 35, 55, 56
Sound: *see also* Aberration (nonrelativistic); Doppler effect (nonrelativistic) **2.5.4**; 27, 29–30, 146
Spacelike: *see* Four-vectors; Space-time intervals
Space-time diagrams, **5.2**; 76, 210
  with oblique axes, 72–73
  and the light cone, 69–71
Space-time intervals, **5.1**; 209–10
  definition, 67
  invariant, 67–68
  lightlike, 68
  spacelike, 68, 72
  timelike, 68–69, 72
Speed: *see also* Light speed in vacuo; Velocity
  closing, 49, 60, 208–09
  invariant, finite, **3.1**; 3, 37, 39, 64, 147
  invariant, infinite (in Newtonian physics), 15, 37, 64
  some typical speeds, 54
  vs velocity, 9
Standard configuration, 14, 53, 141, 175
Stellar aberration: *see* Bradley
Stresses: *see* Acceleration stresses
Systems of particles treated as one particle
  in Newtonian physics, 23–24, 24–25
  relativistically, **11.2**, **11.6.1**; 122, 218–19
  in two-body decay, **11.1**;

Thermodynamics and temperature, 136, 137, 141, 207, 221
Thomas precession, 186
Timelike: *see* Four-vectors; Space-time intervals
Time dilation
  and Doppler effect, 172
  evidence, **4.4**, **6.4**
  from light clock, **3.3.1**
  from Lorentz transformation, **4.2.2**
Trajectories, 70–73

Transformations: *see also* Galilean transformations; Lorentz transformations; Matrix representations
  active and passive, **2.3.4**; 75
"Twin paradox": *see* "Clock paradox"
Two-body decay: *see* Decay
Uncertainty relation: *see* Indeterminacy relation

Units
  atomic, 121
  atomic mass units (amu), 112–13, 116
  conversion of, 117–18
  natural, **Chapter 10**
  of length, and c, 49

Vectors: *see also* Four-vectors; Matrix representation definition, **B.1**
  transformation under rotations, **B.2**, **B.4.1**
Velocity: *see also* Galilean transformations; Lorentz transformations
  combination rule
    Galilean, 15, 29
    Lorentz, 58, 60, 87, 159, 184–85
  differences, 16
  four-velocity, **7.2**; 72, 103, 107
    undefined at light speed, 89, 107
  group, **12.3**; 27–28, 29–30, 158
  phase: *see also* Refractive index
    12.4; 27, 28–30, 36, 49, 158, 159, 184–85
  relative, 88–89, 209, 212
  vs speed, 9

Wave equations, **Appendix D**; 159
Waves: *see* Aberration; Amplitude and phase; Dispersion relations; Doppler effect; Frequency and wave-vector; Galilean transformations; Lorentz transformations; Phase; Plasma; Quantum mechanics (phases); Velocity (group and phase)
Wave-vector: *see* Frequency and wave-vector
World lines: *see* Trajectories

X-ray binary star, 44, 45, 46

Zero-mass: *see also* Black-body radiation; Compton effect; Light speed in vacuo; Photons
  limit, 107
  particles, **9.2.2**; 109–10, 129, 146, 198